NUTRACEUTICALS AND BONE HEALTH

AAP Advances in Nutraceuticals

NUTRACEUTICALS AND BONE HEALTH

Edited by

Deepak Sharma, PhD

Madan Mohan Gupta, PhD

Anil K. Sharma, PhD, MPharm

Raj K. Keservani, PhD, MPharm

Rajesh K. Kesharwani, PhD, MTech

AAP | APPLE ACADEMIC PRESS

First edition published 2024

Apple Academic Press Inc.
1265 Goldenrod Circle, NE,
Palm Bay, FL 32905 USA

760 Laurentian Drive, Unit 19,
Burlington, ON L7N 0A4, CANADA

CRC Press
2385 NW Executive Center Drive,
Suite 320, Boca Raton FL 33431

4 Park Square, Milton Park,
Abingdon, Oxon, OX14 4RN UK

Apple Academic Press exclusively co-publishes with CRC Press, an imprint of Taylor & Francis Group, LLC

Library and Archives Canada Cataloguing in Publication

CIP data on file with Canada Library and Archives

Library of Congress Cataloging-in-Publication Data

CIP data on file with US Library of Congress

ISBN: 978-1-77491-450-2 (hbk)
ISBN: 978-1-77491-451-9 (pbk)
ISBN: 978-1-00341-508-4 (ebk)

ABOUT THE BOOK SERIES: AAP ADVANCES IN NUTRACEUTICALS

Series Editor:
Raj K. Keservani, PhD, MPharm
Faculty of B. Pharmacy, CSM Group of Institutions, Allahabad, India

Rajesh K. Kesharwani, PhD
Nehru Gram Bharati (Deemed to be University), Prayagraj, India

Anil K. Sharma, PhD, MPharm
Department of Pharmacy, School of Medical and Allied Sciences,
GD Goenka University, Gurugram, India

In the modern era, mankind has witnessed a paradigm shift with respect to fundamental eating behavior. The lack of physical workouts and busy schedules at offices and in households have promoted the consumption of junk foods, which eventually results in numerous diseases and disorders since the nutritional content from fast food is inadequate. Plumpness and obesity have become a global health threat. The leading causes of death from most developing nations are noncommunicable diseases such as cardiovascular diseases, cancer, arthritis, osteoporosis, and liver toxicity. Patients suffering from such lifestyle ailments are a bit apprehensive towards prolonged use of costly modern therapeutics, which encourages instead the use of alternative approaches for management of such diseases and disorders.

The emerging sector of the nutraceutical industry encompasses products derived from nature, dietary supplements, and functional foods. Nutraceuticals are used for treatment and prevention of a broad range of diseases, such as the common cold, arthritis, sleep-related disorders, cancers, cardiovascular complications, metabolic disorders, and others.

The research on nutraceuticals is increasing day by day, considering the beneficial effects of food or food supplements in the management of diverse diseases. The issue of paramount concern is standardization and establishment of clinical efficacy of nutraceuticals, which is in fact a challenge for researchers around the globe.

This book series aims at realizing the significance of the variety of nutraceuticals for human well-being. The books in the series will also emphasize the role of dieticians and nutritionists for the prescription of judicious eating habits. The key food components such as carbohydrates, proteins, and lipids as well as micronutrients (vitamins, minerals) are demonstrated to maintain good health, obviating the need for medicines. Thus, nutraceuticals that are indeed derived from food ingredients are believed to be potential alternative therapeutics.

Asymptomatic diseases/disorders necessitate proper diagnosis. Stressed circumstances have been reported to cause weight loss. Workplace stress is a key variable in the genesis and propagation of metabolic disorders. The nuances of the underlying mechanisms are still to be deciphered. Poor lifestyle trends have been attributed along with the consumption of junk foods in several such instances. Sugar-rich carbonated beverages, the use of contraband drugs, and the consumption of liquor in excess are assigned as risk factors. This book series will cover advances and applications in addition to providing the basics of nutraceuticals.

The books in this series will be valuable resources for industry professionals to design and develop quality products for end users. In addition, the book series will be fortified by ideas of innovation, concept building, manufacturing aspects, quality control, and regulatory status of nutraceuticals.

Ostensibly, business has seen a rise in the nutraceutical consumption in recent years. Pure, safe, efficacious products are the need of the hour globally. There are a number of books on nutraceuticals in the market, yet these are only a few books presenting certain aspects of nutraceuticals therein. This book series aims to offer a holistic view of this promising strategy, which is picking up pace with time. This book series will be a unique endeavor to bring manufacturing, research and development, and marketing strategies under a single umbrella.

CURRENT BOOKS IN THE SERIES

- **Micronutrients and Macronutrients as Nutraceuticals**
  Editors: Prakash Chandra Gupta, PhD, Sayan Bhattacharyya, MBBS, MD, Nisha Sharma, PhD, MPharm, Rajesh K. Kesharwani, PhD, and Raj K. Keservani, PhD, MPharm

- **Formulations, Regulations, and Challenges of Nutraceuticals**
  Editors: Tingirikari Jagan Mohan Rao, PhD, Rajesh K. Kesharwani, PhD, Raj. K. Keservani, PhD, and Anil K. Sharma, PhD

- **Food Supplements and Dietary Fibers in Health and Disease**
Editors: Bhushan R. Rane, PhD, MPharm, Raj K. Keservani, PhD, MPharm, Durgesh Singh, DPhil, Nayan A. Gujarathi, PhD, and Ashish S. Jain, PhD
- **Nutraceuticals in Cancer Prevention, Management, and Treatment**
Editors: Raj K. Keservani, PhD, MPharm Bui Thanh Tung, PhD, Sippy Singh, DPhil, and Rajesh K. Kesharwani, PhD
- **Advances in Flavonoids for Human Health and Prevention of Diseases**
Editors: Nisha Sharma, PhD, Deepika Saini, PhD, Rajesh K. Kesharwani, PhD, Prakash Chandra Gupta, and Raj. K. Keservani, PhD
- **The Flavonoids: Extraction and Applications**
Editors: Deepika Saini, PhD, Rajesh K. Kesharwani, PhD, and Raj K. Keservani, PhD, MPharm
- **Immune-Boosting Nutraceuticals for Better Human Health: Novel Applications**
Editors: Urmila Jarouliya, PhD, Raj K. Keservani, PhD, MPharm Rajesh K. Kesharwani, PhD, Virendra K. Patel, PhD, and Adi D. Bharti
- **Applications of Functional Foods in Disease Prevention**
Editors: Raj K. Keservani, PhD, MPharm, and Eknath D. Ahire, MSPharm
- **Plant Metabolites and Vegetables as Nutraceuticals**
Editors: Raj K. Keservani, PhD, MPharm, Bui Thanh Tung, PhD, Rajesh K. Kesharwani, PhD, MPharm, and Eknath D. Ahire
- **Preventive and Therapeutic Role of Vitamins as Nutraceuticals**
Editors: Khemchand R. Surana, Eknath D. Ahire, Raj K. Keservani, PhD, MPharm, and Rajesh K. Kesharwani, PhD
- **Nutraceuticals and Bone Health**
Editors: Deepak Sharma, PhD, Madan Mohan Gupta, PhD, Anil K. Sharma, PhD, MPharm, Raj K. Keservani, PhD, MPharm, and Rajesh K. Kesharwani, PhD
- **Flavonoids as Nutraceuticals**
Editors: Rajesh K. Kesharwani, PhD, MTech, Deepika Saini, PhD, Raj K. Keservani, PhD, MPharm, and Anil Kumar Sharma, PhD
- **Herbals as Nutraceuticals: Their Role in Healthcare**
Editors: Raj K. Keservani, PhD, MPharm, Rajeshwar Kamal Kant Arya, PhD, and Rajesh K. Kesharwani, PhD

- **Nutraceutical Fruits: Overview and Disease Prevention**
  Editors: Raj K. Keservani, PhD, MPharm
- **Nutrigenomics and Nutraceuticals**
  Editors: Raj K. Keservani, PhD, MPharm, Eknath D. Ahire, Shubham J. Khainar, PhD, Sanjay J. Kshirsagar, PhD, and Rajesh K. Kesharwani, PhD
- **Nutraceuticals in Cardiac Health Management**
  Editors: Raj K. Keservani, PhD, MPharm, Rajesh K. Kesharwani, PhD, Praveen Kumar Jain, PhD, Bhushan R. Rane, PhD, and Nayan A. Gujarathi, PhD
- **Nutraceuticals for the Treatment and Prevention of Sexual Disorders**
  Editors: Raj K. Keservani, PhD, MPharm, Sharangouda J. Patil, PhD, and Ivan Aranha, PhD
- **Nutraceuticals in Obesity Management and Control**
  Editors: Raj K. Keservani, PhD, MPharm, and Alka Lohani, PhD
- **Antioxidants as Nutraceuticals**
  Editors: Nayan A. Gujarathi, PhD, Raj K. Keservani, PhD, MPharm, Rajesh K. Kesharwani, PhD, Bhushan R. Rane, PhD, and Yogeeta Sameer Goyal, PhD
- **The Nature of Nutraceuticals: History, Properties, Sources, and Nanotechnology**
  Editors: Rajesh K. Kesharwani, PhD, Prashant Kumar, PhD, and Raj K. Keservani, PhD, MPharm
- **Nutraceuticals in Insomnia and Sleep Problems**
  Editors: Raj K. Keservani, PhD, MPharm, Sayan Bhattacharyya, PhD, and Rajesh K. Kesharwani, PhD
- **Nutraceuticals in Respiratory and Pulmonary Diseases**
  Editors: Deepika Saini, PhD, Rajesh K. Kesharwani, PhD, and Raj K. Keservani, PhD, MPharm
- **Nutraceuticals in Arthritis and Psoriasis**
  Editors: Meenakshi Jaiswal, PhD, Raj K. Keservani, PhD, MPharm Rajesh K. Kesharwani, PhD, and Swati G. Talele, PhD

ABOUT THE EDITORS

Deepak Sharma, PhD
Professor, Institute of Pharmacy, Assam Don Bosco University, Tapesia, Assam. India

Deepak Sharma, PhD, is Professor at the Institute of Pharmacy, Assam Don Bosco University, Tapesia, Assam, India. With a background of classical and instrumental methods of analysis, Dr. Sharma is an expert in the area of analytical method development and validation and stability testing of drug substance and drug products. He has contributed more than 38 research articles in various peer-reviewed and Scopus-indexed journals. He has written one book and one book chapter to date. He is serving various reputed journals as a reviewer. He is also associated as an examiner with various government and private universities. His research interests include analytical methods development, drug therapeutics, pharmacokinetics, and dissolution.

Madan Mohan Gupta, PhD
Associate Professor (Senior Lecturer) in the School of Pharmacy, Faculty of Medical Sciences, The University of the West Indies, St. Augustine, Trinidad & Tobago, WI

Madan Mohan Gupta, PhD, is an Associate Professor (Senior Lecturer) at the School of Pharmacy, Faculty of Medical Sciences, the University of the West Indies, St. Augustine, Trinidad & Tobago, WI. He has more than 18 years of experience in teaching, research, and administration. He has published more than 68 research and review papers. He has written five books and one book chapter. Dr. Gupta has received six national and international awards for excellence in teaching and research. His research interest includes conventional and novel drug delivery system, nutraceuticals, solubility enhancement, taste masking of bitter drugs, and meta-analysis of different types of cancers.

Anil K. Sharma, PhD, MPharm
Assistant Professor, School of Medical and Allied Sciences,
G D Goenka University, Gurugram, India

Anil K. Sharma, PhD, MPharm, is an Assistant Professor in the School of Medical and Allied Sciences at G.D. Goenka University, Gurugram, India. He is an expert in the area of pharmaceutics with a background in drug delivery. He has taught these subjects for nearly 10 years at universities such as the Delhi Institute of Pharmaceutical Sciences and Research, the University of Delhi, and the School of Medical and Allied Sciences, G.D. Goenka University, India. Prior to taking up his current role in 2018, Dr. Sharma served in academic positions such as Lecturer (Pharmaceutics) at the Delhi Institute of Pharmaceutical Sciences and Research, University of Delhi, India. He has published over 28 peer-reviewed papers in the field of pharmaceutical sciences in both national and international journals as well as many book chapters and several edited books. He holds a PhD (Pharmaceutical Sciences) from the University of Delhi, an MPharmacy (Pharmaceutics) from the Rajiv Gandhi Proudyogiki Vishwavidyalaya and a BPharmacy from the University of Rajasthan.

Raj K. Keservani, PhD, MPharm
Associate Professor, Faculty of B.Pharmacy, CSM Group of Institutions, Prayagraj, India

Raj K. Keservani, PhD, MPharm, is working as Associate Professor in the Faculty of B. Pharmacy at the CSM Group of Institutions, Prayagraj, India. He has more than 12 years of academic (teaching) experience from various institutes of India in pharmaceutical education. He has supervised more than 25 undergraduate students for their research work. He has published over 30 peer-reviewed papers in the field of pharmaceutical sciences in national and international journals, six patents, over 70 book chapters, three co-authored books, and over 40 edited books. He is also active as a reviewer for several international scientific journals. His research interests include nutraceutical and functional foods, novel drug delivery systems (NDDS), transdermal drug delivery/drug delivery, health science, cancer biology, and neurobiology. Dr. Keservani graduated with a pharmacy degree from the Department of Pharmacy, Kumaun University, Nainital, Uttarakhand, India. He received his

Master of Pharmacy (MPharm) (specialization in pharmaceutics) from the School of Pharmaceutical Sciences, Rajiv Gandhi Proudyogiki Vishwavidyalaya, Bhopal, India. He earned his PhD degree from IKG Punjab Technical University, Jalandhar, Punjab, India. Dr. Keservani is a life member of the Society of Pharmaceutical Education and Research (SPER).

Rajesh K. Kesharwani, PhD, MTech
Associate Professor, Department of Computer Application,
Nehru Gram Bharati (Deemed to be University), Prayagraj, India

Rajesh K. Kesharwani, PhD, MTech, working as an Associate Professor in the Department of Computer Application at Nehru Gram Bharati (Deemed to be University), Prayagraj, India. He has more than 11 years of research and nine years of teaching experience in various institutes in India, imparting bioinformatics and biotechnology education. He has received several awards, including the NASI-Swarna Jayanti Puruskar by The National Academy of Sciences of India. He has supervised one PhD and more than 20 under- and postgraduate students for their research work and has authored over 49 peer-reviewed articles, 20 book chapters, and 14 edited books with international publishers. He has been a member of many scientific communities as well as a reviewer for many international journals. He has presented many papers at various national and international conferences. Dr. Kesharwani received his PhD from the Indian Institute of Information Technology, Allahabad, and worked at NIT Warangal for two-semester. He has been a recipient of the Ministry of Human Resource recipient of Development (India) Fellowship and a Senior Research Fellowship from the Indian Council of Medical Research, India. His research fields of interest are medical informatics, protein structure and function prediction, computer-aided drug designing, structural biology, drug delivery, cancer biology, nano-biotechnology, and biomedical sciences.

CONTENTS

Contributors .... *xv*

Abbreviations .... *xix*

Preface .... *xxv*

PART I: Bone Health Introduction 1

1. **Introduction to Bone Health and the Role of Nutraceuticals 3**
   Shamsun Nahar Khan

PART II: Beneficial Effects of Nutraceuticals in Bone Disorders 23

2. **Beneficial Aspects of Nutraceuticals in the Management of Osteoporosis 25**
   Sharad Visht and Garima Singh

3. **Management of Fractures via Nutraceuticals 59**
   Arun Soni, Niyati Acharya, and Sanjeev Acharya

4. **Therapeutic Potential of Nutraceuticals in the Management of Scoliosis and Related Complications 83**
   Gunika Gulati, Siddhi Bagwe Parab, Dhruv Sanjay Gupta, and Ginpreet Kaur

5. **Role of Nutraceuticals in the Treatment of Rheumatoid Arthritis 125**
   Devesh U. Kapoor and Manish Jaimini

6. **Nutraceutical Interventions for Effective Management of Paget's Disease 145**
   Suraj Bamankar and Vaishali Y. Londhe

7. **Health Benefits of Nutraceuticals for Gout Patients 175**
   Suresh Kumar, Bhawna Chopra, Rameshwar Dass, and Ashwani K. Dhingra

8. **Role of Nutraceuticals in Prevention and Management of Bursitis 191**
   Pragya Pandey, Savvy Sharma, Utsav Rathore and Madan Mohan Gupta

9. **Carpal Tunnel Syndrome Management through Nutraceuticals 205**
   Ruhul Amin, Faruk Alam, Durgaprasad Kemisetti, Dhrubajoyti Sarkar and Biplab Kumar Dey

PART III: Inflammatory Deformities and Rickets 221

10. Role of Nutraceuticals in Managing Osteomyelitis 223
Joyeta Ghosh, Sudrita Roy Choudhury, and Samarpita Koner

11. Nutraceuticals: New Paradigm for Better Management of Rickets 293
Shyamli Thakur and Luxita Sharma

Index *313*

CONTRIBUTORS

Niyati Acharya
Assistant Professor & Head, Department of Pharmacognosy, Institute of Pharmacy, Nirma University, Ahmedabad, India

Sanjeev Acharya
Assistant Professor & Head, Department of Pharmacognosy, Institute of Pharmacy, Nirma University, Ahmedabad, India

Faruk Alam
Associate Professor, Faculty of Pharmaceutical Science, Assam Down Town University, Panikhaiti, Gandhinagar, Guwahati, Assam, India

Ruhul Amin
Assistant Professor, Faculty of Pharmaceutical Science, Assam Down Town University, Panikhaiti, Gandhinagar, Guwahati, Assam, India

Suraj Bamankar
Scholar, Shobhaben Pratapbhai Patel School of Pharmacy & Technology Management, SVKM's NMIMS, Vile Parle [W], Mumbai, Maharashtra, India

Bhawna Chopra
Associate Professor in Guru Gobind Singh College of Pharmacy, Yamuna Nagar, Haryana, India

Sudrita Roy Choudhury
Research Scholar, Department of Dietetics and Nutrition, NSHM Knowledge Campus-Kolkata, West Bengal, India

Rameshwar Dass
Assistant Professor in Guru Gobind Singh College of Pharmacy, Yamuna Nagar, Haryana, India

Biplab Kumar Dey
Dean, Faculty of Pharmaceutical Science, Assam Down Town University, Panikhaiti, Gandhinagar, Guwahati, Assam, India

Ashwani K. Dhingra
Professor in Guru Gobind Singh College of Pharmacy, Yamuna Nagar, Haryana, India

Joyeta Ghosh
Assistant Professor, Department of Dietetics and Applied Nutrition, Amity University, Kolkata, India

Gunika Gulati
Scholar, Shobhaben Pratapbhai Patel School of Pharmacy and Technology Management, SVKM'S NMIMS, Mumbai, Maharashtra, India

Dhruv Sanjay Gupta
Scholar, Shobhaben Pratapbhai Patel School of Pharmacy and Technology Management, SVKM'S NMIMS, Mumbai, Maharashtra, India

Madan Mohan Gupta
Senior Lecturer (Associate professor), School of Pharmacy, Faculty of Medical Sciences, The University of the West Indies St Augustine Trinidad and Tobago. West-Indies

Manish Jaimini
Prof & Principal, Department of Pharmacy, Maharishi Arvind University, Jaipur, Rajasthan, India

Devesh U. Kapoor
Lecturer (Grant-In-Aid), Dr. Dayaram Patel Pharmacy College, Bardoli, Gujarat, India

Ginpreet Kaur
Associate Professor, Shobhaben Pratapbhai Patel School of Pharmacy and Technology Management, SVKM'S NMIMS, Mumbai, Maharashtra, India

Durgaprasad Kemisetti
Associate Professor, Faculty of Pharmaceutical Science, Assam Down Town University, Panikhaiti, Gandhinagar, Guwahati, Assam, India

Shamsun Nahar Khan
Post Doc Harvard University, USA, MRSC, Professor, Department of Pharmacy, East West University, Dhaka, Bangladesh

Samarpita Koner
Research Scholar, Department of Dietetics and Nutrition, NSHM Knowledge Campus-Kolkata, West Bengal, India

Suresh Kumar
Assistant Professor, Department of Pharmacy, School of Medical and Allied Sciences, GD Goenka University, Gurugram, Haryana, India

Vaishali Y. Londhe
Assistant Professor, Professor, Shobhaben Pratapbhai Patel School of Pharmacy & Technology Management, SVKM's NMIMS, Vile Parle [W], Mumbai, Maharashtra, India

Pragya Pandey
Assistant Professor, Department of Food Technology and Nutrition, School of Agriculture Lovely Professional University, Phagwara, Punjab, India

Siddhi Bagwe Parab
Scholar, Shobhaben Pratapbhai Patel School of Pharmacy and Technology Management, SVKM'S NMIMS, Mumbai, Maharashtra, India

Utsav Rathore
Department of Food Technology and Nutrition, School of Agriculture Lovely Professional University, Phagwara, Punjab, India

Dhrubajoyti Sarkar
Associate Professor, Faculty of Pharmaceutical Science, Assam Down Town University, Panikhaiti, Gandhinagar, Guwahati, Assam, India

Luxita Sharma
Faculty of Health and Allied Sciences, Amity University, Gurgaon, India

Savvy Sharma
Department of Food Technology and Nutrition, School of Agriculture Lovely Professional University, Phagwara, Punjab, India

Garima Singh
Assistant Professor, Faculty of Nutrition and Health, School of Medical and Allied Sciences, G. D. Goenka University, Gurugram, Haryana, India

Arun Soni
SSR College of Pharmacy, Sayli, Silvassa, Dadra, India

Shyamli Thakur
School of Pharmaceutical and Populations Health Informatics, DIT University, Dehradun, Uttarakhand, India

Sharad Visht
Assistant Professor, Faculty of Pharmacy, Tishk International University, Erbil, Kurdistan Region, Iraq

ABBREVIATIONS

5-HTP	5 hydroxytryptophan
AA	arachidonic acid
ADHD	attention deficit-hyperactivity disorder
AHS	allopurinol hypersensitivity syndrome
AIA	adjuvant induced arthritis
AIS	adolescent idiopathic scoliosis
AP	anterior-posterior
AR	anti-resorptive
ASU	avocado-soybean unsaponifiable
ATP	adenosine triphosphate
BF	bone formation
BiFi	bifidobacteria
BIIC	bovine type II collagen
BMD	bone marrow density
BMP	bone morphogenetic protein
CAGR	compound annual growth rate
cAMP	cyclic adenosine monophosphate
CBC	complete blood count
CBD	cannabidiol
CFAA	complete Freund's adjuvant arthritis
CFS	chronic fatigue syndrome
CIA	collagen-induced arthritis
CLPs	common lymphoid progenitors
CNS	central nervous system
COL1A1	collagen type I α1
CRP	C-reactive protein
CS	congenital scoliosis
CTS	carpal tunnel syndrome
CTSG	cathepsin G
CTX	C-terminal telopeptide
Cu	copper
DCs	dendritic cells
DGLA	dihomo-γ-linoleic acid
DHA	docosahexaenoic acid
DHEA	dehydroepiandrosterone
DHEAS	dehydroepiandrosterone-sulfate
DLS	degenerative lumbar scoliosis

DLX5	distal lower homeobox 5
DM	diabetes mellitus
DMARD	disease-modifying antirheumatic drugs
DMS	dexamethasone
DNA	deoxyribonucleic acid
DSHEA	Dietary Supplement Health and Education Act
EC	epicatechin
ECG	epicatechin gallate
EDS	Ehlers-Danlos syndrome
EFSA	European Food and Safety Authority
EGCG	epigallocatechin gallate
EPA	eicosapentaenoic acid
ERK	extracellular Regulated Kinase
ERs	estrogen receptor
FA	fatty acids
FAA	Freund's adjuvant arthritis
FBP	fishbone peptides
FDA	Food and Drug Administration
FGFRs	FGF tyrosine kinase receptors
FGFs	fibroblast growth factors
FHL	fish head lipid
FLSC	fibroblast-like synovial cells
FN	fibrinogen
FSE	Food Supplements Europe
FSSA	Food Safety and Standards Act
FSSAI	Food Safety and Standards Authority of India
GIT	gastrointestinal tract
GLA	gamma-carboxyglutamic acid rich / gamma linolenic acid
Gm	gut microbiota
GMP	good manufacturing practices
GPI	for glycosyl phosphatidylinositol
GPCR	G protein-coupled receptors
GSH	glutathione
HCl	hydrochloric acid
HGF	hepatocyte growth factor
HMGB1	high mobility group box 1
HNPP	hereditary neuropathy with pressure palsy
HPD	hesperidin
HSPG2	heparan sulfate proteoglycan 2
HYAL1	hyaluronoglucosaminidase 1
ICSBP	interferon nonsupervisory factor 8
IEC	import-export certificate
IFN	interferon

IKB	IKappaB kinase
IKK	IκB kinase
IL	interleukin
IL-1β	interleukin-1β
IL-6	interleukin-6
iNOS	inducible nitric oxide synthase
IS	idiopathic scoliosis
IU	international units
JNK	c-Jun N-terminal kinase
LA	linolenic acid
LCFA	long chain fatty acid
LOX	5-lipoxygenase
LP	lipopolysaccharides
MCFA	medium chain fatty acid
MCP	monocyte chemoattractant protein
MCSC	mesenchymal stem cells
M-CSF	macrophage colony-stimulating factors
MDR	multi-drug resistance
MIP	macrophage inflammatory proteins
MMC	mesenchymal multipotent cells
MPO	myeloperoxidase
MR	mannose receptor
MRI	magnetic resonance imaging
MRP8	myeloid-related protein 8
MRSA	methicillin-resistant S. aureus
MSCs	mesenchymal stem cells
MSU	monosodium urate
NAG	naringenin
NF-κB	nuclear factor kappa-B
NK	natural killer
NKC	natural killer cells
NMS	neuromuscular scoliosis
NO	nitric oxide
NPS	natural dietary supplements
NR	nutritional rickets
NRTK	non-receptor tyrosine kinase
NSAIDs	non-steroidal anti-inflammatory drugs
OAT1	organic anion transporter 1
OBs	osteoblasts
OCs	osteoclasts
ODF	osteoclast differentiation factor
OI	osteogenesis imperfecta
OPG	osteoprotegerin

OPGL	osteoprotegerin ligand
PA	posterior-anterior
PDB	Paget's disease of the bone
PES	Pediatric Endocrine Society
PGF	placental growth factor
PKB	protein kinase B
PMN	polymorphonuclear cells
PNS	peripheral nervous system
PPAR	peroxisome proliferator-activated receptor
PPRs	pattern recognition receptors
PRE	prebiotic Inulin
PRO	probiotic
PSSR	physiotherapeutic scoliosis-specific rehabilitation
PTH	parathyroid hormone
PTHrP	parathyroid hormone-related protein
PUFA	polyunsaturated fatty acids
RA	rheumatoid arthritis
RANK/RANKL	receptor activator of nuclear factor kappa- B ligand
RANKL	receptor activator for NFkappaB ligand
RDAs	recommended daily allowances
RNA	ribonucleic acid
ROS	reactive oxygen species
RPL	resveratrol attenuates oligosaccharide
Saa	serum amyloid A
SACs	staphylococcal abscess communities
SCFAc	short-chain fatty acids
SFA	saturated fatty acid
SR	scavenger receptor
SUA	serum uric acid
SYM	symbiotic
T3	triiodothyronine
T4	thyroxine
TACR1	tachykinin receptor 1
TCJ	tart cherry juice
THC	tetrahydrocannabinol
THFA	tetrahydrofolate
TNF	tumor necrosis factor
TNFα	tumor necrosis factor-alpha
TPH	tryptophan hydroxylase
TRANCE	TNF-related activation-induced cytokine
TRP	tryptophan
TSH	thyroid-invigorating chemical
TYMLOS	teriparatide and abaloparatide

UK	United Kingdom
URAT	urate transporter/channel
URAT1	urate transporter 1
UVB	ultraviolet radiation B
VCP	valosin-containing protein
VDRs	vitamin D receptors
VEGF	vascular endothelial growth factor
WOMAC	Western Ontario and McMaster Universities Osteoarthritis Index
XOD	xanthine oxidase
ZA	zoledronic acid

PREFACE

Welcome to the book, "Nutraceuticals and Bone Health." The progression or development of bone is a multifaceted process that requires an endless and ample hoard of nutrients, such as calcium, phosphorus, potassium, protein, vitamin D, magnesium, and fluoride.

This book is written with the aim of providing deep knowledge of how nutraceuticals can be used to uplift bone health, especially in diseased conditions, when we need something more than regular medications. This book aims to provide detailed information about the fundamental aspects of nutrition and how it applies to our own lives. Nourishment is a science-based discipline, so all material included is upheld by thorough, logical examination yet introduced in a reasonable, straightforward manner with a logical foundation.

This book, *Nutraceuticals and Bone Health*, comprises 11 chapters in three sections that cover the introduction, structure, and function of bones, the beneficial effects of nutraceuticals in bone disorders, and inflammatory deformities and rickets.

Part I is titled, Bone Health Introduction, and has one chapter in it. In Chapter 1, Introduction to Bone Health and the Role of Nutraceuticals, the author discusses the structure and function of bones, various disorders related to bone health, and the beneficial nutraceuticals for such disorders written by Shamsun Nahar Khan.

Part II is titled, Beneficial Effects of Nutraceuticals in Bone Disorders, and has eight chapters in it. In Chapter 2, Beneficial Aspects of Nutraceuticals in Management of Osteoporosis," the authors mention that osteoporosis is a metabolic bone disorder characterized by low bone mineral density as well as low bone protein, and the nutraceuticals recognized from dietary agents such as anthocyanin, kaempferol, curcumin, diosgenin, embelin, gambogic acid, genistein, quercetin, resveratrol, zerumbone omega 3, prebiotics, probiotics, and PUFA, are found to be antiosteoporotic, written by Garima Singh and her colleagues.

Flow in Chapter 3, Management of Fractures via Nutraceuticals, presents overall information on factors pertaining to the use of nutraceuticals in the market as bone fracture preventive agents, written by Sanjeev Acharya and associates.

Scoliosis is a very common spinal deformity that affects 2–3% of the population globally and is characterized by the abnormal curvature of the spine discussed by Ginpreet Kaur and her co-worker in Chapter 4, Therapeutic Potential of Nutraceuticals in the Management of Scoliosis and Related Complications.

Rheumatoid arthritis (RA) is a very common autoimmune disorder and forms the basis for Chapter 5, Role of Nutraceuticals in Treatment of Rheumatoid Arthritis, written by Devesh U. Kapoor and Manish Jaimini and reported nutraceuticals to have mammoth potential as a safe and effectual treatment option against RA disorder.

Global regulatory criteria for the pragmatic and safe usage of nutraceuticals can be proposed, given the nutraceutical market's huge development potential as reported by Vaishali Y. Londhe and Suraj Bamankarin in Chapter 6, Nutraceuticals Interventions for Effective Management of Paget's Disease.

Nutraceuticals or natural products, including vegetables, nuts, legumes, fruits, and whole grains, play an important role in the development of hyperuricemia and gout. Xanthine oxidase inhibitors are currently being investigated as potential therapeutics for disorders such as gout, hyperuricemia, ulcers, cancer, and oxidative damage, as written by Suresh Kumar and associates in Chapter 7, Health Benefits of Nutraceuticals in Gout Patients.

Pragya Pandey et al. discuss many nutraceuticals that have been proven to have anti-inflammatory effects, which can, in turn, help in the management of bursitis. Some of the significant nutraceuticals and their role in the prevention and treatment of bursitis is discussed in Chapter 8, Role of Nutraceuticals in Prevention and Management of Bursitis.

Chapter 9, Carpal Tunnel Syndrome Management by Nutraceuticals, written by Faruk Alam and his associates concludes that nutraceuticals play an important role and serve as prophylactic treatment of certain diseases.

Part III is titled, Inflammatory Deformities and Rickets and has one chapter in it.

The inflammatory bone deformities (Osteomyelitis) mainly caused by infection, leading to the destruction or necrosis of bones, can be treated using some specific nutraceuticals discussed by Joyeta Ghosh et al., in Chapter 10, Role of Nutraceuticals in Managing Osteomyelitis.

Chapter 11, Nutraceuticals: New Paradigm for Better Management of Rickets, tells the story of how the justified use of nutraceuticals can make the life of patients suffering from rickets easy and comfortable, written by Shyamli Thakur and her associates.

PART I
Bone Health Introduction

CHAPTER 1

INTRODUCTION TO BONE HEALTH AND THE ROLE OF NUTRACEUTICALS

SHAMSUN NAHAR KHAN

Post Doc Harvard University, USA, MRSC, Professor, Department of Pharmacy, East West University, Dhaka, Bangladesh

ABSTRACT

Bones are the most important and vital parts of in a human body. Calcium and phosphorus are the two most important and fundamental component of the bone growth. Vitamin D3 also known as cholecalciferol, it helps in the absorption of calcium and phosphorus in the body. But still after taking several calcium supplements bone decay and fracture are a very common phenomenon observed in the older age also, much of the time it causes immobility which leads to devastating tragic end of a life as well. Nutraceuticals are the different types of compounds which are extracted from fruits and vegetables and are widely used around the globe for better health and wellbeing. The most important biological properties of nutraceuticals include anti-inflammatory, antioxidant and antidiabetic activities. Some of these compounds are proved to have properties which stimulate or facilitate bone growth by the generation of new bone cells; some these compounds help's to inhibit degeneration of bone. It is essential to take care of the bone with the help of balanced diet and following a healthy life style, Nutraceuticals are always very much important to maintain an excellent bone health and life. In this chapter we discussed structure and function of the bone, various disorders related to bone health and beneficial nutraceuticals.

Nutraceuticals and Bone Health. Deepak Sharma, Madan Mohan Gupta, Anil K. Sharma, Raj K. Keservani, Rajesh K. Kesharwani (Eds.)

1.1 INTRODUCTION

Bones are the most fundamental framework of our body structure and function. Movements and different body postures, bending are always supported by our bones. Bones protect soft tissues and organs of our body in an orchestral fashion. Bones are highly organized and structured organs. Bones are composed of proteins, fibrous collagen, minerals, calcium, phosphorus and other organic and inorganic materials. 99% of the body's calcium is found in the skeletal bones. Healthy structured bones are blessings and need to be taken care of according to its requirement for the children and adult people. Healthy bone development at earlier stage proceeds to have an ideal structured adult human being. There is nutritional and dietary requirement to keep our bones healthy and strong. Major source of balanced nutrition is food but there is alternative nutritional supplements or nutraceuticals are also available in the market.

Nutraceuticals are the compounds commonly extracted from fruits and vegetables, usually these are the compounds with anti-inflammatory, antioxidant and antidiabetic activities, recommended for chronic diseases e.g., osteoporosis, arthritis, cancer and cardiovascular disease management (Ronis et al., 2018). The term "nutraceutical" was introduced by Dr. Stephen DeFelice in 1989 (Brower, 1998). "Nutraceutical" is basically originated from two words "Nutrition" and "pharmaceutics" or" Pharmaceutical," in general, many of the nutritional products are formulated and produced in pharmaceutical industry for the purpose of promoting health (Keservani et al., 2010a, 2010b, 2017, 2020). These products are produced with assurance of quality, safety and efficacy, thus the word "Nutraceuticals" more often used during marketing. "Nutraceuticals" can be defined as food or part of food which has specific role in the disease prevention, cure and mitigation as well. Potential action of nutraceuticals is well explained on the basis of evidence grounded scientific studies along with the pharmacological properties of the ingredients for its specific efficacy. On the other hand, "functional foods" are cooked on the basis of scientific knowledge that its contents are kept functional and can be provided the quantity of the vitamins, fats, carbohydrate, proteins, etc. (Ekta, 2003).

Decaying bone and fracture is a very common phenomenon observed in the older age, much of the time it causes immobility and causes devastating tragic end of a life as well. To prevent bone fracture extra care and proper nutritional maintenances are needed for the old aged people. A well-balanced diet, rich in calcium and vitamin D, exposure in adequate regular sun light and adequate exercise is definitely necessary along with other proportionate trace minerals. Nutraceuticals would be taken by the individual according to the requirement and specifically by the recommendation of a dietician and nutritionist. Without the guidance of nutritionist or physician excessive intake of the nutritional supplement may cause toxicity as well. We all are aware of the importance of calcium intake to prevent the process of decaying bone and osteoporosis in elderly and postmenopausal female population.

1.2 STRUCTURE AND FUNCTION OF BONE

Bones are apparently the sturdiest organ of the human body and it seems to be very stable and static but in reality bones are very much dynamic living tissues and are in the continuous process of being reshaping. General structure of a long bone is presented in Figure 1.1. Both terminal of a bone is observed with a flared outward and known as epiphysis, the frontal part is called proximal epiphysis and the posterior part is called the distal epiphysis.

The middle-centered tubular section of the bone called diaphysis, after the diaphysis, near the end there is a flare outward is called metaphysis which enlarged towards the end contain spongy interior. In the young growing aged people metaphysis is separated from the physis by epiphyseal line which is basically a hyaline cartilage which is also known as (growth plate).

Periosteum layer is present at the most outer surface of the bone; these are composed of fibrous connective tissues. Endosteum is a vascular connective tissue layer lined on the inner surface of the bony wall, its function as to protecting the medullary cavities of bone.

Bones consists of three different highly specialized cells; they are osteoblasts, osteocytes and osteoclasts. Osteoblasts are responsible for bone growth by the generation of new bone cells and secreting collagen which mineralizes the initial bone matrix. The bone matrix requires mineralization which is done by the osteocytes. Osteocytes responsible for the uptake of the minerals from the blood as well reimbursing the minerals to the blood through a process called mineral homeostasis. Osteoclasts assist the dissolution of minerals in to the bone and release and also control release back into the blood. With the direction of osteocytes osteoblasts builds up the bone on the other hand osteoclasts continually break down the bones.

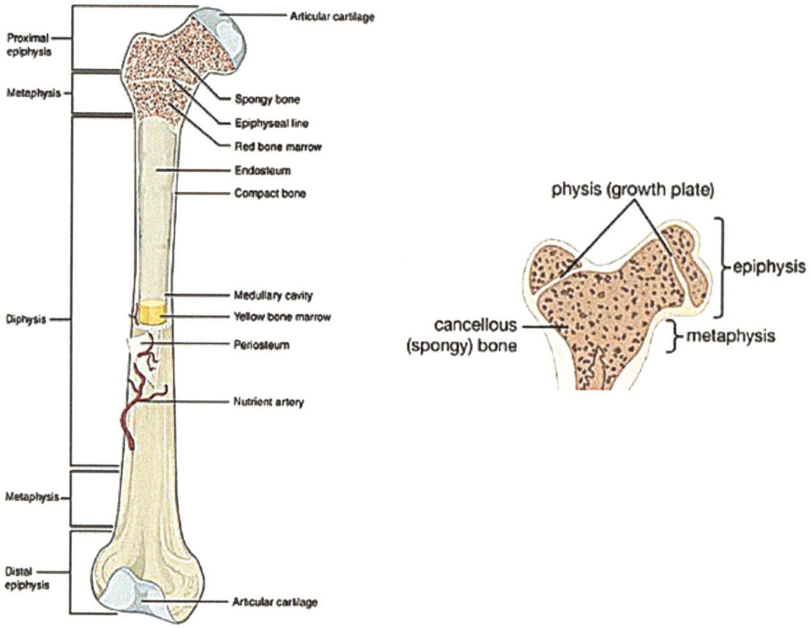

FIGURE 1.1 Structure of bone.

Source: Adapted from Wikidpedia. Image by OpenStax College. https://creativecommons.org/licenses/by/3.0/ and Image on Bone Morphology by Encyclopædia Britannica

Different part of the bones is structurally and compositionally different as the bones are composed of different types of tissues. There are different types of bones observed according to their structures e.g., compact and cancellous bone structure (spongy). The most outer part of the bone which makes the dense and tough outer layer of a bone is known as compact bone and the unit of this bone is known as osteon. The inner layer of the bone is lighter less dense and provides a sponge like structure thus known as spongy bone. Inside the spongy layer of bone there are soft connective tissue prevails, it is responsible for the production of the blood cells of our body.

1.3 MOLECULAR BASIS OF BONE FORMATION, GROWTH AND ROLE OF NUTRACEUTICALS

Osteogenesis refers to the formation of bone at first in embryo and ossification means something turning into bones. There are in general two types of ossification e.g.:

1. Intramembranous ossification
2. Endochondral ossification

1.3.1 INTRAMEMBRANOUS OSSIFICATION

Intramembranous ossification refers to making a bone between fibrous connective tissue membrane and forms flat bones e.g., scull bones. Intramembranous ossification process is presented in the Figure 1.2, in this process there are several factors and signaling pathway observed which are marked as 1, 2 and 3. Wnt/β-catenin is a pathway which is associated with bone cell signaling. The fibroblast growth factors are a group of polypeptides which binds with fibroblast growth factors-tyrosine kinase receptors, and regulate both types of ossification. Curcumin, resveratrol, quercetin, genistein are the nutraceuticals are the most popular and common agents recommended for the healthy bone formation as they stimulate the pre-osteoblast formation at the stage 2.

Mesenchymal cells are commonly found in embryo and generates vascularized primitive connective tissues through condensation process. These stem cells further differentiated into osteogenic cells which are also know as osteoprogenitor cells. Osteoprogenitor cells further differentiated into pre-osteoblasts and eventually mature osteoblasts, a center of ossification is also formed simultaneously.

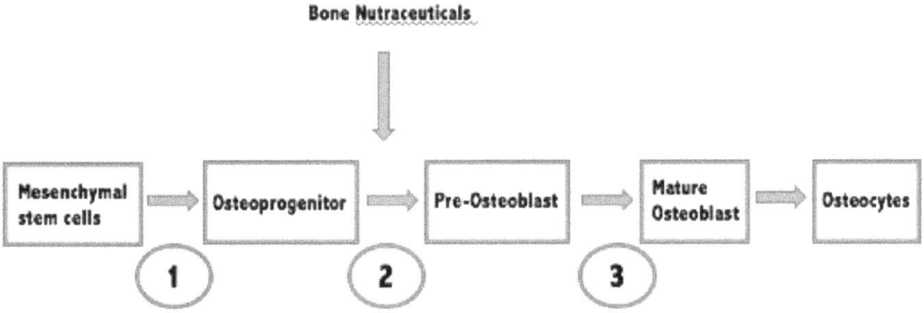

FIGURE 1.2 Intramembranous ossification, 1: Wnt/β-catenin, Fibroblast growth factors; 2: Runx-2, Bone morphogenetic proteins 2 & 4, Wnt/ β-catenin, Fibroblast growth factors, 3: Runx-2 BMP 2 & 4 TGF-β2 Wnt / β-catenin Fibroblast growth factors / fibroblast growth factor receptors.

Osteoblasts secrets an extracellular matrix. Osteoblasts then turn into osteocytes. In this step calcification of osteoids occurs, osteoblasts around outside of the bone forming structure secrets osteods in the middle calcium phosphate and hydroxide combined to form hydroxyappetite needle like crystal structures, it also combines with other minerals salts and ions and give a hard boney structure.

In the nest stage trabecular and periosteum forms where extracellular matrix develops into trabeculae around the blood vessels in a random fashion and forms spongy bone structure, which is also called woven bone (Pandey et al., 2018)

1.3.2 ENDOCHONDRAL OSSIFICATION

Endochondral ossification refers to the bone formation in the cartilage, cartilage serves as a frame work upon which bone formation occurs, eventually cartilage slowly replaced by bones. Examples endochondral ossification are long bones, limbs etc. Ossification of bone processing initiates in the primary ossification center which located in the diphysis of a long bone and secondary ossification center is located in the epiphysis. Chondrocytes, osteoblasts and osteoclasts the cells are involved in the bone formation. At the center of cartilage cells are spontaneously converted in to osteoblasts, and within the diphysis osteoblast lining down bone, and the center is called primary ossification center. In the middle of the diphysis osteoblasts began replacing cartilage with bone. Epiphyseal plates made up of chondrocytes or cartilage making cells, they from cartilage according to the epiphyseal direction and replacing cartilage with bones.

After birth secondary ossification center appears which are located in the epiphysis throughout growth our bones lengthening occurs as long as there is cartilage in between epiphysis and diphysis, eventually with the aging there will be no cartilage then the bone growth will stop.

1.4 ROLE OF CALCIUM, VITAMIN D3 ON THE GROWTH OF BONE

1.4.1 CALCIUM

Calcium and phosphorus are the two most important and fundamental component of the growth of the bones. Requirement of calcium for the adult 1000 to 1,200 mg daily. This required calcium need to be taken from the daily dietary (Keservani et al., 2020) intake of food e.g., egg, milk etc. If the deficiency of calcium cannot be maintained from food in that case supplementary of calcium is provided (Nieves, 2013; IoM 2010).

Calcium supplement is a billion-dollar market in the United States of America. Daily recommended calcium intake is 700–1,200 mg for the post-menopausal older woman, according to the UK national osteoporosis guideline (Li et al., 2018). There are limited calcium rich food source is very limited thus calcium supplements are essential for the patients who are deficient in calcium. Clinical studies showed that high calcium supplements cause enhanced cardiovascular diseases (Reid et al., 2017, Tankeu et al., 2017; Xiao et al., 2013). A few other studies postulated that dietary calcium has better absorption that calcium supplements and thus has a better impact on the bone density compare to the calcium supplements (Booth et al., 2013).

1.4.2 VITAMIN D3

Vitamin D3 is a hormone and also known as cholecalciferol, it helps to absorption of calcium and phosphorus in the body. Daily requirement of vitamin D for an adult person is 800–1,000 IU per day. Synthesis of the Vitamin D3 occurs on the skin of human with the exposure of the sunlight. Specifically, ultraviolet radiation B (UVB radiation) with wavelength 290–315 nm of the sunlight function on the epidermis layer of the skin with 7-dehydrocholesterol and converted to pre-vitamin D3 (Laird et al., 2010). Synthesis of the Vitamin D3 is maximized within a short period of time 10–15 minutes in a bright sunny day depending on the color of the skin which is depending upon the concentration of the melanin. In the presence of the higher concentration of melanin (darker skin) in the skin is responsible for the absorption and scattering of the UVB radiation thus it reduces conversion of 7-dehydrocholesterol to the precursor of vitamin D3. On the other hand, lighter skin is able to have rapid synthesis of the pre-vitamin D3.

Over exposure in the sun light is the cause of skin cancer as well, thus it is often challenging to have a potential exposure of the sunlight for the vitamin D3 synthesis (Bonila et al., 2014). Vitamin D3 is responsible for the bone growth and mineralization on the bone.

1.5 ESSENTIAL MINERALS AND TRACE MINERALS FOR THE BONE HEALTH

Mineralization is one of the most important factor for the growth of the bone, predominantly magnesium (Mg) and zinc (Zn) are essential for organic bone matrix synthesis along with potassium, magnesium (Mg). Besides these there are trace minerals e.g., boron, copper (Cu) iron, zinc (Zn), Manganese (Mn) and selenium are also very important component for the bone growth and development as well, these trace minerals are used as co factor for different enzymes for the metabolism (Saltman & Strause, 1993). Milk is considered as the most important dietary source of minerals, studies showed that milk itself contains more than 20 essential minerals which present at different concentration (Cashman, 2006).

1.6 OSTEOPOROSIS AND BONE LOSS

The word osteoporosis originated from Greek word "Osteon" means bone and "porosis" means porous. Osteoporosis is a process of decaying bones or loss of bones and cause ease of fracture, its commonly found with the woman with menopause. It is a very slow progressing disease due to the dysregulation of pro-inflammatory cytokines. Osteoporosis has complex etiology and is considered to have multiple factors e.g., genetic, nutritional, environmental and individual hormonal level leading to this disease.

Osteoporosis progresses without any significant sign and symptoms e.g., pain, swelling etc. Woman whose bone density tested T scores of -2.5 or lower values are suspected to have bone factures and need to be under medication by healthcare provider.

Vitamin D and calcium are the most important nutraceuticals recommended for osteoporosis for 14–50 years old, pregnant/lactating woman 600 IU is the most commonly recommended dose along with fortified milk intake. Replacing lost of estrogen and hormone therapy helps to combat osteoporosis. Ca, Cu, Mn and Zn are the essential mineral nutrients for the proper mineral bone density maintenance in postmenopausal women (Saltman & Strause, 1993).

1.6.1 ROLE OF ESTROGEN IN THE OSTEOPOROSIS

Estrogen is a sex hormone and it is very much important for the bone health of as it activates the activity of osteoblasts, during menopause the estrogen level of the blood reduced and thus the bone formation is not accomplished. Estrogens plays a crucial role in regulating bone homeostasis and preventing bone loss in the postmenopausal patients.

There are two types estrogen receptor e.g., (ERs), ERα (OMIM133430) and ERβ (OMIM601663), they belong to the superfamily of nuclear receptor. Both of the receptors are found with polymorphism which has significant contribution the, there is a strong correlation of these polymorphism and to the pathogenesis of osteoporosis. Polymorphisms of ERα and ERβ originated from their pharmacogenomic implications which is able to provide a better guideline for the treatment and medicine prescription. Estrogen replacement therapy and selective ER modulators are the two important approaches for the treatment of osteoporosis in the post menopausal patients (Gennari et al., 2005). Both types of estrogen receptors are expressed in osteoblasts, osteoclasts, and bone marrow (Bland 2000, Compston, 2001).

1.6.2 DEHYDROEPIANDROSTERONE (DHEA)

One of the most important precursor to androgens and estrogens respectively in male and female is Dehydroepiandrosterone (DHEA). Several clinical studies demonstrated reduced serum levels of dehydroepiandrosterone-sulfate (DHEAS) is found in both elderly male and female. This product is available in the market with claiming as anti-aging remedy which helps to increase muscle. DHEA is capable of inhibiting osteoclastic bone resorption.

1.6.3 ROLE OF CALCITONIN ON THE OSTEOPOROSIS

A clinical case study reported by Gur and his colleagues in 2002. The patients were selected with post menopausal osteoporosis (70 patients) and a control group (30 patient) who did not have osteoporosis. They observed with a lower levels of serum Mg, copper (Cu), and Zn in the patient group but not in the control group. Gür and his colleagues demonstrated the role of calcitonin therapy in the regulation of Mg, Cu and Zn in patients with postmenopausal osteoporosis with comparison of a control group. Different preparation of calcitonin are commercially available, e.g., calcitonin-salmon (Fortical® and Micalcin®). These are synthetic hormones which help to reduce the chances of spine fractures but not hip or other types of bone fractures. Most commonly these hormones are administered by the injectable routs and it can be used by the inhalation as well.

1.7 RECOMMENDED DRUGS AGAINST BONE LOSS BY USFDA

There are many medicines are approved by the USFDA as the treatment of bone loss as follows, but most of them are reported with some dangerous and uncomfortable side effects as follows (Pandey et al., 2018):

- **Estrogen:** Estrogen is a sex steroid and already discussed its necessity in the previous section. Estrogen is used as the hormone replacement therapy, it inhibits the osteoclast development. It has some side effects e.g., Endometrial cancer, stroke.

- **Estrogen Derivatives:** Estrogen derivatives are also approved by the USFDA e.g., Estrogen. It prevents bone loss by the inhibition of osteoclast apoptosis. Side effects of this derive are Breast cancer Nausea, diarrhea, cramps.
- **Estrogen Mimic Agents:** There are the medicines which mimic the effects of estrogen e.g., Raloxifene which inhibits osteoclast development, reported side effects are leg cramps, hot flashes.
- **Peptide Hormone:** Calcitonin is a peptide hormone which is produced by the parafollicular cells. Calcitonin is involved in to regulate levels of calcium and phosphate in the blood, opposing the action of parathyroid hormone. Its mechanism of action is through inhibition of osteoclasts. Reported side effects of calcitonin are diarrhea. The reposted side effects of calcitonin are Nausea, skin redness.
- **Peptides:** There are two peptide molecules e.g., Teriparatide and Abaloparatide (TYMLOS) approved by the USFDA. Both acts as to induce bone formation. Common side effects are observed as Headache, hypercalcemia. Teriparatide also causes diarrhea.
- **Bisphosphonates:** Zoledronic acid/ibandronate/ Alendronate/ risedronate are belong to the Bisphosphonates group of drugs. They potentially prevent bone loss by the inhibition of osteoclast. They showed side effects e.g., GI toxicity, weight loss, bone pain, low calcium levels.

Compare to the nutraceuticals, most of these medicines are associated with hazardous side effects thus natural products were prioritized over these medicines, and created a billion-dollar markets all over the world.

1.8 ROLE OF NUTRACEUTICALS IN BONE REMODELING

There are natural products from dietary or plant sources e.g., embelin, diosgenin, reseveratrol, gambogic acid, genistein, plumbagin, quercetin, butein, cardamonin, coronarin D, curcumin, zerumbone.

These nutraceuticals have the intrinsic capacity of modulation of the cell signaling pathways. They are capable to reverse or slow down osteoporosis in postmenopausal woman. The most important advantages of these compounds are that they do not show any side effects as they are obtained from natural sources.

Lin et al., and his research group demonstrated and correlated that the polyphenol class of components in the food enhances bone mineral density and bone metabolism. This study was conducted on the post-menopausal women, the women were divided into three different groups and the groups were respectively administered three different plants extracts e.g., I) anti-resorptive (AR) combination of pomegranate fruit (*Punica granatum* L.) and grape seed (*Vitis vinifera* L.) extracts, II) a bone formation (BF) combination of quercetin (*Dimorphandra mollis* Benth) and III) licorice (*Glycyrrhiza-glabra* L.) extracts. This research group used microarray analysis to evaluate association of these extracts with the bone health and compared with the gene library. The study showed that the anti-resorptive (AR) combination down regulated the genes which are responsible for the induction of bone resorption e.g., tachykinin receptor 1 (TACR1) and cathepsin G (CTSG), anti-resorptive (AR) combination also found to up regulate the genes responsible for the extracellular matrix development including heparan sulfate proteoglycan 2 (HSPG2) and hyaluronoglucosaminidase 1 (HYAL1). On the other hand the BF combination showed up- regulation of bone morphogenetic protein 2 (BMP-2) and COL1A1 (collagen type I α1) genes, these genes are linked to bone and collagen formation while down-regulating genes linked to osteoclastogenesis (Lin et al., 2016). These nutraceuticals are inexpensive; show no side effect while still possessing anti-inflammatory properties (Pandey et al. 2018).

1.8.1 POPULAR NUTRACEUTICALS MAINTAIN BONE HEALTH

Several other natural compounds are recommended as anti-osteoporosis potential e.g., apigenin, betullinic acid, emodin, eugenol, fisetin, gambogic acid, catechingallate, caffeic acid, curcumin and calebin A, coronarin D, luteolin, obovatol, lycopene, oleanolic acid, plumbagin, reseveratrol, rosmarinic acid, silymarin, thiocolchicoside, wedelolactone, withanolides, zerumbone, zyflamend, gossypin, guggulsterone, γ-tocotrienol, isodeoxyelephantopin, harpagoside. Carotenoids are another group of phytonutrients. These are plant pigments, belong to the tetraterpenoids class of natural products. They are basically fat-soluble and considered to be antioxidants. The most common source of carotenoids are vegetables e.g., carrots, sweet potatoes, spinach, kale. Biophosphates are used for the collard greens, papaya, bell peppers, and tomatoes. There are four carotenoids (beta-carotene, alpha-carotene, gamma-carotene). Table 1.1 presented popular and widely used nutraceuticals with their natural sources.

TABLE 1.1 Structures of Nutraceuticals and Their Sources from Plant Origin

Nutraceuticals (Natural Compounds) Recommended for Osteoporosis	Natural Sources
Acetoxychavicol acetate	Thai ginger
Capsaicin	Red chili
Embelin	Vidanga
Indole 3-carbinol	Broccoli
Boswellic acid	Salaiguggul
Celastrol	Thunder God vine
Fisetin	Straberries
Lycopene	Tomatoes
Plumbagin	Plaumbago
Withnolide	Tomatilo
Berberin	Oregone grape
Deguelin	Pea
Guggulsterone	Giggul gum
Resveratrol	Grapes
Quercetin	Oak forest
Zerumbone	Asian Ginger
Betuliniac acid	Holy basil
Butein	Chinese lacquer
Escin	Horse chestnut
Gambogic acid	False mangosteen

1.9 PHYTOESTROGENS CLASS OF NUTRACEUTICALS

Phytoestrogens are nutraceuticals isolated from different types of plants as well from dietary soy-derived products. Phytoestrogens are natural products found in plant, these compounds have pharmacological function similar to estrogen e.g., agonist or antagonistic effects. Phytoestrogens differ in their affinity for estrogen receptor alpha versus estrogen receptor-beta. There are three most common phytoestrogens e.g., daidzein, glycitein and genistein (glycine max), they belong to the isoflavone class of nutraceutical. Isoflavone is known inhibitors of several transcription factors which are associated with bone loss e.g., NF-κB, AP-1, cytokines such as IGF-1 and TGFβ. Isoflavone also found to upregulate Smad5, Runx2 (Dai et al., 2013).

Different group of scientists showed that dietary supplementation of soy product can be beneficial in the prevention of bone loss.

Naturally they can be obtained from different plants species and as well as from the soybeans and soy products, chickpeas and legumes. For the convenience of the patients many of the phytoestrogens are not are also purified and concentrated into high-dose dietary supplements. Asian diets are usually more rich in isoflavone class of compounds (30 to 60 mg/d) compare to the American diets (1–3 mg/d) (Chun et al., 2009, Messina et a., 2006). Thus now a day's American dietary guideline added soy products in place of dairy or proteins.

1.10 FISH HEAD LIPIDS DIETARY SUPPLEMENTS

Along with the nutraceuticals from plant sources, there are many studies showed that fish head lipids also prevent osteoporosis. Zhu and his colleagues recently showed that lipids extracted from the head of *Tilapia nilotica* in the osteoporosis rats. The osteoporosis rat model was prepared by ovariectomization. The composition of this fish head lipid (FHL) was neutral lipids (NL, 77.84%), phospholipids (PL, 11.86%), and glycolipids (GL, 6.47%). This study demonstrated that this fish head can improve bone microstructure, increase bone mineral density, and decrease bone resorption. The proposed antiosteoporotic mechanism is the changed microbial communities, proinflammation factors, serum metabolites, and metabolic pathways. FHL reduced the content of *Alistipes* in the gut and resulted the enhanced quantity of beneficial bacteria such as *Oscillospira, Roseburia,* and *Dubosiella.* These caused declining proinflammation and changing the metabolites and metabolic pathways such as arachidonic acid metabolism and primary bile acid metabolism. FHL improved bone loss through reducing inflammation and changing the metabolites and metabolic pathways such as arachidonic acid metabolism and primary bile acid metabolism, etc., by altering gut microbiota (Zhu et al., 2022).

1.11 QUALITY ASSURANCE OF NUTRACEUTICALS

Quality control of nutraceuticals are very much important, as the pure and extracted compounds can cause serious side effects. Some examples and toxicity is presented here.

Among the most widely recommended, marketed and consumed nutraceuticals includes isoflavone and flavonoid (Keservani and Sharma, 2014; Keservani et al., 2016a, 2016b) compounds from naturally occurring plant sources such as polyphenols group of compounds e.g., resveratrol isolated from red grapes, anthocyanins compounds from berries, flavonols from dark chocolate, catechins from tea. These compounds are still not reported with significant toxicities, but epigallocatechin gallate which is the active catechol isolated from green tea extract reported against liver injury and causing enhance oxidative stress (Mazzanti et al., 2009). The isoflavone compounds in their purified form were able to produce uterine hypertrophy or malformation of reproductive organs, inhibit and androgen hormone production, reduce fertility, reduce testis size, and stimulate estrogen-dependent tumor growth, these evidences indicating health risks following hormone replacement therapy in postmenopausal women. It is a report of endometriosis in women having isoflavone nutraceuticals (Mahady et al., 2003), since it is a clear indication of estrogenicity, thus there is a possibility of increased risk of estrogen causing cancer among the consumers.

As these isoflavone compounds are purified and concentrated, they are able to present in high plasma concentration on the other hand when isoflavones are taken from the soy food sources along with mixtures of proteins and hundreds of compounds and found less toxicity than the purified isoflavones (Ronis et al., 2017).

KEYWORDS

- **anti-resorptive**
- **dehydroepiandrosterone**
- **dehydroepiandrosterone-sulfate**
- **estrogen receptor**
- **teriparatide and abaloparatide**
- **ultraviolet radiation**

REFERENCES

Atbinici, H., Sipahioglu, S., Aksoy, N., Baykara, I., & Isikan, U. E., (2015). Effects of salmon calcitonin treatment on serum and synovial fluid bone formation and resorption markers in osteoporosis patients. *Acta Orthop. Traumatol. Turc., 49*(2), 160–165.

Aubin, J. E., & Liu, F., (1996). The osteoblast lineage. In: Bilezikian, J. P., Raisz, L. G., & Rodan, G. A., (eds.), *Principles of Bone Biology* (pp. 51–67). Academic Press, San Diego.

Bandeira, L., Lewiecki, E. M., & Bilezikian, J. P., (2006). Pharmacodynamics and pharmacokinetics of oral salmon calcitonin in the treatment of osteoporosis. *Expert Opin. Drug Metab. Toxicol., 12*(6), 681–689.

Beekman, K. M., Veldhuis-Vlug, A. G., Den, H. M., Maas, M., Oleksik, A. M., Tanck, M. W., Ott, S. M., et al., (2017). *The Effect of Raloxifene on Bone Marrow Adipose Tissue and Bone Turnover in Post-menopausal Women with Osteoporosis.* http://dx.doi.org/10.1016/j.bone.2017.10.011.

Binkley, N., Bone, H., Gilligan, J. P., & Krause, D. S., (2014). Efficacy and safety of oral recombinant calcitonin tablets in postmenopausal women with low bone mass and increased fracture risk: A randomized, placebo-controlled trial. *Osteoporos. Int., 25*(11), 2649–2656.

Bland, R., (2000). Steroid hormone receptor and action on bone. *ClinSci, 98*, 217–240.

Bonilla, C., Ness, A. R., Wills, A. K., Lawlor, D. A., Lewis, S. J., & Smith, G. D., (2014). Skin pigmentation, sun exposure and vitamin D levels in children of the Avon longitudinal study of parents and children. *BMC Public Health, 14*, 597.

Booth, A., & Camacho, P., (2013). A closer look at calcium absorption and the benefits and risks of dietary versus supplemental calcium. *Postgrad Med., 125*(6), 73–81.

Bragdon, B., Oleksandra, M., Sven, S., Daniel, K., Joanne, J., & Anja, N., (2011). Bone morphogenetic proteins: A critical review. *Cellular Signaling, 23*(4), 609–620.

Brower, V., (1998). Nutraceuticals: Poised for a healthy slice of the healthcare market? *Nat. Biotechnol., 16*, 728–731.

Canalis, E., Economides, A. N., & Gazzerro, E., (2003). Bone morphogenetic proteins, their antagonists, and the skeleton. *Endocr. Rev., 24*, 218–235.

Caplan, A. I., (1991). Mesenchymal stem cells. *J. Orthop. Res., 9*, 641–650.

Cashman, K. D., (2006). Milk minerals (including trace elements) and bone health. *Int Diary J., 16*(11), 1389–1398.

Chun, O. K., Chung, S. J., & Song, W. O., (2009). Urinary isoflavones and their metabolites validate the dietary isoflavone intakes in US adults. *J. of American Diet Assoc., 109*, 245–254.

Compston, J. E., (2001). Sex steroids and bone. *Physiol. Rev., 81*, 419–447.

Cummings, S. R., Ferrari, S., Eastell, R., Gilchrist, N., Beck, J. J. E., McClung, M., Roux, C., et al., (2018). Vertebral fractures following discontinuation of denosumab: A post-hoc analysis of the randomized placebo-controlled freedom trial and its extension. *J. Bone Miner. Res., 33*(2), 190–198.

Das, S., & Crockett, J. C., (2013). Osteoporosis – a current view of pharmacological prevention and treatment. *Drug Des. Devel. Ther., 7*, 435–448.

Ducy, P., & Karsenty, G., (1995). Two distinct osteoblast-specific cis-acting elements control expression of a mouse osteocalcin gene. *Mol. Cell Biol., 15*, 1858–1869.

Ducy, P., Starbuck, M., Priemel, M., Shen, J., Pinero, G., Geoffroy, V., Amling, M., & Karsenty, G., (1999). A Cbfa1-dependent genetic pathway controls bone formation beyond embryonic development. *Genes Dev., 13*, 1025–1036.

Ducy, P., Zhang, R., Geoffroy, V., Ridall, A. L., & Karsenty, G., (1997). Osf2/Cbfa1: A transcriptional activator of osteoblast differentiation. *Cell, 89*, 747–754.

Fernandez-Garcia, D., Munoz-Torres, M., Mezquita-Raya, P., De La Higuera, M., Alonso, G., Reyes-Garcia, R., Ochoa, A. S., et al., (2008). Effects of raloxifene therapy on circulating osteoprotegerin and RANK ligand levels in post-menopausal osteoporosis. *J. Endocrinol. Investig., 31*(5), 416–421.

Gao, J., Gao, C., Li, H., Wang, G. S., Xu, C., & Ran, J., (2017). Effect of zoledronic acid on reducing femoral bone mineral density loss following total hip arthroplasty: A meta-analysis from randomized controlled trails. *Int. J. Surg., 47*, 116–126.

Gennari, L., Merlotti, D., Paola, V. D., Calabrò, A., Becherini, L., Martini, G., & Nuti, R., (2005). Human genome epidemiology (huge) review. *American Journal of Epidemiology, 161*, 307–320.

Goetz, L. G., Mamillapalli, R., Devlin, M. J., Robbins, A. E., Majidi-Zolbin, M., & Taylor, H. S., (2017). Cross-sex testosterone therapy in ovariectomized mice: Addition of low-dose estrogen preserves bone architecture. *Am. J. Physiol. Endocrinol. Metab., 313*(5), E540–E551.

Gomes-Filho, J. E., Wayama, M. T., Dornelles, R. C., Ervolino, E., Yamanari, G. H., Lodi, C. S., Sivieri-Araujo, G., Dezan-Junior, E., & Cintra, L. T., (2015). Raloxifene modulates regulators of osteoclastogenesis and angiogenesis in an oestrogen deficiency periapical lesion model. *Int. Endod. J., 48*(11), 1059–1068.

Greenspan, S. L., Vujevich, K., Britton, C., Herradura, A., Gruen, G., Tarkin, I., Siska, P., Hamlin, B., & Perera, S., (2018). Teriparatide for treatment of patients with bisphosphonate-associated atypical fracture of the femur. *Osteoporos. Int., 29*(2), 501–506.

Grigg, A., Butcher, B., Khodr, B., Bajel, A., Hertzberg, M., Patil, S., D'Souza, A. B., et al., (2017). An individualized risk-adapted protocol of pre- and post-transplant zoledronic acid reduces bone loss after allogeneic stem cell trans- plantation: Results of a phase II prospective trial. *Bone Marrow Transplant, 52*(9), 1288–1293.

Grigoriadis, A. E., Heersche, J. N. M., & Aubin, J. E., (1988). Differentiation of muscle, fat, cartilage, and bone from progenitor cells present in a bone-derived clonal cell population: Effect of dexamethasone. *J. Cell Biol., 106*, 2139–2151.

Gür, A., Colpan, L., Nas, K., Cevik, R., Saraç, J., Erdoğan, F., & Düz, M. Z., (2002). The role of trace minerals in the pathogenesis of postmenopausal osteoporosis and a new effect of calcitonin. *J. Bone Mineral Metabolism, 20*(1), 39–43.

Inada, M., Yasui, T., Nomura, S., Miyake, S., Deguchi, K., Himeno, M., Sato, M., et al., (1999). Maturational disturbance of chondrocytes in Cbfa1-deficient mice. *Developmental Dynamics: An Official Publication of the American Association of Anatomists, 214*, 279–290.

Jang, W. G., Kim, E. J., Kim, D. K., Ryoo, H. M., Lee, K. B., Kim, S. H., Choi, H. K., & Koh, J. T., (2012). BMP2 protein regulates osteocalcin expression via Runx2- mediated Atf6 gene transcription. *J. Bio. Chem., 287*(2), 905–915.

Jang, W. G., Kim, E. J., Lee, K. N., Son, H. J., & Koh, J. T., (2011). AMP- activated protein kinase (AMPK) positively regulates osteoblast differentiation via induction of Dlx5-dependent Runx2 expression in MC3T3E1 cells. *Biochem. Biophys. Res. Commun., 404*, 1004–1009.

Javed, A., Barnes, G. L., Jasanya, B. O., Stein, J. L., Gerstenfeld, L., Lian, J. B., & Stein, G. S., (2001). Runthomology domain transcription factors (Runx, Cbfa, and AML) mediate repression of the bone sialoprotein promoter: Evidence for promoter context-dependent activity of Cbfa proteins. *Mol. Cell Biol., 21*, 2891–2905.

Kalra, E. K., (2003). Nutraceutical-definition and introduction. *AAPS PharmSci., 5*(3), 27, 28.

Kaneko, T., Okamura, K., Yonemoto, Y., Okura, C., Suto, T., Tachibana, M., Tamura, Y., Inoue, M., & Chikuda, H., (2017). Short-term daily teriparatide in patients with rheumatoid arthritis. *Mod. Rheumatol.*, 1–6.

Kelvin, L., Li, K., Xia-Fang, W., Li, D. Y., Chen, Y. C., Zhao, L. J., Liu, X. G., Guo, Y. F., et al., (2018). The good, the bad, and the ugly of calcium supplementation: A review of calcium intake on human health. *Clic. Interven. Aging, 13*, 2443–2452.
Keservani, R. K., & Sharma, A. K., (2014). Flavonoids: Emerging trends and potential health benefits. *Journal of Chinese Pharmaceutical Sciences, 23*(12), 815.
Keservani, R. K., Kesharwani, R. K., Sharma, A. K., Vyas, N., & Chadoker, A., (2010b). Nutritional supplements: An overview. *International Journal of Current Pharmaceutical Review and Research, 1*(1), 59–75.
Keservani, R. K., Kesharwani, R. K., Vyas, N., Jain, S., Raghuvanshi, R., & Sharma, A. K., (2010a). Nutraceutical and functional food as future food: A review. *Der Pharmacia Lettre, 2*(1), 106–116.
Keservani, R. K., Sharma, A. K., & Kesharwani, R. K., (2016a). Nutraceutical and functional foods for cardiovascular health. In: *Food Process Engineering* (pp. 257–278, 291–312). Apple Academic Press, CRC Press. ISBN: 9781771884020.
Keservani, R. K., Sharma, A. K., & Kesharwani, R. K., (2016b). Medicinal effect of nutraceutical fruits for the cognition and brain health. *Scientifica*, 10. Article ID 3109254. doi: 10.1155/2016/3109254.
Keservani, R. K., Sharma, A. K., & Kesharwani, R. K., (2017). An overview and therapeutic applications of nutraceutical and functional foods. *Recent Advances in Drug Delivery Technology*, 160–201. ISBN:9781522507543.
Keservani, R. K., Sharma, A. K., & Kesharwani, R. K., (2020). *Nutraceuticals and Dietary Supplements: Applications in Health Improvement and Disease Management*. CRC Press. ISBN: 9781771888738.
Khalid, A. B., & Krum, S. A., (2016). Estrogen receptors alpha and beta in bone. *Bone, 87*, 130–135.
Kim, E. J., Cho, S. W., Shin, J. O., Lee, M. J., Kim, K. S., & Jung, H. S., (2013). Ihh and Runx2/Runx3 signaling interact to coordinate early chondrogenesis: A mouse model. *PloS One, 8*, e55296.
Kim, I. S., Otto, F., Zabel, B., & Mundlos, S., (1999). Regulation of chondrocyte differentiation by Cbfa1. *Mechanisms of Development, 80*, 159–170.
King, J. A., Marker, P. C., Seung, K. J., & Kingsley, D. M., (1994). BMP5 and the molecular, skeletal, and soft-tissue alterations in short ear mice. *Dev Biol., 166*, 112–122.
Komori, T., (2006). Regulation of osteoblast differentiation by transcription factors. *J. Cell. Biochem., 99*, 1233–1239.
Komori, T., Yagi, H., Nomura, S., Yamaguchi, A., Sasaki, K., Deguchi, K., Shimizu, Y., et al., (1997). Targeted disruption of Cbfa1 results in a complete lack of bone formation owing to maturational arrest of osteoblasts. *The Cell., 89*, 755–764.
Laird, E., Ward, M., McSorley, E., Strain, J. J., & Wallace, J., (2010). Vitamin D and bone health; potential mechanisms. *Nutrients, 2*, 693–724.
Langdahl, B. L., Libanati, C., Crittenden, D. B., Bolognese, M. A., Brown, J. P., Daizadeh, N. S., Dokoupilova, E., et al., (2017). Romosozumab (sclerostin monoclonal antibody) versus teriparatide in postmenopausal women with osteoporosis transitioning from oral bisphosphonate therapy: A randomized, open- label, phase 3 trial. *Lancet, 390*(10102), 1585–1594.
Lange, U., Classen, K., Muller-Ladner, U., & Richter, M., (2017). Weekly oral bisphosphonates over 2 years prevent bone loss in cardiac transplant patients. *Clin Transplant. 31*(11), e13122.

Lappe, J., Kunz, I., Bendik, I., Prudence, K., Weber, P., Recker, R., & Heaney, R. P., (2013). Effect of a combination of genistein, polyunsaturated fatty acids and vitamins D3 and K1 on bone mineral density in postmenopausal women: A randomized, placebo-controlled, double-blind pilot study. *Cochrane Central Register of Controlled Trials, 52*(1), 203–215.

Lee, M. H., Kim, Y. J., Kim, H. J., Park, H. D., Kang, A. R., Kyung, H. M., Sung, J. H., et al., (2003). BMP-2- induced Runx2 expression is mediated by Dlx5, and TGF- 1 opposes the BMP-2-induced osteoblast differentiation by suppression of Dlx5 expression. *J. Biol. Chem., 278*, 34387–34394.

Lengner, C. J., Drissi, H., Choi, J. Y., Van, W. A. J., Stein, J. L., Stein, G. S., & Lian, J. B., (2002). Activation of the bone-related Runx2/Cbfa1 promoter in mesenchymal condensations and developing chondrocytes of the axial skeleton. *Mechanisms of Development., 114*, 167–170.

Lin, Y., Kazlovaa, V., Ramakrishnan, S., Murray, M. A., Fast, D., Chandra, A., & Gellenbeck, K. W., (2016). Bone health nutraceuticals alter microarray mRNA gene expression: A randomized, parallel, open-label clinical study. *Phytomedicine, 23*, 18–26.

Lindsay, R., Cosman, F., Lobo, R. A., Walsh, B. W., Harris, S. T., Reagan, J. E., Liss, C. L., Melton, M. E., & Byrnes, C. A., (1999). Addition of alendronate to ongoing hormone re- placement therapy in the treatment of osteoporosis: A randomized, controlled clinical trial. *J. Clin. Endocrinol. Metab., 84*(9), 3076–3081.

Mahady, G., Parrot, J., Lee, C., Yun, G., & Dan, A., (2003). Botanical dietary supplement use in peri- and postmenopausal women. *Menopause, 10*, 65–72.

Mandema, J. W., Zheng, J., Libanati, C., & Perez, R. J. J., (2014). Time course of bone mineral density changes with denosumab compared with other drugs in post- menopausal osteoporosis: A dose-response-based meta-analysis. *J. Clin. Endocrinol. Metab., 99*(10), 3746–3755.

Martin, J. J. R., Kim, B. P., & James, W., (2018). Adverse effects of nutraceuticals and dietary supplements *Annu. Rev. Pharmacol. Toxicol. 58*, 583–601.

Mazzanti, G., Menniti-Ippolito, F., Moro, P. A., Cassetti, F., & Raschetti, R., (2009). Hepatotoxicity from green tea: A review of the literature and two unpublished cases. *Eur. J. Clin. Pharmacol., 65*, 331–341.

Meier, C., Uebelhart, B., Aubry-Rozier, B., Birkhauser, M., Bischoff-Ferrari, H. A., Frey, D., Kressig, R. W., et al., (2017). Osteoporosis drug treatment: Duration and management after discontinuation. A position statement from the SVGO/ASCO. *Swiss Med. Wkly., 147*, w14484.

Messina, M., Nagata, C., & Wu, A. H., (2006). Estimated Asian adult soy protein and isoflavone intakes. *Nutr. Cancer, 55*, 1–12.

Monda, V., Lupoli, G. A., Messina, G., Peluso, R., Panico, A., Villano, I., Salerno, M., et al., (2017). Improvement of bone physiology and life quality due to association of risedronate and anastrozole. *Front. Pharmacol., 8*, 632.

Murakami, T., Saito, A., Hino, S., Kondo, S., Kanemoto, S., Chihara, K., Sekiya, H., et al., (2009). Signaling mediated by the endoplasmic reticulum stress transducer OASIS is involved in bone formation. *Nat. Cell Biol., 11*, 1205–1211.

Nakamura, Y., Suzuki, T., & Kato, H., (2017). Denosumab significantly improves bone mineral density with or without bisphosphonate pre-treatment in osteoporosis with rheumatoid arthritis: Denosumab improves bone mineral density in osteoporosis with rheumatoid arthritis. *Arch. Osteoporos, 12*(1), 80.

Nakatsukasa, K., Koyama, H., Ouchi, Y., Sakaguchi, K., Fujita, Y., Matsuda, T., Kato, M., Konishi, E., & Taguchi, T., (2018). Effect of denosumab administration on low bone mineral density (T-score -1.0 to -2.5) in postmenopausal Japanese women receiving adjuvant aromatase inhibitors for non-metastatic breast cancer. *J. Clinical Onco., 36*(15), e12536.

National Academy of Sciences IoM, (2010). Dietary reference intakes for calcium and vitamin D. IoM, over 2 years prevent bone loss in cardiac transplant patients. *Clin. Transpl., 31*(11).

Nieves, J. W., (2013). Skeletal effect of nutrients and nutraceuticals beyond calcium and vitamin D. *Osteopores Int., 24*, 771–786.

Otto, F., Thornell, A. P., Crompton, T., Denzel, A., Gilmour, K. C., Rosewell, I. R., Stamp, G. W., et al., (1997). Cbfa1, a candidate gene for cleidocranial dysplasia syndrome, is essential for osteoblast differentiation and bone development. *Cell, 89*, 765–771.

Owen, M., Pittenger, M. F., Mackay, A. M., Beck, S. C., Jaiswal, R. K., Douglas, R., Mosca, J. D., et al., (1999). Multilineage potential of adult human mesenchymal stem, cells. *Science, 284*(5411), 143–147.

Pandey, M. K., Gupta, S. C., Karelia, D., Gilhooley, P. J., Shakibaei, M., & Aggarwal, B. B., (2018). Dietary nutraceuticals as backbone for bone health. *Biotech Adv., 36*(6), 1633–1648.

Reginster, J. Y., Al-Daghri, N. M., & Bruyere, O., (2017). Abaloparatide comparator trial in vertebral endpoints (ACTIVE) confirms that abaloparatide is a valuable addition to the armamentarium against osteoporosis. *Expert. Opin. Pharmacother., 18*(17), 1811–1813.

Reid, I. R., Birstow, S. M., & Bolland, M. J., (2017). Calcium and cardiovascular disease. *Endocrinol. Metab., 32*(3), 339–349.

Richard, N. W., Jordan, G., Zhongliang, W., Youlin, D., Min, Q., Michael, P., Qian, Z., et al., (2014). Bone morphogenetic protein (BMP) signaling in development and human diseases. *Genes & Diseases, 1*, 87–105.

Rosen, V., Cox, K., & Hattersley, G., (1996). Bone morphogenetic protein. In: Bilezikian, J. P., Raisz, L. G., & Rodan, G. A., (eds.), *Principles of Bone Biology* (pp. 661–671). Academic Press, San Diego, CA.

Saltman, P. D., & Strause, L. G., (1993). The role of trace elements in osteoporosis, *JA College of Nutri., 12*(4), 384–389.

Sapir-Koren, R., & Livshits, G., (2017). Postmenopausal osteoporosis in rheumatoid arthritis: The estrogen deficiency-immune mechanisms link. *Bone, 103*, 102–115.

Sharma, A., & Pradeep, A. R., (2012). Clinical efficacy of 1% alendronate gel as a local drug delivery system in the treatment of chronic periodontitis: A randomized, controlled clinical trial. *J. Periodontol., 83*(1), 11–18.

Shimizu, T., Tanaka, T., Kobayashi, T., Kudo, I., Nakatsugawa, M., Takakura, A., Takao-Kawabata, R., & Ishizuya, T., (2017). Sequential treatment with zoledronic acid followed by teriparatide or vice versa increases bone mineral density and bone strength in ovariectomized rats. *Bone Reports, 7*, 70–82.

Shin, K., Park, S. H., Park, W., Baek, H. J., Lee, Y. J., Kang, S. W., Choe, J. Y., et al., (2017). Monthly oral ibandronate reduces bone loss in Korean women with rheumatoid arthritis and osteopenia receiving long-term glucocorticoids: A 48-week double-blinded randomized placebo-controlled investigator-initiated trial. *Clin. Ther., 39*(2), 268–278 (e262).

Shirley, M., (2017). Abaloparatide: First global approval. *Drugs, 77*(12), 1363–1368.

Son, H. E., Kim, E. J., & Jang, W. G., (2018). Curcumin induces osteoblast differentiation through mild-endoplasmic reticulum stress-mediated such as BMP2 on osteoblast cells. *Life Sceinces, 193*, 34–39.

Soung Do, Y., Talebian, L., Matheny, C. J., Guzzo, R., Speck, M. E., Lieberman, J. R., Speck, N. A., & Drissi, H., (2012). Runx1 dose-dependently regulates endochondral ossification during skeletal development and fracture healing. *J. Bone and Mineral Res., 27*, 1585–1597.

Southmayd, E. A., & De Souza, M. J., (2017). A summary of the influence of exogenous estrogen administration across the lifespan on the GH/IGF-1 axis and implications for bone health. In: *Growth Hormone & IGF Research* (pp. 2–13).

Streicher, C., Heyny, A., Andrukhova, O., Haigl, B., Slavic, S., Schuler, C., Kollmann, K., et al., (2017). Estrogen regulates bone turnover by targeting RANKL expression in bone lining cells. *Sci. Rep., 7*(1), 6460.

Suzuki, T., Sukezaki, F., Shibuki, T., Toyoshima, Y., Nagai, T., & Inagaki, K., (2018). Teriparatide administration increases periprosthetic bone mineral density after total knee arthroplasty: A prospective study. *J. Arthroplasty, 33*(1), 79–85.

Tandon, M., Chen, Z., & Pratap, J., (2014). Runx2 activates PI3K/Akt signaling via mTORC2 regulation in invasive breast cancer cells. *Breast Cancer Research: BCR, 16*, R16.

Tankeu, A. T., Ndip-Agbor, V., & Noubiap, J. J., (2017). Calcium supplementation and cardiovascular risk: A rising concern. *J. Clin. Hypertens., 19*(6), 640–646.

Taylor, S. M., & Jones, P. A., (1979). Multiple new phenotypes induced in 10T1/2 and 3T3 cells treated with 5-azacytidine. *Cell, 17*, 771–779.

Tella, S. H., Kommalapati, A., & Correa, R., (2017). Profile of abaloparatide and its potential in the treatment of postmenopausal osteoporosis. *Cureus, 9*(5), e1300.

Von-Metzler, I., Krebbel, H., Kuckelkorn, U., Heider, U., Jakob, C., Kaiser, M., Fleissner, C., et al., (2009). Curcumin diminishes human osteoclastogenesis by inhibition of the signalosome-associated I kappaB kinase. *J. Cancer Res. Clin. Oncol., 135*(2), 173–179.

Watanabe, R., Fujita, N., Takeda, S., Sato, Y., Kobayashi, T., Morita, M., Oike, T., et al., (2016). Ibandronate concomitantly blocks immobilization-induced bone and muscle atrophy. *Biochem. Biophys. Res. Commun., 480*(4), 662–668.

Welch, R. D., Jones, A. L., Bucholz, R. W., Reinert, C. M., Tjia, J. S., Pierce, W. A., Wozney, J. M., & Li, X. J., (1998). Effect of recombinant human bone morphogenetic protein-2 on fracture healing in a goat tibial fracture model. *J. Bone Miner. Res., 13*, 1483–1490.

Wu, G., Xu, R., Zhang, P., Xiao, T., Fu, Y., Zhang, Y., Du, Y., et al., (2018). Estrogen regulates stemness and senescence of bone marrow stromal cells to prevent osteoporosis via ERbeta-SATB2 pathway. *J. Cell. Physiol., 233*(5), 4194–4204.

Xiao, Q., Murphy, R. A., Houston, D. K., Harris, T. B., Chow, W. H., & Park, Y., (2013). Dietary and supplemental calcium intake and cardiovascular disease mortality: The national institutes of health-AARP diet and health study. *JAMA Intern. Med., 173*(8), 639–646.

Yamaguchi, A., & Kahn, A. J., (1991). Clonal osteogenic cell lines express myogenic and adipocytic developmental potential. *Calcif. Tissue Int., 49*, 221–225.

Yamaguchi, A., Komori, T., & Suda, T., (2000). Regulation of osteoblast differentiation mediated by bone morphogenetic proteins, hedgehogs, and Cbfa1. *Endocr. Rev., 21*(4), 393–411.

PART II

Beneficial Effects of Nutraceuticals in Bone Disorders

CHAPTER 2

BENEFICIAL ASPECTS OF NUTRACEUTICALS IN THE MANAGEMENT OF OSTEOPOROSIS

SHARAD VISHT[1] and GARIMA SINGH[2]

[1]Assistant Professor, Faculty of Pharmacy, Tishk International University, Erbil, Kurdistan Region, IRAQ.

[2]Assistant Professor, Faculty of Nutrition and Health, School of Medical and Allied Sciences, G.D. Goenka University, Gurugram, Haryana, India

ABSTRACT

Osteoporosis is a metabolic bone disorder characterized by low bone mineral density as well as low bone protein, which affects the quality life of the senile population. Additionally, change in the composition of bone microarchitecture as well as the change in bone protein led to porousness of bones and eventually higher fracture risk. This is very much common in postmenopausal women because of their decreased reproductive hormone. Some medications are being used to reduce the symptoms of osteoporosis; although, these medications are having long term side effects like carcinogenicity etc. Components from fruits and vegetables are a source of protective nutrients for the maintenance of bone health. The nutraceuticals recognize from dietary agents such as anthocyanin, kaempferol, curcumin, diosgenin, embelin, gambogic acid, genistein, quercetin, reseveratrol, zerumbone omega 3, prebiotics, probiotics and PUFA, can found to be anti-osteoporotic. These nutraceuticals are having anti-inflammatory properties with no side effects. Category of Nutraceuticals are very wide

Nutraceuticals and Bone Health. Deepak Sharma, Madan Mohan Gupta, Anil K. Sharma, Raj K. Keservani, Rajesh K. Kesharwani (Eds.)

in terms of bone health. Calcium is not an only mineral required for bone health, some other minerals like phosphorous, magnesium, copper and zinc also play a vital role in bone formation and repair. Same in the case with vitamins for example vitamin C and vitamin K2, they also support the bone health. It has been documented that three p's are quite helpful for treating and reducing the symptoms of osteoporosis i.e., poly unsaturated fatty acids probiotics and prebiotics. Nutraceuticals good for bone health are also called anti-osteoporotic compounds.

2.1 INTRODUCTION

Over 200 million individuals worldwide suffer from osteoporosis and poor bone mass. It is caused by a reduction of bone mass, which results in weak and porous bones, as well as an increased risk of fractures of weight-bearing bones. If it continues for an extended period of time, it may result in height loss, stooped posture, humpback, and other bone abnormalities, as well as extreme discomfort. Osteoporosis is a bone disease that is more prevalent in the elderly and postmenopausal women with low estrogen levels. It is caused by a disruption in the normal process of bone development, which is a continual cycle of bone creation and degradation. Though pharmacological therapy for osteoporosis is usually associated with long-term adverse effects, nutraceuticals are an excellent alternative for reducing osteoporotic symptoms. This chapter discusses how including nutraceuticals such as minerals, herbs, phytochemicals, and dairy products into the diet might benefit osteoporotic patients by promoting bone health. Nutrition has a critical role in the prevention and treatment of osteoporosis. Indeed, a balanced diet rich in calcium (1,000 mg/day), vitamin D (400–600 IU), and the recommended levels of protein, magnesium, and vitamin K may significantly benefit bone health. Although some kinds of nutraceuticals such as anthocyanin, kaempferol, curcumin, gambogic acid, genistein, quercetin, reseveratrol, EPA, ALA, DHA, prebiotics, and probiotics all benefit bone health (Keservani et al., 2010a, 2010b, 2017, 2020).

2.2 PHYSIOLOGY OF BONE FORMATION AND REMODELING

Bones are not stationary constructions inside the human body; they continually change over the course of time. This skeletal change is alluded to as bone reconstruction, and it effectively protects the underlying parts of the skeletal framework while additionally contributing metabolically deep down's calcium and phosphorus statement. Reconstruction speeds up the method involved with demineralizing old or harmed bone and mineralizing new bone. Osteoclasts and osteoblasts are the essential cells associated with the demineralization and mineralization phases of bone redesigning; however, osteocytes additionally add to this interaction. Hormonal signs (calcitonin and parathyroid chemical) advance the activity of these cells, particularly osteoclasts, and the equivalent is valid for osteoblasts. This transaction of bone shaping cells and chemicals advances the course of bone redesigning.

2.2.1 THE CELLS OF BONE FORMATION: THE MAJOR PLAYERS

The course of bone arrangement starts during the beginning phases of fetal turn of events and is reliant upon the collaboration of two unmistakable cell genealogies. Mineralization is completed by osteoblastic cells, which begin in the mesenchymal tissue. Mesenchymal foundational microorganisms might form into a scope of unmistakable cell types, including osteoblasts, chondrocytes, myocytes, and adipocytes. Osteoclastic cells are answerable for bone demineralization and bone misfortune. They are plummeted from the hematopoietic genealogy, which includes an assortment of platelet types found in the bone marrow (Cosman, 2018).

2.2.1.1 OSTEOCLASTS

Osteoclasts are the cells dependable of demineralization process. Demineralization is additionally a significant interaction for bone redesigning as it starts bone breakdown responses subsequently bone cycle would go on in an appropriate channel. When the combination of osteoblastic cells and osteoclastic forerunner cells has begun, the subsequent multinucleated osteoclast connects deep down surface and initiate demineralization. This bone framework is made out of an inorganic piece of calcium phosphate precious stones and a natural part made out of collagen, proteoglycans, and glycoproteins. The demineralization interaction implies inorganic mixtures are returning to blood by means of bones. Then, at that point, it is accepted that one more stage is directed by mononuclear cells of macrophage genealogy proceed to corrupt and store natural material by means of blood to start the mineralization stage.

2.2.1.2 OSTEOBLASTS

These cells are significant for the production of the bone framework by keeping new collagen and minerals. Osteoblast initiates the mineralization interaction and in the long run changes into an osteocyte or kicks the bucket ordinarily (apoptosis).

2.2.1.3 OSTEOCYTES

Osteocytes are the most bountiful type of bone cell. They add to bone improvement by conveying signs to neighboring osteocytes in light of bone pressure. Liquid stream inside the bone is additionally constrained by osteocytes in light of mechanical pressure, thus cell flagging might happen because of changes in liquid stream. These phones take an interest in an interaction known as mechanotransduction, which changes mechanical powers over to biochemical signs. Osteocytes convey this sign and educate adjoining cells on the most proficient method to adapt to and adjust to mechanical pressure (Anagnostis et al., 2018) (Figure 2.1).

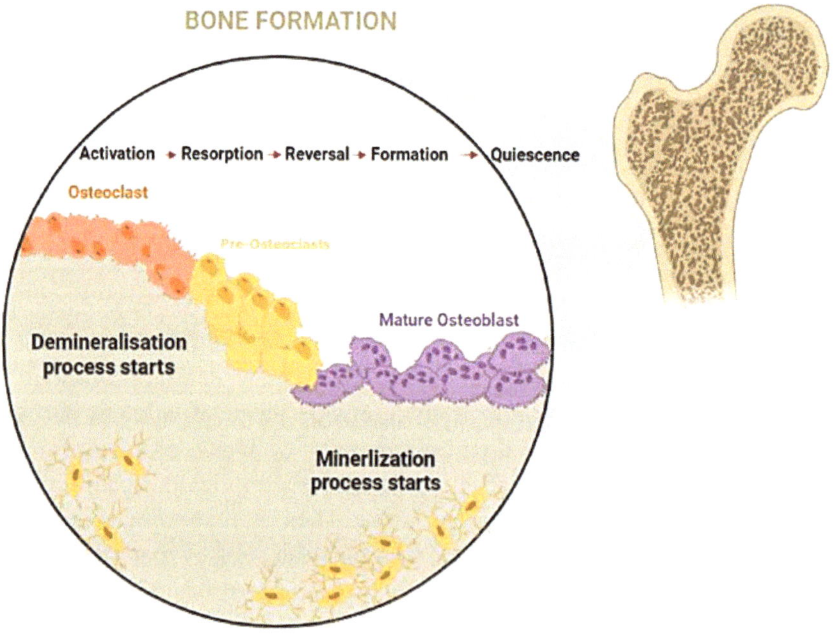

FIGURE 2.1 Process of bone formation.

2.2.2 HORMONAL IMPACT ON BONE FORMATION (MECHANISM)

Chemicals are the body's substance couriers. Our body is a natural instrument that imparts in synthetic terms. Chemicals are answerable for keeping up with the body's balance by kneading every cell and invigorating substance processes (digestion) in light of the body's requirements. With regards to bone homeostasis, a solid hormonal pathway is expected to support bone wellbeing.

2.2.2.1 PARATHYROID HORMONE (PTH)

PTH is a polypeptide chemical delivered by the parathyroid organ's fundamental cells. PTH works in various ways, including supporting calcium assimilation through expanding vitamin D ingestion, advancing calcium reabsorption by renal tubules, and demineralizing bones. It is discharged in light of a decrease in blood calcium levels. The chemical has a physiological negative criticism instrument that is impacted by the blood calcium level. At the point when the plasma calcium fixation diminishes, the parathyroid organ distinguishes this and deliveries PTH chemical, which follows up on osteoclast cells through the blood. Osteoclasts keep up with homeostasis by delivering calcium into the blood. On the other hand, raised blood calcium levels confine PTH discharge. This in the long run outcomes in expanded blood calcium, which restrains the development of PTH by the parathyroid organ's fundamental cells. This is only one piece of the calcium homeostasis jigsaw, since PTH additionally follows up on the kidneys and digestive organs to control calcium and phosphate levels (Bakr et al, 2019).

2.2.2.2 OESTROGEN

An absence of estrogen brings about expanded bone demineralization, an interaction in which bone breakdown surpasses bone creation, bringing about a deficiency of bone mass. It is discharged by the ovaries. As indicated by creature research, estrogen might change neighborhood factors that administer osteoblast and osteoclast forerunners. Estrogen can possibly restrain the turn of events and capacity of interleukin-6, subsequently forestalling bone deterioration. Furthermore, it is imagined that osteoclast movement ascends without even a trace of estrogen, bringing about a more noteworthy level of bone turnover (Gallagher et al., 2018).

2.2.2.3 CALCITONIN

Calcitonin is a polypeptide chemical discharged by thyroid para-follicular cells, frequently known as C cells, in response to raised calcium levels. Calcitonin invigorates the calcification interaction and represses bone demineralization, as its name suggests. Calcitonin is remembered to serve a basic capacity in calcium homeostasis and skeletal turn of events. Be that as it may, calcitonin is utilized restoratively to treat osteoporosis (Kankre et al., 2018).

2.2.2.4 GROWTH HORMONE

Development hormone (GH), a peptide chemical delivered by the pituitary organ, invigorates bone creation and demineralization through insulin-like development factors. It is frequently alluded to as the expert chemical because of its impact on most of cell digestion. GH supports osteoblast multiplication and movement straightforwardly and in a roundabout way through IGF, however it additionally invigorates osteoclastic bone demineralization action; yet, the aggregate effect of this double action favors bone redesigning.

2.2.2.5 THYROID HORMONE

Thyroid-invigorating chemical (TSH), thyroxine (T4), and triiodothyronine (T3) instigate chondrocyte multiplication and osteoblastic movement in the epiphyseal plate of long bones. The pace of bone turnover is diminished or expanded in hypothyroidism and hyperthyroidism, individually. The greatness of bone turnover is an aftereffect of T3/impact T4's on the amount and movement of osteoblasts and osteoclasts. For example, thyrotoxicosis' raised metabolic state brings about improved osteoblast capacity and osteoclast amount and movement, bringing about expanded demineralization (Wang et al., 2018) (Figure 2.2).

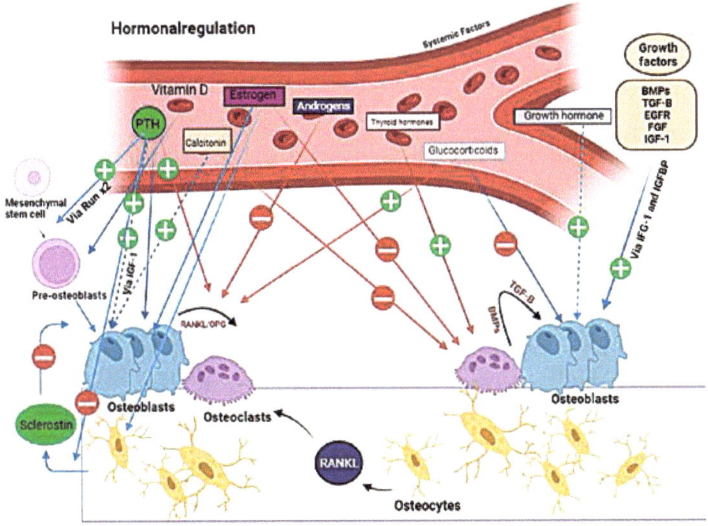

FIGURE 2.2 Role of hormones in bone health.

2.3 RELATED FACTORS RESPONSIBLE FOR OSTEOPOROSIS

2.3.1 AGE FACTOR

The possibility creating osteoporosis ascends with age attributable to the disintegration of bones. After the age of 30, the speed at which your bone tissue corrupts and is consumed by the body progressively ascends, while the rate at which new bone is framed step by step declines. In ladies, bone misfortune is all the quicker and frequently happens after month-to-month feminine cycles stop, when the chemical estrogen creation stops (for the most part between the ages of 45 and 55). Men's bone debilitating ordinarily starts progressively, around the age of 45 to 50, when testosterone creation declines. Ladies' bones are frequently more modest and lighter than men. As an outcome, ladies foster osteoporosis at a far higher rate than guys. Osteoporosis ordinarily clearly affects people until they arrive at the age of 60 or more established, since bone proteins are additionally decreased in osteoporosis.

2.3.2 GENETIC FACTOR

If a person's mother, father, or brother has been diagnosed with osteoporosis or has fractured bones as a result of a minor incident, that person is at a higher risk of developing osteoporosis. Although several variables such as improved bone-friendly food intake and physical activity negate this.

2.3.3 LOW CALCITRIOL LEVELS

Lack of vitamin D contributes fundamentally to the debilitation of the mineralization period of bone creation, bringing about an expanded amount of nonmineralized osteoid being supplanted by bone. Lack of vitamin D has been displayed to detrimentally affect calcium digestion, osteoblastic movement, framework solidification, bone redesigning, and bone thickness (Hejazi et al., 2020). Lack of vitamin D is associated with auxiliary hyperparathyroidism and sped up bone turnover. Vitamin D deficiency has been connected to an expanded gamble of osteoporosis (lips et al., 2006).

2.3.4 HORMONAL FACTOR

Vitamin D deficiency contributes significantly to the impairment of the mineralization phase of bone production, resulting in an increased quantity of nonmineralized osteoid being replaced by bone. Vitamin D deficiency has been shown to have a detrimental effect on calcium metabolism, osteoblastic activity, matrix ossification, bone remodeling, and bone density (Hejazi et al., 2020). Vitamin D deficiency is connected with secondary hyperparathyroidism and accelerated bone turnover. Vitamin D insufficiency has been linked to an increased risk of osteoporosis (lips et al., 2006).

2.3.5 HYPERPROTEINOSIS

Unnecessarily high protein in the eating regimen with no actual work prompts filtering of calcium from the unresolved issues the homeostasis of the body. Protein rich eating routine is acidic in nature as it can make the blood acidic. To keep up with the PH of blood, calcium emerges from issues that remains to be worked out and additional calcium will discharge out from the kidneys subsequently bones will become permeable. Kidney is the fundamental organ dependable to discharge out the additional calcium from the body. It has been seen in the cases who have taken outrageous non-veggie lover diet with no work out having a low mineral thickness. With high protein diet it is compulsory to keep up with the actual work status to stay away from early osteoporosis.

2.3.6 OTHER FACTORS

Smokers lose bone thickness more quickly than non-smokers. Utilization of a lot of liquor might debilitate bone development and increment the gamble of falling. For guys, over the top liquor utilization is characterized as multiple beverages each day; for ladies, it is characterized as more than one beverage each day. View pictures of normal cocktails. By applying the muscles and bones against gravity, weight-bearing exercises, for example, strolling, running, step climbing, moving, or lifting loads keep up with bones solid and sound. Exercise might assist you with keeping up with your equilibrium and diminish your possibility falling. People who are slender or have a minuscule casing are more defenseless to foster osteoporosis. Be that as it may, being overweight seriously jeopardizes individuals for fostering extra significant clinical issues, like sort 2 diabetes, hypertension, and coronary supply route infection (CAD). Aside from that, certain ailments, like hyperthyroidism or hyperparathyroidism, taking specific prescriptions like corticosteroids for a drawn-out timeframe, having specific medical procedures, for example, having your ovaries eliminated before menopause, over the top consuming less calories or having a dietary problem, for example, anorexia nervosa, as well similar to a female competitor with inconsistent feminine cycles because of low muscle to fat ratio, all add to bone diminishing.

2.4 NUTRACEUTICALS

2.4.1 MINERALS

2.4.1.1 MAJOR MINERALS

2.4.1.1.1 Calcium

Osteoporosis is more inclined to shape in the individuals who consume inadequate calcium. A lack of persistent of calcium adds to the advancement of osteoporosis. Lack of calcium is related with diminished bone thickness, early bone misfortune, and an expanded gamble of breaks. Calcium has for quite some time been perceived as basic and fundamental for the wellbeing and conservation of bones. Calcium is a basic part of bone, with the skeleton containing close to 100% of the body's calcium holds. Furthermore, the calcium in our bones fills in as a store for calcium in the blood, which is expected for solid neurons and muscles.

The amount of calcium we really want fluctuates all through our life-times. In our young adult years, we really want additional calcium because of the quick development of our bones. The body's ability to assimilate calcium diminishes with age, which is one reason seniors additionally need expanded portions. A portion of around 1000mg each day is prompted for a solid grown-up. To forestall calcium deficiency by food alone, it is important to comprehend the significant calcium sources. Ragi (364 mg/100g), poppy seeds (1372 mg/100g), milk (105 mg/100ml), and dark gingelly seeds (1664 mg/100g) are for the most part magnificent calcium sources (Gopalan et al, 2020).

2.4.1.1.2 Phosphorous

Phosphorus is a mineral that records for 1% of a singular's complete body weight. It is the body's second generally bountiful mineral. It is viewed as in the body's cells as a whole. Most of the body's phosphorus is situated during the bones and teeth. Phosphorus is a basic supplement for human wellbeing since it plays out a scope of physiological capacities. These incorporate underlying capacities, since phosphorous is a critical part of cell layers as phospholipids and furthermore adds to the film's trustworthiness and penetrability. Furthermore, phosphorus is associated with metabolic exercises including energy, corrosive/base equilibrium, and intracellular cell flagging. 85% of the phosphorus in the body is situated in bones, though 15% is available in delicate tissues and generally 1% in extracellular liquid. Adenosine triphosphate is the synthetic equation for ATP (one protein and three phosphorus). Like lack of calcium, phosphorus inadequacy brings about rickets and hindered development in youngsters, as well as osteomalacia (grown-up rickets) in grown-ups. Notwithstanding, dietary phosphorus deficiency is very unprecedented in people, attributable to the way that it is found in an expansive scope of food varieties. Phosphorus deficiency happens just within the sight of starvation, malabsorption condition, or ineffectively oversaw parenteral nourishment. Along these lines, most of occurrences of phosphorus deficiency are brought about by an issue with kidney work, explicitly diminished phosphorus cylindrical reabsorption. There are a few renal infections that are described by low plasma phosphate and a low cylindrical maximal phosphorus reabsorption rate. These remember hereditary irregularities that outcome for strangely high measures of phosphorus-related chemicals like parathyroid chemical (PTH) or in renal cylindrical phosphate reabsorption inadequacies. Oral phosphate supplements are fundamental for these patients' treatment. With the exception of these specific patient classifications, most people can help a satisfactory measure of phosphorus through a scope of suppers. Phosphorus and calcium collaborate in the gastrointestinal framework and restrain each other's assimilation, which is the reason it is suggested that the dietary calcium and phosphorus proportions be 1:1; any other way, a more noteworthy amount of one of them would repress the ingestion of the other. Calcitonin and parathyroid chemical, similar to calcium, keep up with phosphorus homeostasis. On the off chance that phosphorus admission is unreasonable, parathyroid chemical invigorates the renal tubules to decrease reabsorption, keeping phosphorus from expanding in the blood and keeping up with the proportion. Albeit most of people fall into this proportion because of their low calcium and high phosphorus consumes less calories.

Expanded phosphorus utilization, particularly when joined with a lack of calcium, corrupts bone wellbeing through debilitating calcium assimilation (Krishnarao & associates, 1972). It has been accounted for that eliminating the parathyroid organ forestalled an expansion in bone breaking down in light of a high-phosphorus diet. As of late, the bone framework protein Osteoid was found as another part interceding the sped-up bone breakdown reaction to a high phosphorus consumption, as well as the way that unreasonable dietary phosphorus detrimentally affects bone network all through improvement (Draper et al., 1983).

Inorganic phosphate augmentations in handled food varieties are entirely bioavailable and have been exhibited to supply a critical amount of phosphorus to the eating routine. Cola drinks are a critical wellspring of added inorganic phosphate, which has been connected to bone wellbeing consistently. Cola drinks incorporate a high centralization of phosphoric corrosive, which is immediately assimilated and is frequently intoxicated between suppers. Furthermore, cola beverages might supersede milk in the eating regimen, bringing about a diminishing in calcium utilization and an expansion in inorganic phosphorus consumption.

2.4.1.1.3 Magnesium

Magnesium is put away in the bone framework at a centralization of around 60%. 33% of magnesium is viewed as on cortical bone, either on the outer layer of hydroxyapatite or inside the precious stone's hydration shell (Alfrey et al., 2012). Magnesium, similar to calcium and phosphorus, is a basic mineral in the mineralization interaction. It goes about as a repository for interchangeable magnesium, which is fundamental for the support of sound extracellular centralizations of absolute cations (emphatically charged particles) (Jehnen et al., 2012). How much magnesium on the outer layer of the bone is corresponding to the amount of magnesium in the blood, as occurs in persistent renal problems (Cunningham et al., 2012). Most of bone magnesium is in all probability stored as a fundamental part of the apatite precious stone and filters out during bone demineralization. Aside from its underlying capacity in precious stones, magnesium is expected for the development and endurance of every single living cell, including osteoblasts, osteoclasts, and osteocytes. Magnesium is expected for an assortment of physiological cycles. To start, magnesium is fundamental for ATP, the essential wellspring of energy in cells. Furthermore, magnesium is a cofactor for many chemicals associated with the production of lipids, proteins, and nucleic acids. Magnesium's positive charge assists with keeping up with cell layers. Furthermore, it goes about as a calcium adversary (Iseri et al., 1984) and fills in as a sign transducer. Subsequently, obvious changes in magnesium homeostasis affect cell and tissue exercises.

Controlling magnesium homeostasis is by all accounts basic for bone wellbeing. Magnesium in low and high portions inconveniently affects the bones. Lack of magnesium causes osteoporosis straight by slowing down precious stone arrangement and bone cell work, and in a roundabout way by obstructing the parathyroid organ's discharge and activity of parathyroid chemical and by empowering a poor-quality provocative reaction. At the point when the blood magnesium level is expanded, mineralization issues are found. (2013) (Sara et al.). Roughly 60% of complete magnesium is put away in bone. On cortical bone, 33% of skeletal magnesium is viewed as on the outer layer of hydroxyapatite or in the hydration shell encompassing the precious stone (Alfrey et al., 2012). It goes about as a repository for interchangeable magnesium, which is important to keep the cation fixations in the extracellular space at sound levels (Jahnen et al., 2012). How much magnesium on the outer layer of the bone is corresponding to the degree of magnesium in the blood, since surface bone magnesium ascends with expanding magnesium in persistent renal problems. Most of bone magnesium is apparently stored as a significant part of the apatite precious stone, and its filtering happens because of bone framework demineralization. Magnesium is expected for an assortment of metabolic cycles. Magnesium is expected for the amalgamation of adenosine triphosphate, the essential wellspring of energy in cells. Furthermore, magnesium is a cofactor for many chemicals associated with cell lipid, protein, and nucleic corrosive digestion. Magnesium balances out cell layers and starts film potential because of its positive charge (cation). Furthermore, it goes about as a calcium adversary and a sign transducer. Notwithstanding the way that calcium and phosphorus have a critical influence in the mineralization interaction, obvious changes in magnesium homeostasis affect bone arrangement.

2.4.1.2 MICRO MINERALS

2.4.1.2.1 Copper

Copper is a basic follow metal associated with an assortment of metabolic exercises, including enzymatic responses, nucleic corrosive amalgamation, cancer prevention agent protection instruments, iron digestion, and immunological reaction. Copper inadequacy might bring about debilitated bones, strange cholesterol levels, and cardiovascular infection (Bost et al., 2016). Copper's essential job is to frame catalysts that move electrons (oxidases) to diminish sub-atomic oxygen, as is expected to keep up with the equilibrium of energy digestion at the cell level (Kubiak et al., 2020). Among these compounds is lysyl oxidase, which utilizes lysine and hydroxylysine (the two of which are found in ligament) as substrates to make the cross-joins fundamental for the formation of connective tissues, including new bone creation. Various *in vitro* examinations have shown that copper advantageously affects the cells that administer bone digestion. Copper particle has been displayed to forestall osteoclastic breakdown (Li et al., 2007), though different writers have shown that copper's advantageous impacts are portion subordinate. Copper was displayed to improve the reasonability and advancement of osteoblastic cells at low portions, however was viewed as cytotoxic at bigger fixations. Furthermore, copper advances mesenchymal foundational microorganism improvement, which brings about bone creation (Ding et al., 2014). Dietary copper inadequacy might bring about a diminishing in how much copper in our bodies and, subsequently, in the movement of specific chemicals. Copper deficiency has physiological ramifications, including connective tissue irregularities, which add to skeletal issues (Turnlund, 1999). It has been shown that a copper-insufficient eating regimen restrains the amalgamation of cross connections in collagen and elastin from different tissues, bringing about expanded bone porosity. On the off chance that the dietary admission of copper is inadequate, this outcomes in diminished mechanical strength. It is plainly obvious that copper deficiency thoroughly debilitates the metabolic course of collagen amalgamation (Jonas et al., 1993).

2.4.1.2.2 Iron

Iron is a basic minor element since it takes an interest in oxygen transport and starts an assortment of metabolic enzymatic activities in the body, including collagen creation and vitamin D digestion. The relationship among iron and bone wellbeing depends on a few clinical observational investigations in iron-over-burden people who had bone misfortune as an outcome of inappropriate iron assimilation. Iron deficiency has been demonstrated to significantly affect bone digestion since iron conveys oxygen, which is required by each cell for metabolic exercises. For this reason, this supplement is basic for cell endurance, including bone cells. Persistent lack of iron harms bone and improves the probability of creating bone-related messes; subsequently, complete recuperation from sickliness and its counteraction ought to be urged to upgrade generally speaking wellbeing, including bone wellbeing (Laura Toxqui & Pilar Vaquero, 2015).

2.4.1.2.3 Zinc

Zinc is a basic minor component for calcium and vitamin D to be assimilated appropriately and arrive at the cells where they act to make bones. In spite of the fact that calcium is the essential mineral associated with bone creation, zinc is additionally present during the bones, representing a minuscule extent of the bones. Zinc is expected by bone cells for an assortment of metabolic cycles associated with bone redesigning. Zinc initiates chemicals that go about as impetuses in streamlining bone metabolic pathways; subsequently, zinc is basic for solid bone mineralization by adding to bone wellbeing. Calcium isn't the sole mineral expected for bone creation; a few minor components like zinc, copper, and boron additionally assume a part. Zinc has been displayed to invigorate osteoblastic bone development and mineralization. This metal straightforwardly initiates the compound answerable for improving bone protein amalgamation in cells and energizes the production of other cell proteins. Zinc has been exhibited to support record factor quality articulation, bringing about an expansion in the movement of osteoblast cells, which brings about the mineralization interaction. Furthermore, zinc has been exhibited to safeguard mature osteoclasts from debasement by forestalling the development of osteoclast-like cells from bone marrow cells and instigating apoptotic cell passing in mature osteoclasts. Zinc restrains the atomic component (NF) receptor, which is answerable for the excitement of osteoclastic movement. Zinc-subordinate chemicals have been exhibited to support the movement of osteoblastic and osteoclastic cells in light of bone wellbeing and to be associated with cell record. Zinc supplementation has been related with an expansion in bone mass (Masayoshi Yamaguchi, 2009).

2.4.2 VITAMINS

2.4.2.1 ASCORBIC ACID

Ascorbic corrosive is basic for solid gums and bones. Ascorbic corrosive is expected for the amalgamation of collagen, which goes about as a concrete for connective tissues and furthermore fills in as the reason for bone mineralization. Ascorbic corrosive is an unsaturated polyhydroxy liquor that has a place with the classification of unsaturated polyhydroxy alcohols. A water-dissolvable supplement all plants and most creatures, except for people, incorporate (Sun et al., 2018). Overdosage of ascorbic corrosive might cause moderate side effects like looseness of the bowels and queasiness, since the overabundance is dispensed with by means of the pee (Sun et al., 2018). Poisonousness is ordinarily stayed away from since the kidneys are incredibly skillful of discharging it by means of the pee. Scurvy is a clear cut gathering of side effects brought about by a lack of ascorbic corrosive. Scurvy shows itself in various ways on the bone, including bone breaking down, osteonecrosis, bone misfortune, and obsessive cracks (Smirnoff et al., 2001). In addition to the fact that it is a cancer prevention agent nutrient these days, however it additionally has an influence in bone digestion, quality guideline, and insusceptible framework initiation (Finck et al., 2014).

2.4.2.2 K2 (MENAQUINONE)

Vitamin K is accessible in two structures: nutrient K1 (phylloquinone) and nutrient K2 (phylloquinone) (menaquinone). Vitamin K is available in dull green verdant food varieties, for example, spinach and cruciferous vegetables like broccoli, spinach, and Brussel sprouts. Nutrient K2 is produced in the lower gastrointestinal framework by microorganisms. Furthermore, it is available in follow levels in grass-took care of creature meats and dairy items, and in higher fixations in specific cheeses and a couple of matured food varieties, for example, natto, a Japanese soybean item, curd, and yogurt (Mark Richens, 2019). Nutrient K2 is additionally characterized into various subtypes, the most critical of which are menaquinone 4 and menaquinone 7.

One of its most basic positions is to keep up with the body's calcium testimony. At the end of the day, it instigates bone calcification and safeguards veins and kidneys against calcification. It has been shown that the elements of phylloquinone and menaquinone are exceptionally unmistakable, which is the reason they should be delegated particular supplements. A few examinations have shown that nutrient menaquinone (MK-4) restrained vein calcification, however phylloquinone didn't. Menaquinone pills have been displayed to advance bone and cardiovascular wellbeing, while phylloquinone has no such impact. Menaquinone-rich creature sources incorporate grass-took care of dairy items, egg yolks, liver, and other organ meats. Vitamin K is a fat-dissolvable nutrient, and that implies it is insufficient in low-fat and lean creature items and isn't assimilated successfully on a low-fat eating regimen. While creature food varieties incorporate the Menaquinone-4 subtype, matured food varieties like sauerkraut, natto, curd, yogurt, and miso have a higher centralization of the more drawn-out subtypes, to be specific Menaquinone – 5 to Menaquinone – 14. The job of Vitamin K2 is all around displayed in Figure 2.3, where it is expressed, that Menaquinone is answerable for diminishing vascular calcification and advancing bone calcification.

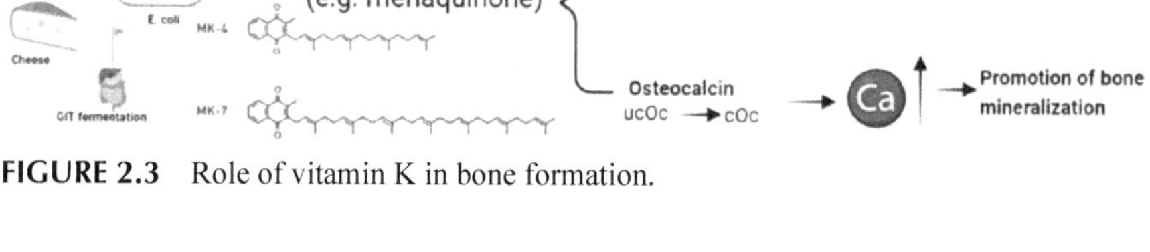

FIGURE 2.3 Role of vitamin K in bone formation.

2.4.2.3 1, 25- DIHYDROXYCHOLECALCIFEROL

For the greater part of the populace, bright B light with a frequency of 290–315 nm is the essential wellspring of vitamin D creation (Holick et al., 2004). In the top epidermis of the skin, bright B radiation changes over 7-dehydrocholestrol to pre-cholecalciferol by photolysis of the B ring structure followed by isomerization. Because of these underlying changes, the particle leaves the plasma layer and enters the extracellular space, in the long run arriving at the blood vessels, where it is limited by the vitamin D restricting protein and conveyed to the liver (Holick et al., 2004). In the late spring, ideal vitamin D age happens following 10–15 minutes of bright B openness (Diamond et al., 2005). In spite of the fact that there are different factors that impact the amalgamation of cholecalciferol in the skin during UVB openness, like skin pigmentation, since melanin shade slows down Vitamin D3 age. Our bodies have a homeostatic instrument that guarantees Vitamin D isn't poisonous, i.e., UVB openness starts the development of dormant photoproducts, for example, tachysterol and lumisterol, which effectively affect calcium digestion (Holick et al., 2003) and forestall hypervitaminosis D brought about by sun openness.

Topographical area is a basic component in deciding the effectiveness of vitamin D amalgamation, since combination is point subordinate, or relying upon the sunlight-based pinnacle point. As scope rises, the amount of light arriving at the world's surface in the frequency range 290–315 nm diminishes because of an improved course of transversion and ozone assimilation. Along these lines, at 35° scope, vitamin D creation is occasional, with the most significant levels occurring throughout the late spring and practically zero development happening throughout the colder time of year. Furthermore, way of life, ecological, and physiological factors could influence creation. Sunscreen use, time spent outside, and the wearing of full apparel all significantly affect UVB openness, as do ecological factors, for example, contamination, overcast cover, and the presence of ozone (Matsuoka et al., 1987). For instance, a sun defensive component of 15 in sunscreen, when applied accurately, diminishes vitamin D development (Matsuoka et al., 1987). Other physiological factors, for example, skin tone and progress in years additionally sway vitamin D age, since more obscure skin has a more noteworthy melanin fixation thus needs more bright B openness time to make a comparative amount of vitamin D as lighter skin (Matsuoka et al., 1987). With expanding age, the amount of nutrient D3 forerunner cholesterol in the skin diminishes fundamentally, bringing about a lessening in vitamin D amalgamation. Various investigations have shown that people beyond 70 four years old produce up to multiple times less cutaneous amalgamation than those matured 20 years (Maclaughlin et al., 1985). Vitamin D is additionally present in little levels in the food, though scarcely any dietary sources normally contain enough of the nutrient to contribute fundamentally to suggested day to day prerequisites. Nutrient D2 (ergocalciferol) and nutrient D3 are the two types of vitamin D found in the eating regimen (cholecalciferol). Nutrient D2 is acquired from plants and some organism. As recently expressed, nutrient D3 is orchestrated from 7-dehydrocholesterol, a type of cholesterol found in skin cells. Nutrient D3 is acquired in the eating routine from creature items like slick fish, fish oils, eggs and dairy items. Most of preliminaries and meta-investigations demonstrate that a portion of vitamin D of 800 IU/day in blend with an adequate admission of calcium is ideal, albeit a few examinations recommend a significantly more noteworthy advantage at higher admissions.

2.4.3 PHYTOCHEMICALS

2.4.3.1 RED CLOVER

An enhancement produced using red clover might assist with safeguarding ladies' bones against early disintegration related with maturing and osteoporosis. Various investigations have shown that ladies who take a red clover-determined dietary enhancement containing isoflavones (Promensil) have significantly less bone osteolysis of the spine than different ladies. Furthermore, the enhancement appeared to have preventive advantages on hip and different bones. Isoflavones contained in plant food varieties, for example, soybeans and red clover have been supported as an option in contrast to menopausal chemical treatment for osteoporosis counteraction. Estrogen inadequacy is a key gamble factor for osteoporosis, and ladies are reluctant to use menopausal chemical treatment to forestall or get the sickness due the dangers associated with it, notwithstanding the way that the impacts of soy or other isoflavone supplements have been restricted and brief.

2.4.3.2 ALFALFA

Horse feed (*Medicago sativa*), some of the time known as lucerne, is a perpetual herbaceous plant having a place with the vegetable family Fabaceae. Horse feed (*Medicago sativa*) is a plant that certain individuals use to enhance their eating regimens with calcium, potassium, phosphorus, iron, and nutrients A, C, E, and K.

Horse feed is most frequently developed for dairy cattle feed. It is frequently utilized as an embellishment in people and appears to restrain cholesterol assimilation in the stomach.

Horse feed is utilized to treat over the top cholesterol, diabetes, acid reflux, and an assortment of different infections.

Horse feed has a high centralization of phytonutrients, like proteins, minerals, and nutrients. It has the most nutrients and minerals of any plant in the world. Minerals like calcium, copper, iron, magnesium, phosphorus, potassium, selenium, sodium, zinc, manganese, boron, chromium, and silicon, as well as nutrients like vitamin A, L-ascorbic acid, vitamin D, vitamin K, vitamin E, pantothenic corrosive, folic corrosive, thiamin, riboflavin, niacin, pentothenic corrosive, pyridoxal phosphate, and choline, are fundamental

Horse feed additionally has a high centralization of amino acids. It contains 14 amino acids, eight of which are significant for human wellbeing. Leucine, phenylalanine, tryptophan, threonine, isoleucine, methionine, valine, and lysine are the eight fundamental amino acids found in horse feed. Because of this blend of supplements, horse feed is one of the most powerful restorative plants on earth.

Horse feed incorporates a high centralization of vitamin K, which supports the creation and upkeep of bones by communicating with vitamin D and glutamic corrosive to initiate osteocalcin, a protein chemical expected for both bone creation and support. The blend of these three supplements is basic for bone development, since the body can't involve calcium in its entire without every one of the three. L-ascorbic acid advances bone arrangement by invigorating the improvement of bone-shaping cells. Furthermore, a few investigations demonstrate that the cancer prevention agent properties of L-ascorbic acid might assist with safeguarding bone cells from hurt. As indicated by a few explorations, vitamin an is by all accounts to a great extent advantageous for bone wellbeing. The Alfalfa plant, to a limited extent because of its phytoestrogenic activity, may assist with keeping up with bone thickness and diminish menopausal side effects. Horse feed has a high centralization of phytoestrogens. Phytoestrogens are exceptionally interesting phytoestrogens. These are non-steroidal phenolic compounds having a steric structure practically identical to that of steroidal estrogens. Their similarity to estrogen receptors empowers them to tie to them and display estrogenic and additionally hostile to estrogenic activities. Isoflavones, lignans, and coumestans are the most predominant phytoestrogens. Phytoestrogens are exceptional natural mixtures that have been displayed in a few investigations to upgrade human wellbeing in an assortment of ways, including the aversion of menopausal side effects, osteoporosis, bosom and prostate malignant growth, and cardiovascular infection. In spite of the fact that there are extra factors influencing how much phytoestrogens in Alfalfa like hereditary variables (plant genotype), ecological circumstances (temperature, precipitation, stickiness, and daylight), biotic elements (contagious infections), phases of development (early vegetative or late blossom), standard period of horse feed, plant parts (leaves, stems, and blossoms), the executives' methodologies (cutting recurrence), and safeguarding technique. Genistein is the essential phytoestrogen found in horse feed, trailed by kaempferol and coumestrol (Tucak et al., 2020).

2.4.3.3 HORSETAIL

Since horsetail contains silicon, a mineral fundamental for bone wellbeing, it has been suggested as a treatment for osteoporosis (diminishing bone). As indicated by an Italian examination, ladies who took horsetail dry concentrate or Osteosil calcium 270 mg two times day (a horsetail/calcium blend utilized in Italy to treat osteoporosis and breaks) saw an expansion in bone thickness. Horsetail is a relative of tremendous, tree-like plants that prospered 400 million years prior during the Paleozoic age. Horsetail, a close to cousin of the greenery, is a perpetual weed that fills in districts of Europe, Asia, the Middle East, and North America. The plant is a perpetual (it returns consistently), with empty stems and branches that initially look like asparagus.

Horsetail's bone-transforming properties have been ascribed for the most part to its high silica fixation. No other plant has as quite a bit of this mineral. Horsetail has the most noteworthy centralization of this mineral.

Silica, which is found in bones, helps construct ligament by advancing collagen amalgamation, advances bone thickness, and helps in calcium assimilation from the small digestive system. *Equisetum arvense* (normal horsetail) is one such spice that is prescribed for osteoporosis treatment because of its exceptionally high silica fixation in the plant world. Silica is fundamental for the assimilation and utilization of calcium, as well as the amalgamation of collagen (establishing protein). Furthermore, auxiliary metabolites like flavonoids (quercetin, kaempferol, luteolin, and apigenin) and triterpenoids (oleanolic corrosive, betulinic corrosive, and ursolic corrosive) have been displayed to safeguard against osteoporosis-related bone misfortune (Badole & Kotwal, 2004).

2.4.3.4 ONIONS

Free revolutionaries make harm our bones, making them become aggravated and distorted. Quercetin is a powerful cancer prevention agent and mitigating substance that safeguards the body against free extreme harm. Furthermore, they smother the osteoclastosis interaction. The development and separation of bone breakdown cells (osteoclasts) is restrained, keeping specific osteoclasts from creating and separating bone. As an outcome, bone mass is kept up with and, surprisingly, expanded. Furthermore, onions are wealthy in bioavailable sulfur compounds, which the body need to incorporate glutathione, our essential intracellular cancer prevention agent, and to stay away from unreasonable homocysteine development. Homocysteine is a sulfur-containing insignificant amino corrosive that oxidizes collagen in the bone and supply routes and is additionally provocative in nature. In members given onion, critical modifications in the degrees of soluble phosphatase (ALP), free revolutionaries, absolute cancer prevention agent limit (TEAC), and various cell reinforcements were distinguished. Furthermore, it was shown that supplementation with onion juice improved the BMD of three postmenopausal ladies respectably. Utilization of onion juice advantageously affects bone misfortune and BMD by means of expanding cancer prevention agent movement, thus might be recommended for the therapy of various bone-related sicknesses, including osteoporosis (Law et al., 2016).

2.4.3.5 GREEN TEA

Consuming green tea and additionally gulping green tea bioactive substances might assist matured ladies and men with forestalling bone misfortune, consequently bringing down their gamble of osteoporotic breaks. Tea, produced using the dried leaves of the Camellia sinensis plant of the Theaceae family, is a well-known refreshment with a three-billion-kilogram yearly result. Green tea is fabricated by drying new leaves at high temperatures (by fricasseeing or simmering) to inactivate the oxidizing catalysts. Green tea is an unoxidized/unfermented refreshment that incorporates different polyphenolic compounds alluded to as catechins or tea polyphenols. These mixtures incorporate epigallocatechin gallate (EGCG), epicatechin gallate (ECG), epicatechin (EC), and epigallocatechin gallate (EGCG) (EGC). EGCG is the most predominant catechin and has accumulated the best review interest. Catechins in tea make up generally 30% to 40% of the extractable solids in dried green tea leaves (Keservani & Sharma, 2014; Keservani et al., 2016a, 2016b; Shen et al., 2009) (Figure 2.4).

2.4.3.6 BERRIES

Berries are notable for their high cancer prevention agent content and related medical advantages. Cancer prevention agents are particles that have an

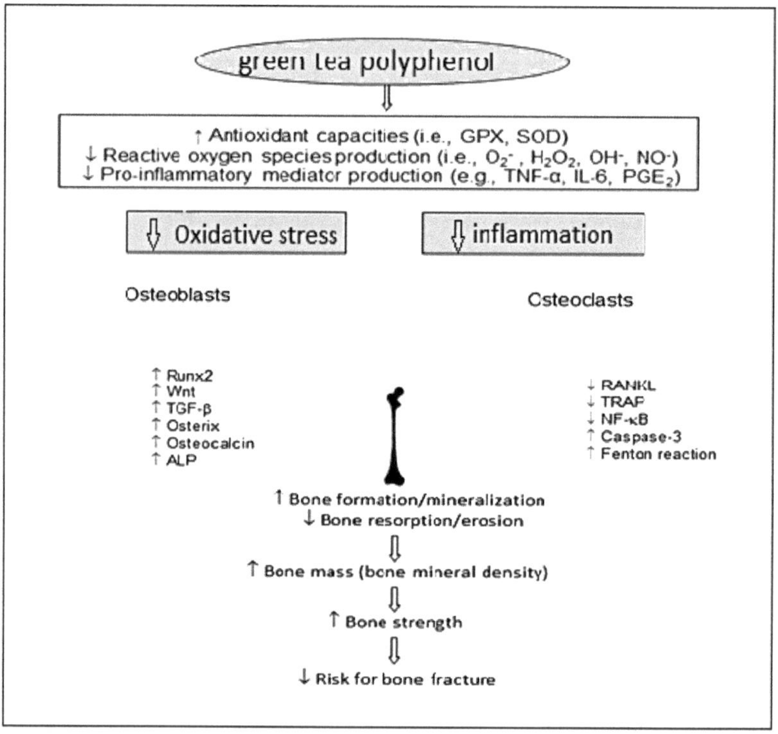

FIGURE 2.4 Role of green tea polyphenols in bone health.

Source: Adapted from Huang et al. (2020)

additional an electron and move it to electron-insufficient radicles. Our body has an assortment of regular cancer prevention agent chemicals that every cell might incorporate in light of the body's requirements. Furthermore, berries have mitigating impacts (Paredes et al., 2010). Berries incorporate an assortment of cancer prevention agents, the most bountiful of which are anthocyanins. Anthocyanins are a water-dissolvable flavonoid shade that reaches in shading from red to purple. Anthocyanins, otherwise called anthocyanidins, are a gathering of synthetic substances that incorporate pelargonidin, cyanidin, delphinidin, and petunidin (Krenn et al., 2007). Anthocyanins are quite possibly the most remarkable phenolic cancer prevention agent (Slatnar et al., 2012). Anthocyanin-rich food varieties have been connected to a diminished gamble of coronary illness, hyperglycemia, and hyperlipidemia, as well as certain types of malignant growth insurance and bone wellbeing (Gupta et al., 2012). With maturing, blood cancer prevention agent movement diminishes, oxidative pressure increments, and bone mass abatements. Ordinarily, the body's regular cell reinforcement protection framework forestalls the unreasonable production of radicles (exceptionally responsive without electron particles) by making chemicals, for example, superoxide-dimutase, catalase, and glutathione (Krenn et al., 2007).

2.4.3.7 RESVERATROL

Resveratrol, a polyphenol observed normally in red grapes and berries, has been displayed to work as a phytoestrogen. Once more, since it is estrogen agreeable, it is viewed as bone amicable also. It has been shown that it benefits both the foundational and cerebral circulatory frameworks, apparently by means of initiating endothelial estrogen receptors. Resveratrol additionally appears to have an anti-osteoporotic impact *in vitro* and *in vivo*, most outstandingly in ovariectomized rodent models that look like postmenopausal osteoporosis instigated by estrogen hardship. Customary supplementation with 75 mg of resveratrol two times day to day might postpone bone misfortune in the lumbar spine and femoral neck, which are successive break areas in postmenopausal ladies without unmistakable osteoporosis. 2020 (Wong et al.). Various investigations have shown that trans (resveratrol) significantly regulates marks of bone digestion. Trans-resveratrol may help with bone development by invigorating osteoblast movement and multiplication and separation (Mizutani et al., 1998).

2.4.3.8 SOY

Soy suppers incorporate fluctuating measures of phytoestrogens, which are powerless plant-based steroid synthetic substances that impersonate estrogen in the female body. Estrogen safeguards the bones; postmenopausal estrogen inadequacy is related with bone misfortune. All things considered, one of the most widely recognized treatments for postmenopausal bone misfortune was chemical substitution treatment, in which estrogen is synthetically assimilated into the circulatory system and cells. Worries about the numerous serious wellbeing outcomes related with long haul chemical substitution treatment have started interest in elective treatments and anti-osteoporotic approaches. Various examination have shown a connection between soy diet consumption and diminished osteoporosis side effects.

Utilization of soy food varieties may in a roundabout way work on bone strength by conveying top notch protein and enhancing consumes less calories with bone-defensive isoflavones. As indicated by review, a maximum usage of creature based protein corrupts bone wellbeing and increments calcium discharge in the pee (Barzel & Massey, 1998), since this is exceptionally acidic in nature, however a high admission of soy protein brings calcium discharge interestingly, down to meat and dairy protein (Zemel, 1988). A few ongoing investigations have exhibited that soybean isoflavones are advantageous for forestalling or keeping up with bone homeostasis (Poulsen & Kruger, 2008), while others have shown blended consequences for the potential wellbeing impacts of soy food varieties and isoflavones (Messina et al. 2009; Lampe, 2009; Tempfer et al. 2007). Furthermore, it has been noted in most of studies that phytoestrogens, especially those got from soy, have significant anti-osteoporotic properties and the capacity to mitigate osteoporosis side effects.

2.4.3.9 PRUNE

At the point when a plum is dried, it becomes prune, which is ordinarily found in European countries. Just the firm-fleshed natural product (plum) of *Prunus domestica* types might be dried into prunes since it has a higher centralization of dissolvable solids and doesn't mature during the drying system. Prunes incorporate starches, dietary fiber (Keservani et al., 2020), protein (Keservani et al., 2015; Singh et al., 2017b), vitamin (Singh et al., 2017a) K, B nutrients, and an assortment of fundamental minerals. As far as nourishment, most of information demonstrates that dried plum (*Prunus domestica* L.) is a compelling superfood for forestalling and turning around bone and underlying misfortune. Ongoing exploration exhibits dried plum's convenience in forestalling and rectifying bone misfortune brought about by estrogen chemical inadequacy. As indicated by other examination, postmenopausal ladies who recently ingested a critical number of dried plums day to day safeguarded a more significant level of bone mineral thickness than the individuals who didn't. This might be attributable to the bioactive synthetic substances found in dried plums, which additionally have bone-defensive properties. As indicated by most of exploration, dried plum in its entire is an exceptionally successful utilitarian or superfood for forestalling bone deterioration in postmenopausal females, with the potential for dependable anti-osteoporotic benefits (Arjmandi et al., 2017).

2.4.4 MILK AND MILK PRODUCTS

Calcium's impact on bone mineral thickness is generally settled (Caroli et al., 2011). An adequate calcium consumption related to bone-accommodating food varieties advances bone mineral thickness (BMD) and safeguards the older against bone misfortune and osteoporotic cracks (Rizzoli et al., 2010). The most noteworthy wellspring of dietary calcium is milk and dairy food varieties, which additionally supply vitamin D (when dairy items are strengthened), protein, phosphorus, potassium, and other bone-accommodating large scale and micronutrients (Caroli et al., 2011). Dairy items might be the best wellspring of calcium in the eating routine attributable to their high calcium content, high calcium assimilation rate because of the presence of casein protein and amino acids, and modest expense. Furthermore, dairy items incorporate the most noteworthy centralizations of protein, calcium, magnesium, potassium, zinc, and phosphorus of any food. Meta-investigations of a few clinical exploration on calcium supplementation and expanded dietary dairy items in youngsters have demonstrated the positive impacts of calcium and dairy items on bone mineral thickness all through the developing period (Huncharek et al., 2008). Expanded dietary calcium/dairy item utilization is connected with an ascent in complete body and lumbar spine BMC in youngsters with low benchmark admissions. Calcium-enhanced dinners significantly expanded bone mass in prepubertal young ladies and young men (Bonjour et al, 1997), and this impact persevered for 1–3 years after calcium supplementation was suspended (Chevelley et al., 2005). There should be no doubt that dairy items, as a magnificent wellspring of calcium, incredibly advance bone wellbeing that endures a more drawn out measure of time.

2.4.5 THREE P'S

2.4.5.1 PUFA

Unsaturated fats might have a basic capacity in the science and wellbeing of bones (Watkins et al., 2001). It has been displayed in creature models that dietary polyunsaturated unsaturated fat (PUFA) deficiency brings about serious osteoporosis (Kruger & Horrobin, 1997). *In vitro* and creature studies have shown the constructive outcomes of PUFAs, especially n-3 unsaturated fats, on bone digestion (Watkins et al., 2003; Poulsen et al., 2007; Salari et al., 2008). By upgrading calcium assimilation, dietary PUFAs might assist with keeping up with calcium homeostasis and backing bone wellbeing (Kruger and Horrobin, 1997; Poulsen et al., 2007). The phone instruments of the activities of PUFAs on bone include adjustment of unsaturated fat metabolites like prostaglandins, resolvins and protectins, a few flagging pathways, cytokines and development factors (Kruger et al., 2010). Prostaglandin E2, produced from n-6 PUFAs, is the significant prostaglandin viewed as in bone. It is a critical advertiser of bone resorption yet additionally advances bone development at low portions (Watkins et al., 2001). N-3 PUFAs might influence bone digestion well in more ways than one: they decrease amalgamation of prostaglandin E2, produce less provocative prostanoids, smother proinflammatory cytokines and bring about lipid middle people (for instance, resolvins), which restrain aggravation instigated bone resorption (Kruger et al., 2010). Diet is a basic way of life part that can possibly adjust hazard and help in the counteraction of osteoporosis (Albertazzi & Coupland, 2002). Utilization of fat, both immersed and unsaturated, has been related to bone misfortune. Among the numerous sorts of lipids, polyunsaturated unsaturated fats (PUFAs) are acquiring consideration for their different capacities in osteoporosis counteraction and treatment (Mangano et al., 2013). Omega-3 unsaturated fats (n-3 FA) and omega-6 unsaturated fats (n-6 FA) are the important antecedents of eicosanoids (flagging particles), which control intracellular flagging and cell-to-cell associations (Albertazzi & Coupland, 2002). It has been recommended that eicosanoids, which are delivered from omega-6 unsaturated fats, increment cell multiplication and aggravation, which might detrimentally affect bone resorption. Eicosanoids got from omega-3 unsaturated fats have mitigating properties, restraining bone resorption (Casado-Diaz et al. 2013). As found in Figure 2.5, DHA (docosahexaenoic corrosive) and EPA (eicosapentaenoic corrosive) smother the amalgamation of such synthetic substances that invigorate osteoclastic bone resorption, subsequently debilitating bone wellbeing.

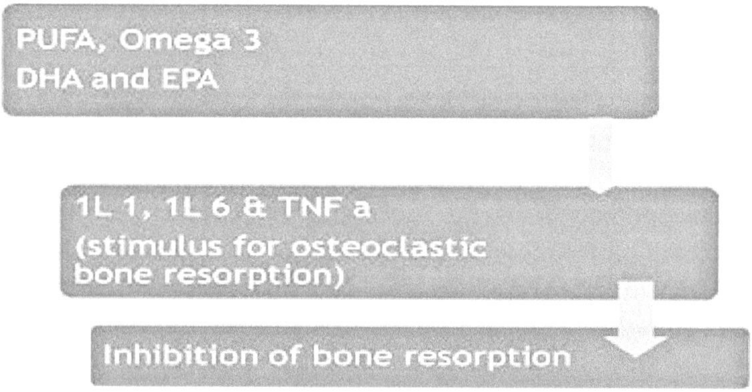

FIGURE 2.5 Role of PUFA in bone health.

2.4.5.2 PROBIOTICS

Probiotics are living microscopic organisms, for example, *Lactobacillus* and *Bifidobacteria* that advance sound wellbeing. Presently, it has been shown unequivocally that probiotics are exceptionally valuable to our bodies in an assortment of ways. These are answerable for gastrointestinal maturation and the development of short chain unsaturated fats, which have an assortment of wellbeing benefits when consumed in little sums. Short chain unsaturated fats are promptly assimilated, offer quick energy, and are not processed with bile. Furthermore, research has shown that probiotics are basic for insusceptibility. Studies have shown unequivocally that people who use probiotic merchandise consistently have a more noteworthy degree of insusceptibility than the individuals who consume less prebiotic things. Probiotics additionally play a part in bringing down the gamble of creating sicknesses like liver infection, provocative entrail illness, cardiovascular illness, and diabetes. As recently referenced, stomach microscopic organisms play a basic capacity in the insusceptible framework. Ongoing review has additionally uncovered that stomach microscopic organisms might safeguard against bone misfortune. Osteoporosis is a typical bone infection that is associated with age and postmenopausal status. Probiotics forestall bone misfortune brought about by estrogen inadequacy, diabetes, or glucocorticoid treatment by restraining osteoclast bone resorption and advancing osteoblast bone development. They support 25-hydroxyvitamin D levels in people, change calcium admission and assimilation, and insignificantly diminish bone misfortune in more established postmenopausal ladies, all at a greatness practically identical to that announced with calcium and vitamin D supplementation. Probiotics are found in matured dairy items, which might assist with keeping a sound calcium balance, stay away from auxiliary hyperparathyroidism, and slow the age-related ascent in bone resorption and misfortune (Rene Rizzoli and Emmanuel Biver, 2020). One potential effect of probiotics on bone wellbeing is that stomach microscopic organisms work as antecedents of nutrients, for example, K2 and B12, which are expected for calcium homeostasis and bone creation (R. D. Hancock and R. Viola, 2001). (Estate et al., 1995). Furthermore, microscopic organisms make short chain unsaturated fats that bring down the PH and work on mineral assimilation by their solubilization. Phytate is an antinutrient that represses the assimilation of practically all minerals. Probiotics produce a catalyst called phytase, which once again introduces minerals that have been diminished by phytates, making them available, like calcium. In this sense, probiotics are normally anti-osteoporotic.

2.4.5.3 PREBIOTICS

Prebiotics are normally happening and orchestrated substances found in plant-based consumes less calories. In exceptionally straightforward terms, they are the supplements for advantageous microscopic organisms, or these parts give a substrate to gastrointestinal microflora to mature. These are frequently oligosaccharides and polysaccharides (long chain glucose polymer) or dissolvable food filaments that the stomach microscopic organisms digest. This would advance gastrointestinal microflora advancement and work on the wellbeing of the digestive probiotic culture, since microorganisms would not be able to flourish (Gibson et al., 2010). Prebiotic substances are progressively being used by drug firms nowadays, attributable to their advantageous wellbeing benefits. Oligosaccharides, for example, inulin-type fructans, galacto-oligosaccharides found in heartbeats, and beta-glucans found in oats are completely perceived filaments that have been displayed to increment maturation by the gastrointestinal microbiota and to have conceivable prebiotic activity. Curiously, these oligosaccharides got from different food classifications have for quite some time been perceived for their capacity to advance lactobacillus and bifidobacteria improvement (Slavin J. 2013). Prebiotic-rich food varieties like vegetables, tubers, and grains advance the development of gastrointestinal microflora, subsequently diminishing destructive microorganisms. Entire heartbeats, chicory root, leeks, Jerusalem artichokes, asparagus, garlic, onion, wheat, oats, and soybeans all contain prebiotic filaments. Bananas additionally contain follow levels of inulin, a fructose-rich prebiotic.

Oligosaccharides, which are unpalatable non-nature, are viewed as the most encouraging prebiotics for bone wellbeing right now. Lactulose, galacto-oligosaccharides, fructo-oligosaccharides, oligo-fructose, inulin, and beta-glucan are a couple of models. These possibly advantageous dietary parts, concentrates, and manufactured compounds are accessible in a scope of polymerization degrees (DP), which alludes to how much sugar monomers in each chain. Oligosaccharides are the most bountiful class of prebiotics, with each chain containing somewhere in the range of four and ten sugar monomers. Furthermore, disaccharides (lactitol, lactulose, and so forth) and longer-chain particles (10–60 monomers for every chain, for example, inulin and long-chain fructo-oligosaccharides, have prebiotic properties (Roberfroid et al., 2010). Prebiotic research has detonated in prominence during the most recent quite a few years. No matter what the prebiotic parts talked about above, other bioactive filaments and prebiotics have been distinguished for their potential medical advantages. Fructo-oligosaccharides, inulin, galacto-oligosaccharides, lactose subordinates, and dissolvable maize fiber are the most proficient prebiotics for bone wellbeing. Prebiotics are anti-osteoporotic in nature since they further develop calcium assimilation from the lower digestive system parcel by diminishing the PH. This is achieved by microbial maturation and hydrolysis processes that outcome in the production of short chain unsaturated fats. Expanded colonic corrosiveness is accepted to restrain calcium's chelation with contrarily charged mixtures, for example, phytates and oxalates. Calcium is let out of these particles, expanding its bioavailability for assimilation and, eventually, bone calcification. Prebiotic filaments have been demonstrated to improve the centralization of short chain unsaturated fats (acetic acid derivation, propionate, butyrate, isobutyrate, valerate, and isovalerate) in the colon, consequently advancing bone wellbeing (Weaver et al., 2010) (Figure 2.6).

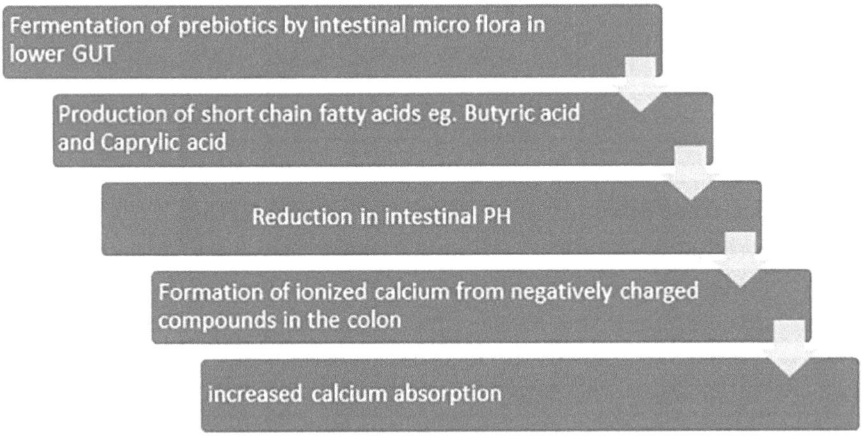

FIGURE 2.6 Role of prebiotics in bone health.

CONCLUSION

Our bone strength is a proxy for our level of aging. Individuals with excellent bone health seem more fit and youthful than those with poor bone health. While the formation and maintenance of bones are dependent on a variety of variables, nutraceuticals work in synergy with appropriate physical activity. These Nutraceuticals enhance bone mineralization through a variety of metabolic mechanisms. Several are intricately related. Everyone is now aware of the need of calcium and vitamin D as the two primary components necessary for healthy bone health. Although study has shown that in addition to these two molecules, additional components such as manganese, copper, zinc, iron, vitamin C, and vitamin K2 are critical. In fact there are so many Functional foods that increase the absorption of calcium for example prebiotics and probiotics but very few population knows about this fact. Additionally, several researches discovered that including nutraceuticals such as phytoestrogens into the diets of postmenopausal females improves bone mineral density, which is critical for overall health. Certain critical facts, such as a high protein diet with little exercise, are related with weak bones. Although fiber reduces calcium absorption, as a prebiotic, it boosts the formation of short chain fatty acids by gut microflora fermentation, which improves mineral absorption. Superfoods such as alfalfa, prunes, berries, soy, and onions contain phytochemicals that help maintain a healthy bone mineral density. This chapter discusses these facts about nutraceuticals and their impact in osteoporosis in detail. One may stay clear of this ailment by eating a balanced diet and supplementing with anti-osteoporotic nutraceuticals.

KEYWORDS

- **anti-osteoporotic**
- **nutraceuticals**
- **osteoblast**
- **osteoclast**
- **remodeling**

REFERENCES

Albertazzi, P., & Coupland, K., (2002). Polyunsaturated fatty acids. Is there a role in postmenopausal osteoporosis prevention. *Maturitas, 42*(1), 13–22.

Alfrey, A. C., & Miller, N. L., (1973). Bone magnesium pools in uremia. *J. Clin. Investig.,52*, 3019–3027.

Baek, K. H., Oh, K. W., Lee, W. Y., et al., (2010). Association of oxidative stress with postmenopausal osteoporosis and the effects of hydrogen peroxide on osteoclast formation in human bone marrow cell cultures. *Calcified Tissue International, 87*(3), 226–235.

Bennett, T., Desmond, A., Harrington, M., et al., (2000). The effect of high intakes of casein and casein phosphopeptide on calcium absorption in the rat. *British Journal of Nutrition, 83*(6), 673–680.

Bost, M., Houdart, S., Oberli, M., Kalonji, E., Huneau, J. F., & Margaritis, I., (2016). Dietary copper and human health: Current evidence and unresolved issues. *J. Trace Elem Med Biol., 35*, 107–115.

Campbell, J. M., Fahey, Jr. G. C., & Wolf, B. W., (1997). Selected indigestible oligosaccharides affect large bowel mass, cecal and fecal short-chain fatty acids, pH and microflora in rats. *Journal of Nutrition, 127*(1), 130–136.

Casado-Díaz, R., Santiago-Mora, G., Dorado, & Quesada-Gómez, J. M., (2013). The omega-6 arachidonic fatty acid, but not the omega-3 fatty acids, inhibits osteoblastogenesis and induces adipogenesis of human mesenchymal stem cells: potential implication in osteoporosis. *Osteoporosis International, 24*(5), 1647–1661.

Cunningham, J., Rodriguez, M., & Messa, P., (2012). Magnesium in chronic kidney disease stages 3 and 4 and in dialysis patients. *Clin. Kidney J., 5*, i39–i51.

Da Silva, F., & Williams, R., (1991). Copper: Extra cytoplasmic oxidases and matrix formation. In: Da Silva, F., & Williams, R., (eds.), *The Biological Chemistry of the Elements: The Inorganic Chemistry of Life* (pp. 388–399). Clarendon Press: Oxford, UK.

Ding, H., Gao, Y. S., Wang, Y., Hu, C., Sun, Y., & Zhang, C., (2014). Dimethyloxaloylglycine increases the bone healing capacity of adipose-derived stem cells by promoting osteogenic differentiation and angiogenic potential. *Stem Cells Dev., 23*, 990–1000.

Draper, H., Sie, T. L., & Bergan, J., (1972). Osteoporosis in aging rats induced by high phosphorus diets. *J. Nutr., 102*(9), 1133–1141.

Genuis, S. J., & Schwalfenberg, G. K., (2007). Picking a bone with contemporary osteoporosis management: Nutrient strategies to enhance skeletal integrity. *Clinical Nutrition, 26*(2), 193–207.

Gilman, J., & Cashman, K. D., (2006). The effect of probiotic bacteria on transepithelial calcium transport and calcium uptake in human intestinal-like caco-2 cells. *Current Issues in Intestinal Microbiology, 7*(1), 1–6.

Hagiwara, H., Hiruma, Y., Inoue, A., Yamaguchi, A., & Hirose, S., (1998). Deceleration by angiotensin II of the differentiation and bone formation of rat calvarial osteoblastic cells. *Journal of Endocrinology, 156*(3), 543–550.

Hamilton-Miller, J. M. T., (2004). Probiotics and prebiotics in the elderly. *Postgraduate Medical Journal, 80*(946), 447–451.

Hancock, R. D., & Viola, R., (2001). The use of micro-organisms for L-ascorbic acid production: Current status and future perspectives. *Applied Microbiology and Biotechnology, 56*(5, 6), 567–576.

Harris, D., (1997). Copper. In: O'Dell, B., & Sunde, R., (eds.), *Handbook of Nutritionally Essential Mineral Elements* (pp. 231–273). Marcel Dekker: New York, NY, USA.

Huang H-T, Cheng T-L, Lin S-Y, Ho C-J, Chyu JY, Yang R-S, Chen C-H, Shen C-L. (2020) Osteoprotective Roles of Green Tea Catechins. *Antioxidants, 9*(11):1136. https://doi.org/10.3390/antiox9111136

Huttunen, M. M., Tillman, I., Viljakainen, H. T., Tuukkanen, J., Peng, Z., Pekkinen, M., et al., (2007). High dietary phosphate intake reduces bone strength in the growing rat skeleton. *J. Bone Miner Res., 22*(1), 83–92.

Institute of Medicine (US) Panel on Micronutrients. (2001). *Dietary Reference Intakes for Vitamin A, Vitamin K, Arsenic,Boron, Chromium, Copper, Iodine, Iron, Manganese, Molybdenum, Nickel, Silicon, Vanadium*. National Academies Press: Washington, DC, USA. Available from: https://www.ncbi.nlm.nih.gov/books/NBK222310/ doi: 10.17226/10026.

Iseri, L. T., & French, J. H., (1984). Magnesium: Nature's physiologic calcium blocker. *Am. Heart J., 108*, 188–193.

Jahnen-Dechent, W., & Ketteler, M., (2012). Magnesium basics. *Clin. Kidney J.,* (5), i3–i14.

Jonas, J., Burns, J., Abel, E. W., Cresswell, M. J., Strain, J. J., & Paterson, C. R., (1993). Impaired mechanical strength of bone in experimental copper deficiency. *Ann. Nutr. Metab., 37*, 245–252.

Kararli, T. T., (1995). Comparison of the gastrointestinal anatomy, physiology, and biochemistry of humans and commonly used laboratory animals. *Biopharmaceutics and Drug Disposition, 6*(5), 351–380.

Keservani, R. K., & Sharma, A. K., (2014). Flavonoids: Emerging trends and potential health benefits. *Journal of Chinese Pharmaceutical Sciences, 23*(12), 815.

Keservani, R. K., Kesharwani, R. K., Sharma, A. K., Vyas, N., & Chadoker, A., (2010b). Nutritional supplements: An overview. *International Journal of Current Pharmaceutical Review and Research, 1*(1), 59–75.

Keservani, R. K., Kesharwani, R. K., Vyas, N., Jain, S., Raghuvanshi, R., & Sharma, A. K., (2010a). Nutraceutical and functional food as future food: A review. *Der Pharmacia Lettre, 2*(1), 106–116.

Keservani, R. K., Sharma, A. K., & Jarouliya, U., (2015). Protein and peptide in drug targeting and its therapeutic approach. *ArsPharmaceutica (Internet), 56*(3), 165–177.

Keservani, R. K., Sharma, A. K., & Kesharwani, R. K., (2016a). Nutraceutical and functional foods for cardiovascular health. In: *Food Process Engineering* (pp. 257–278, 291–312). Apple Academic Press, CRC Press. ISBN: 9781771884020.

Keservani, R. K., Sharma, A. K., & Kesharwani, R. K., (2016b). Medicinal effect of nutraceutical fruits for the cognition and brain health. *Scientifica,* 10. Article ID 3109254. doi: 1 0.1155/2016/3109254.

Keservani, R. K., Sharma, A. K., & Kesharwani, R. K., (2017). An overview and therapeutic applications of nutraceutical and functional foods. *Recent Advances in Drug Delivery Technology*, 160–201. ISBN:9781522507543.

Keservani, R. K., Sharma, A. K., & Kesharwani, R. K., (2020). "*Nutraceuticals and Dietary Supplements: Applications in Health Improvement and Disease Management*. CRC Press. ISBN: 9781771888738.

Krishnarao, G., & Draper, H., (1972). Influence of dietary phosphate on bone resorption in senescent mice. *J. Nutr., 102*(9), 1143–1145.

Kubiak, K., Klimczak, A., Dziki, Ł., Modranka, R., & Malinowska, K., (2010). Influence of copper (II) complex on the activity of selected oxidative enzymes. *Pol. Merkur. Lekarski., 28*, 22–25.

Lan, G. Q., Abdullah, N., Jalaludin, S., & Ho, Y. W., (2002). Efficacy of supplementation of a phytase-producing bacterial culture on the performance and nutrient use of broiler chickens fed corn-soybean meal diets. *Poultry Science, 81*(10), 1522–1532.

Li, B., & Yu, S., (2007). *In vitro* study of the effects of copper ion on osteoclastic resorption in various dental mineralized tissues. *Chin. J. Stomatol., 42*, 110–113.

Li, F. Y., Chaigne-Delalande, B., Kanellopoulou, C., Davis, J. C., Matthews, H. F., Douek, D. C., Cohen, J. I., et al., (2011). Second messenger role for Mg2+ revealed by human T-cell immunodeficiency. *Nature, 475*, 471–476.

Lopez, H. W., Coudray, C., Levrat-Verny, M. A., Feillet-Coudray, C., Demigné, C., & Rémésy, C., (2000). Fructooligosaccharides enhance mineral apparent absorption and counteract the deleterious effects of phytic acid on mineral homeostasis in rats. *Journal of Nutritional Biochemistry, 11*(10), 500–508.

Ma, Y. F., Stimpel, M., Liang, H., Pun, S., & Jee, W. S. S., (1997). Impact of antihypertensive therapy on the skeleton: Effects of moexipril and hydrochlorothiazide on osteopenia in spontaneously hypertensive ovariectomized rats. *Journal of Endocrinology, 154*(3), 467–474.

Mangano, K. M., Sahni, S., Kerstetter, J. E., Kenny, A. M., & Hannan, M. T., (2013). Polyunsaturated fatty acids and their relation with bone and muscle health in adults. *Current Osteoporosis Reports, 11*(3), 203–212.

Matar, C., Amiot, J., Savoie, L., & Goulet, J., (1996). The effect of milk fermentation by *Lactobacillus helveticus* on the release of peptides during *in vitro* digestion. *Journal of Dairy Science, 79*(6), 971–979.

Ouwehand, A. C., Salminen, S., & Isolauri, E., (2002). Probiotics: An overview of beneficial effects. *Antonie Van Leeuwenhoek, 82*(1–4), 279–289.

Roberfroid, M. B., (2000). Prebiotics and probiotics: Are they functional foods? *American Journal of Clinical Nutrition, 71*(6), 1682s–1687s.

Singh, P., Kesharwani, R. K., & Keservani, R. K., (2017a). Antioxidants and vitamins: Roles in cellular function and metabolism. In: *Sustained Energy for Enhanced Human Functions and Activity* (pp. 385–407). Academic Press. ISBN:9780128054130.

Singh, P., Kesharwani, R. K., & Keservani, R. K., (2017b). Protein, carbohydrates, and fats: Energy metabolism. In: *Sustained Energy for Enhanced Human Functions and Activity* (pp. 103–115). Academic Press. ISBN:9780128054130.

Sjögren, K., Engdahl, C., Henning, P., Lerner, U. H., Tremaroli, V., & Lagerquist, M. K., (2012). The gut microbiota regulates bone mass in mice. *Journal of Bone and Mineral Research, 27*(6), 1357–1367.

Turnlund, J., (1999). Copper. In: Shike, M., & Ross, A., (eds.), *Modern Nutrition in Health and Disease* (pp. 241–252). Williams & Wilkins: Baltimore, MD, USA.

Villa, M. L., Marcus, R., Delay, R. R., & Kelsey, J. L., (1995). Factors contributing to skeletal health of postmenopausal Mexican-American women. *Journal of Bone and Mineral Research, 10*(8), 1233–1242.

Weaver, C. M., Martin, B. R., Story, J. A., et al., (2010). Novel fibers increase bone calcium content and strength beyond efficiency of large intestine fermentation. *J. Agric. Food. Chem., 58*, 8952–8957.

Weber, P., (1999). The role of vitamins in the prevention of osteoporosis—A brief status report. *International Journal for Vitamin and Nutrition Research, 69*(3), 194–197.

Wyshak, G., (2000). Teenaged girls, carbonated beverage consumption, and bone fractures. *Arch. Pediatr. Adolesc. Med., 154*(6), 610–613.

Wyshak, G., & Frisch, R. E., (1994). Carbonated beverages, dietary calcium, the dietary calcium/phosphorus ratio, and bone fractures in girls and boys. *J. Adolesc. Health, 15*(3), 210–215.

CHAPTER 3

MANAGEMENT OF FRACTURES VIA NUTRACEUTICALS

ARUN SONI,[1] NIYATI ACHARYA,[2] and SANJEEV ACHARYA[2]

[1]*SSR College of Pharmacy, Sayli, Silvassa, India.*

[2]*Institute of Pharmacy, Nirma University, Ahmedabad India.*

Bone fracture cases are estimated to be 1.5 million each year globally. The chances of fracture are widely seen in aged people and women. Distal radius, proximal femur, ankle, proximal humerus, and metacarpal bones are areas that's more prone to develop fracture. As nutrition is a major factor for upbringing bone growth and density until the 13th year of life, its strength maintained in bone is acquired in the rest of the years. The revolution from infant stage to adult stage requires an adequate Vitamin D and calcium rich diet for bone rigidity and bone mass density to overcome fractures and bone loss. As the saying goes 'prevention is better than cure,' the focus is on tackling nutrition deficiencies related to the skeletal system rather than curing bone cracks and osteoporosis. Here's when nutraceuticals or dietary supplements come into picture. It is made up of two words nutrients and pharmaceuticals which defines it to be the candidate from either food or natural segment that has preventive effects for diseases and other nutritional health benefits. Nutraceuticals are exceptional healthcare elements that are neither really pharmaceutical products nor a part of dietary constituents. It's a widely accepted concept for lowering fracture symptoms and bone density loss. Now-a-days various nutrients are implied as therapeutic agents for bone related disorders. This chapter sheds light on the bone care by means of nutraceutical products to show therapeutic action with no side effects like that of pharmaceutical drugs. It deals with classes of nutrients based on source and chemical present. Also, gives ideas about bone composition, function and

Nutraceuticals and Bone Health. Deepak Sharma, Madan Mohan Gupta, Anil K. Sharma, Raj K. Keservani, Rajesh K. Kesharwani (Eds.)

bone formation, and the resorption process as a normal physiologic function of the body. It explains in detail about types of nutraceutical products that are applied in bone restoration and presents overall information on bone development in the body and factors pertaining to use of nutraceuticals in the market as bone fracture preventive agents.

3.1 INTRODUCTION

Bone is the significant framework which is responsible for support and protection to the body. Bone physiology not only functionalizes for attaining the body's posture but also build ability to withstand external shock conditions, bone-tissue attachments, protection of vital organs and storage of nutrients like calcium and phosphate (Rajput, et al, 2019). Calcium is majorly deposited in bone tissue with a total of 99%. Bone includes necessary substances like collagen mucopolysaccharide that helps in reducing rigidity and on the other hand minerals like calcium phosphate assures the strong bone structural integrity and maintains cells in development of bone. Bone density makeup signalizes from the embryonic stage and attains its peak up to adolescent age (Karpouzos, DIamantis, et al., 2017). Circumstances like pregnancy, inheritance, asthmatic annotations, anorexia, disease refers to distorted bone density levels. Osteoporosis is one of the leading causes for bone fracture and tendency for fracture risk in a person with osteoporosis is higher compared to others (Jani, 2019).

Bone remodeling is the process of regeneration of new bone tissue by replacing aged bone tissue. Remodeling is worked by elements: Osteoblasts and Osteoclasts. Releasing minerals from bone is characteristic of osteoclasts (Qureshi & Pisulkar, 2020). However, mineralizing of the bone minerals after initial replacements of it from bone tissue is carried out by osteoblasts. Hampering the procedure of remodeling can withhold many skeletal health issues. Reduction in any one of the factors such as bone density, bone mineral mass has a great influence on bone health. Widely faced concern is low bone mass which is due to age, race, body size, gender, genetics are inevitable factors and other avoidable factors such as hormonal imbalance, exhaustive life pattern, improper diet and habits of smoking and alcoholism. (Jani, 2019).

Since earlier decades different drug therapies are being used for management of bone related ailments. Unsatisfactorily, these drugs came up with several adverse events, side effects, hypersensitivity reactions, late fracture restoration, GI disturbances, etc. So, a need arose to develop a product which would render nutritional reconstitution with safe and efficient application. An innovation that was regarded as one of the prognoses to this problem was introduction of nutraceuticals for depletion of bone related consequences with the advantage of no prescription consumption. Nutraceuticals (Keservani et al., 2010a, 2010b, 2017, 2020) refers to nutrients and pharmaceuticals in one place, basically it is food which contains essential nutrients, vitamins (Singh et al., 2017a) and minerals for betterment of health crisis and supplies these components to areas of deficiency to tackle any disease (Nieves, 2021; Roberts & Drissi, 2020).

Bone health requires inculcating a balanced diet with proper nutrients, minerals, vitamins in day-to-day lifestyle. For prevention and cure of skeletal health condition, the key aspect is correct nutrition. For ideal bone functioning and inhibiting osteoporosis, proper and definite nutrition is important (Sale & Eliot Sale, 2019).

This data provides with types of nutraceuticals used in healing a fracture and process of restoring bone matrix by aid of nutraceuticals recently available till date (Sale & Eliot Sale, 2019).

3.2 REVOLUTION IN NUTRACEUTICALS DEVELOPMENT

Studies have shown that elder person mostly female with 60 years of age and male with 70 years. of age are prone to risk of hip fracture. In 2007, it was assessed that over 200 million female population suffered from osteoporosis from which aged 60 contributes 1/10th of the habitat and aged 70 & 80 constitutes 1/5th and 2/5th population respectively. According to data of 2017, 8.9 million cases of bone fracture have seen annually pertaining to bone depletion. US ratio depicted 53 million people suffering from bone loss Around 3 million fracture incidences would be noted till 2025 because of bone deficiency (Nwosu & Ubaoji, 2021).

Perception of nutraceuticals made an entry 3000 years ago. As per the statement of Hippocrates in 460–377 BC, food was regarded to be used as medicine and medicine to be used as food. As necessity is key to invention, USA food developers started application of iodine in salt which proved beneficial against goiter. "Nutraceuticals" was appealed by Stephen DeFelice in 1989, the chairman and founder of Foundation for innovation in medicine, Cranford, New Jersey. For skeletal health in the late 20th century research was done to show milk as a suggestive diet. In 1863 Louis Pasteur developed techniques of pasteurization which killed bacteria, fungi, and yeast in milk by heating it above 161°F and thus increasing shelf life of milk. By 2016, the dairy industry grew to provide quality milk items and enhanced milk production (Jani, 2019).

Countries like England, Japan have already considered nutraceuticals a hub for dietary supplements. In the UK, France, Germany, diet was given utmost importance than physical fitness or persistence for gaining good health. Canada implemented formulation of nutraceutical in the form of medicinal forms rather than considering it as food. India has visualized nutraceutical as a product which is more related to plant origin, herbal extract, organic form for mitigating normal physiology. As per reports of 2019, nutraceutical showed rapid growth in industrial scale with predicted growth rate of 7.5%. The worldwide market is expected to increase sales from 241 billion $ in 2019 to 373 billion $ in 2025 (Nwosu & Ubaoji, 2021).

The impact of nutrition on bone health has been comprehensively measured, and various foods and nutrients have been studied for their potential effects on bone fragility. Fortification of food with the micronutrients which are beneficial for bone health and structure has opened a new avenue in the development of Nutraceutical products for better bone health. The effects of micronutrients like magnesium and some of the vitamins like Vitamin K have been reported to alter the bone metabolism and bone strength. Several clinical trial results have shown benefits of calcium-vitamin D-fortified foods with respect to normal food equivalents nutrient, in randomized controlled design. Successful evidence was obtained in postmenopausal women and old aged population in food fortification with calcium and vitamin D in the levels of vitamin D status, prevention of secondary hyperparathyroidism and reduction of accelerated bone turnover. (Whiting, et. al., 2016).

3.3 CLASSIFICATION OF NUTRACEUTICALS

Nutraceutical is a widespread concept that is acceptable within the category as a food supplement with numerous health benefits and nutritional values in food. Controlling any disorder and symptoms by means of biological treatment regards to generating a good health perspective. The reason for implementing nutraceuticals is to improve health condition, and prevention of disorders by modification of daily diet with postulated long-term benefits. The characteristics of nutraceuticals can be evaluated on the basis of their natural origin, pharmacological properties, phytochemicals and chemical composition of the manufactured goods. Different groups of nutrients are done in order to classify them into i) Dietary supplement, ii) Functional food, iii) Medicinal food and iv) Pharmaceuticals (Figure 3.1).

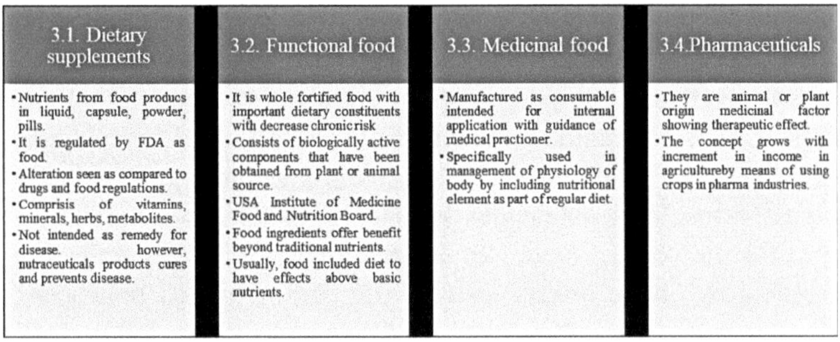

FIGURE 3.1 Determines the classification of nutraceuticals (Jani, 2019).

Source: Adapted from Jani, 2019.

3.4 BONE CHARACTERISTICS

3.4.1 *STRUCTURE AND ANATOMY*

The human bone is made up of 206 bones for an adult. While, the overall bone count at birth is 270. Further division of bone is done into 2 primary regions. Axial part includes 88 bones with rib cage, vertebral column and skull. The Appendicular region contains 126 bones and shoulder girdle, pelvic girdle and upper and lower boundaries. A healthy bone resembles a honeycomb like structure when seen in microscopic view while on the other hand eroded bone appears with increased diameter of pores in bone. Bones are classified based on their shape and size with varying functions. According to their size, there are 4 types of bones: long bone, short bone, irregular bone and flat bone. 1) long bones- humor, femur, radii, ulna, metacarpals, tibiae, fibulae, metatarsals and phalanges.2) short bones- carpal, tarsals, patella, sesamoid bones.3) flat bones- skull, mandible, scapulae, sternum and ribs. 4) Irregular bones- vertebrae, sacrum, coccyx and hyoid bone. Longest bone in the human species is the femur which lies in the thigh region and extends up to the knees. Irregular bones can be seen in the wrist and can carry out movement of hand. Pair of 12 ribs are bent, flat bones present on either side of the thoracic cavity to protect lungs and heart. The long bone has 2 parts: epiphysis and diaphysis. Diaphysis is a tubular structure situated between proximal and distal ends of bone; medullary cavity is hollow region in diaphysis filled with yellow marrow. While epiphysis is the either end of long bone (Karpouzos, et al., 2017).

3.4.2 COMPOSITION

Bone comprises 50–70% of minerals which is hydroxyapatite crystal. Organic matrix in bone contributes 20–40% with 90% of total bone protein that is called type 1 collagen and some other proteins. Water content in some bones is upto 10%. Bone also comprises lipid constituents. The bone matrix consists of 98% of organic part with collagen type 1, non-collagenous part and proteoglycans. The Remaining 2% of organic matter consists of cells like osteoblast, osteoclasts, osteocytes and 10% of total bone volume. Osteoblasts are bone forming cells, osteocytes are mature bone cells, osteoclasts are cells that break down bone matrix, osteoid are unmineralized bone matrix. The inorganic part of the matrix consists of calcium and phosphorus. Chemical composition of bone consists: 39% calcium which involves 99% of the body. 0.2% of potassium that acquires 4% in body calcium. Sodium content is 0.7% and has 35% concentration in the body. Magnesium is present in 0.4% and total of it in the body in 50%. Content of carbonate present in bone is 9.7% and contributes to 80% in the body's carbonate concentration. Whereas, phosphate in bone counts for 17% with 99% of body's phosphate content (Pandey et al., 2018).

3.4.3 TYPES AND PARTS OF BONES.

Bones are differentiated on the basis of size into

1. **Long Bones:** They are compact bones with the presence of little marrow and located in bones of limbs.
2. **Short Bones:** mainly comprises cancellous bones that have layering of cortical bone characterized by cube shaped, small structure and rough appearance.
   E.g. – Hand bones, tarsals of feet.
3. **Flat Bones:** Specified as thin layer curved bone with outer sheet of cancellous bone filmed by cortical bone. Located in the axial skeleton.
   E.g. – skull, ribs, scapula.
4. **Irregular Bones:** different characteristics which are not seen in other bones. Formed by cancellous bone covered by cortical bone.
   E.g. – Pelvis, vertebrae.
5. **Sesamoid Bone:** spherical bone developed from tendons
   E.g. – patella

Based on the architecture of bone it classifies into cortical and trabecular bones. 1) Cortical bone is present in 80% of the total bone of the body and stronger than trabecular bone. It is 10% porous and forms a concentric circle pattern. It sub classifies as Periosteum, Intracortical area and Endosteum. 2) Trabecular / Spongy is 20% part of total bones. It is 50 – 90% porous and is the main tissue of the long bone. But acquires ten times the volume and surface of the volume and surface of the cortical bone. Also called as cancellous bone, it is lighter, less dense, and flexible than compact bone.

Different parts of bones are cortical, cancellous, periosteum, marrow. The Bones of the skeleton are divided into appendicular (limbs, shoulder, pelvic girdle) and axial (skull, vertebral column, thoracic cage) (Jani, 2019).

3.4.4 FORMATION AND FUNCTIONS

Long bone functionalizes for supporting weight and movement. Flat bones and irregular bones act as protective elements for vital organs and tissues. Sesamoid bone protects tendons from stress and wearing. Bone not only gives the body an overall shape but also supports movement, protects, maintains mineral homeostasis, provides blood cell formation and triglyceride formation. It allows backing of muscle movement and helps for extension, adduction, abduction and other actions. Osteocytes are present for engulfing toxic material and heavy compounds for detoxification. Works maintaining the pH levels because bones can alter salt composition (Figure 3.2).

a. Support and movement: bones adhere to ligament and tendons offering a framework of skeleton to coordinate muscles, ligaments and tendons.
b. Protection: bones provide protective layering for soft organs. E.g., Cranium surrounds brain, vertebral column is around spinal cord, rib cage covers lungs and heart and pelvis protects genitourinary system.
c. Most minerals are stored in bones from which 99% is calcium, 85% is phosphate and 50% magnesium. It helps in stabilizing the homeostasis of minerals in blood and bones which are released when required.
d. Blood cell formation takes place in the red bone marrow region from stem cells. Bone marrow is responsible for making 2 million RBCs every sec. In newborn only red bone marrow is present and by the time they replace it by yellow bone marrow. By adult stage the red bone marrow becomes half and remains 30% by old age.

e. Triglyceride storage takes place in adipose tissue which is part of yellow bone marrow, an energy reserve of the body (Rajput, et al., 2018).

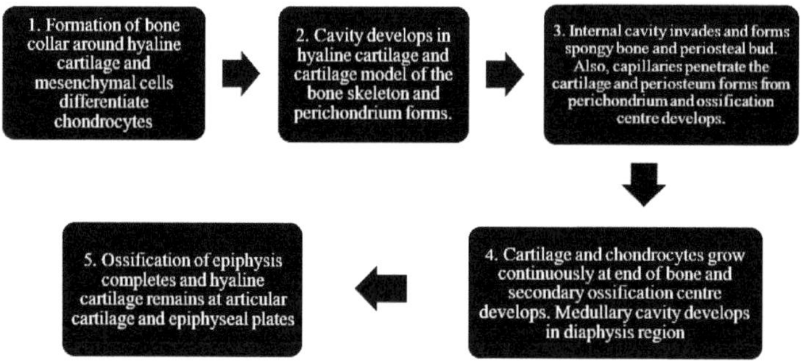

FIGURE 3.2 Denotes process of bone forming.

3.5 BONE DISEASE AND FRACTURE

3.5.1 MECHANISM OF OSTEOPOROSIS AND FRACTURE

- **Osteoporosis:** Osteoporosis is made of two words, 'osteo' that is bone and 'porosis' that refers to pores. Range of breaking of bone is higher than bone formation which further leads to porous bone with low bone density eventually leading to fracture. Characterized by thinning and weakening of bone. A bone comprises a hard layer that is cortical bone and a soft layer of trabeculae, promising for supportive function. However, cortical bone is formed by a number of circular osteons with centrally situated haversian canal for passage of blood. It is surrounded by lamellae which happens to be mostly collagen and calcium phosphate. A bone is replaced by a new formed bone in several years, where spongy bone requires 3 to 4 years and compact bone needs 10 years. Bone remodeling is a process of resorption of aged bone and formation of new bone. Bone resorption works by breaking of bone by osteoclasts while bone formation acts by developing bone by osteoblasts. Remodeling depends on levels of calcium ions that are regulated by PTH, Vitamin D and calcitonin. Condition of osteoporosis is seen when bone mass lowers due to increased action of osteoclasts as compared to osteoblasts. Increase risk for fracture and bone fragility because of widened diameter of haversian canal, thinning of cortical part, lesser number of trabeculae, low estrogen levels, decreased calcium, alcoholism, drugs, smoking, heparin, L- Thyroxin, inactivity of body. Also, diseases like Turner's syndrome, Cushing syndrome, hyperprolactinemia, diabetes mellitus, and Klinefelter syndrome contribute to osteoporosis (Foger-Samwald., 2020).

- **Fracture:** It is a condition of bone where complete or partial collapse of a part of the skeleton. Occurrence is seen when the external force or trauma established on the bone is higher than the bone's stress bearing capacity. Various risk factors associated with it are depleted Vitamin D levels, use of glucocorticoids, alcoholism, smoking groups as avoidable factors whereas unavoidable factors are age, osteogenesis, malabsorption of calcium and Vitamin D. Increased reabsorption leads to bone fracture and osteoporosis shows decreased bone density. Types of fractures: 1. Simple fracture (closed) – here bone breaks but skin remains unharmed. 2. Compound fracture (open) – It occurs when bone tears ligaments and muscles to penetrate out of skin. 3. Greenstick fracture- It has bone broken on one side and bent on either. 4. Impacted fracture- when a piece of bone wedges in another bone. 5. Comminuted fracture- multiple fragments of bones. 6. Spiral fracture- Fracture that turns out to be in spiral projection happens due to twisting action.

There are fractures based on features: vertebral fracture is most common and occurs in weak bones of the spine. Femoral neck fracture, distal radius fracture seen in post-menopausal women.

Bone healing process is initiated by 1. Inflammatory phase which is a response to a stress condition by sending an immune cell to a traumatic location where it removes cell debris and damaged tissue. 2. Reparative phase where the body triggers osteoblasts to form callus that is new bone tissue that would connect the end of fractured bone. 3. Remodeling phase that replaces callus with regular bone so healing is complete.

Complications caused by fractures are damaging of surrounding blood vessels which can lead to bleeding, altering sensation due to nerve damage and tear of ligaments and tendons, fat embolism by long bone fracture, compartment syndrome, delay repair, bone deformity, instability, joint stiffness (Pandey et al., 2018).

3.5.2 BONE REMODELING

The process of bone remodeling is a multistage process which is presented in the Figure 3.3.

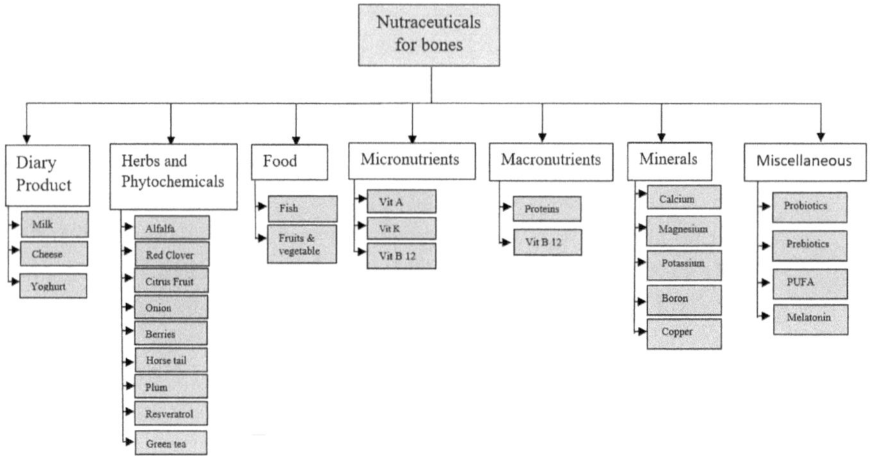

FIGURE 3.3 The process of bone remodeling.

3.6 NUTRACEUTICAL USED IN BONE FRACTURE

3.6.1 CLASSIFICATION BASED ON SOURCE

Classification of nutraceuticals used in bone disorders is shown in Figure 3.4 (Karpouzos et al., 2017).

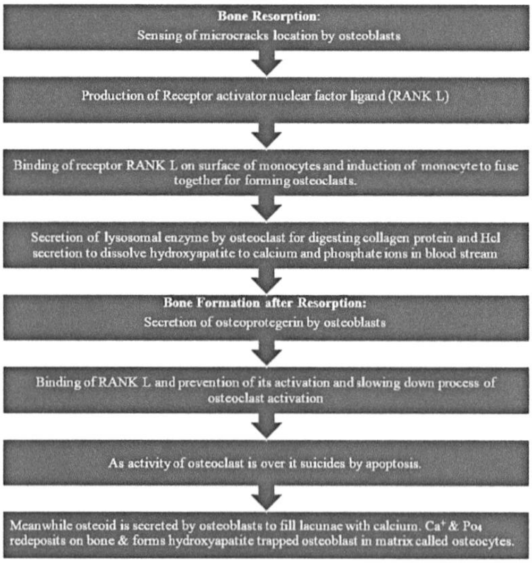

FIGURE 3.4 Classification of nutraceuticals used in bone disorders.

3.6.1.1 MINERALS

Calcium: Calcium is the ion that total ups for 1.9% of body's weight. Thus, it is the fifth widely available compound in the body besides hydrogen, carbon, oxygen and nitrogen. It is stored in teeth and bones from 99% of total calcium content and enhances mechanical strength. Calcium helps improve bone density. Study has referred to calcium tablets to decrease fracture in age women. Therapeutic drugs available for osteoporosis but efficient calcium uptake was required. Bisphosphonate developed attraction for hydroxyapatite crystals and distorted bone resorption activity which ultimately leads to higher bone mass by means of bone tissue. For adequate content of calcium in bone, FDA accepted teriparatide drug against osteoporosis that was a parathyroid hormone human recombinant. The study showed successful bone mineral density elevations. For good bone health it becomes necessary to regulate calcium amount in bone for which supplement is a fundamental element to treat osteoporosis. Diet including milk, cheese, soybean, yogurt, leafy vegetables (spinach) tackle the urge of bone calcium. Figure 3.5 shows food rich in calcium with concentration (Nataranjan et al., 2019).

Sr. No.	Food	Nutrient	Concentration (mg)
1.	Milk	Calcium	290
2.	Canned milk	Calcium	329
3.	Yoghurt	Calcium	320
4.	Feta cheese	Calcium	160
5.	Gruyere	Calcium	300
6.	Parmesan	Calcium	414
7.	Cheddar	Calcium	200
8.	Mozzarella	Calcium	207
9.	Cod liver oil	Calcium	400
10.	Salmon	Vit D	360
11.	Orange juice	Vit D	50
12.	Tuna fish	Vit D	200
13.	Milk	Vit D	103-105
14.	Egg yolk	Vit D	20
15.	Beef liver	Vit D	15
16.	Swiss cheese	Vit D	12

FIGURE 3.5 Food rich in calcium with concentration.

Source: Adapted from Nataranjan et al., 2019

- **Magnesium**: It plays a vital role in carrying out the physiology function of the body. From the total body's magnesium, 60% is stored in bone on the surface of hydroxyapatite crystal surface. Magnesium regulates PTH levels, helps to prevent acidosis and forms apatite crystals. It resembles interest in ATP and performs as a cofactor for enzymes. In the case of osteoporosis due to deficiency of magnesium the apatite crystal enlarges and cannot handle pressure. Scarcity of magnesium lowers PTH that is important for bone and calcium function. It produces inflammatory responses on behalf of bone improvement by inflammatory cytokines. Also heals acidosis and shields the bone framework from degradation. Magnesium is obtained from grains, seeds, green vegetables, etc. A research based on magnesium content denoted that Bone Marrow Density (BMD) level by magnesium in white race increased compared to black race people. Data found higher BMD in white females and males. Increase evidence of wrist fracture was seen in postmenopausal females treated with maximum magnesium amount. Scientific study done to check integrated hip bone mineral content by magnesium oxide supplement was done in a randomized trial proved successful in rising bone mass (Pandey et al., 2018).
- **Potassium**: Helps to store bone calcium and regulates pH. It neutralizes the activity produced by food like meat. Potassium maintains basic surroundings so decreases the need of calcium and shortens the requirement of salts in the skeletal system. Foods with higher amounts of potassium like spinach, sweet potato, apricots, and bananas assist in countering osteoporosis. Research on premenopausal women with magnesium supplement displayed significant BMD hype in femoral neck bones and lumbar spine (Jani, 2019).
- **Boron**: It is naturally gained from prunes, resins, dried apricots and avocados. Helps in retaining bone density by preventing calcium depletion and bone demineralization. Human studies were done to see the result of boron on bone density. It was seen that urinary calcium outflow decreased by 40% and magnesium lowered by 33%. Both calcium and magnesium are needed for bone health. Boron leads to estrogen, Vitamin D stability and half-life enhancement with avoidance of bone demineralization (Nataranjan et al., 2019).
- **Copper**: Copper removes free radicals and cross links collagen. Herbs and phytochemicals. Majorly used in bone protection from

outer damage and endogenous substances. Copper activates osteoclasts as it is a cofactor for antioxidant enzymes and gives out free radicals in the body. Also, it activates lysyl oxidase to crosslink lysine in elastin and collagen. The percentage daily intake should be 0.9 mg/day in adults for bone deficiency. Copper deficiency rats were studied for efficiency of copper supplements on bone fragility which resulted in lowering of torque (Nieves, 2021).

3.6.1.2 DAIRY PRODUCTS

Osteoporosis can be overcome by use of calcium rich milk and milk product intake in daily diet. Moreover, substances like proteins, lipids, sodium, zinc, phosphate are more dominant than calcium. Constituents like milk basic proteins (MBP), lactoferrin, and phospeptide are functional elements that are valuable in bone healing. Basic milk protein consists of Transforming growth factor, vascular endothelial growth factor and Insulin-like growth factor that are active molecules. MBP works by increasing differentiation and spread of osteoblasts and inhibits resorption and generation of osteoclasts. MBP has cystatin C, helps in calcium release from bone by inhibiting cysteine protease. This occurs in the presence of thrombin, Interleukin-1, prostaglandin E2. MBP raises osteocalcin, which is a non-collagenous protein of bone. Effect of MBP were studied on menopausal women. As a result, an increase in bone mass density was seen in the MBP group with 1.21%, Placebo group with 0.66% increase in BMD. Lactoferrin is a multifunctional element of milk, it is a glycoprotein that denotes a strong action on bone. It causes depression in osteoclast cells and rapid multiplication of itself with no cell death of osteoblast. Lactoferrin provides bone building because of low density lipoprotein (Karpouzos et al., 2017; Rajput et al., 2018).

3.6.1.3 FOOD

Fish like sardine, swordfish, and salmon are source of Vitamin D and way safe to decrease changes of hip fracture by boosting Vitamin D levels. Research has proved for 33% hip fracture risk cutoff by fish diet. Fish are high in calcium content and omega-3 fatty acids like EPA and DHA. Omega-3 fatty acid reduces PGE-2 and IL-1 concentration. Lab studies determine the activating osteolysis process. Prevention of hip fracture is associated with consuming DHA and EPA rich fish products. Vegetables and fruits are full of nutrients like zinc, iron, potassium, magnesium which is prone to raised bone density levels in both sexes. Various nutrients like Vitamin C and niacin proves helpful in increased bone formation of arms. Nutrients like phosphorus, folic acid, potassium, magnesium and Vitamin C proteins recover bone loss and improve mineral density in the bone of the neck. Vitamin C is associated with bone collagen hydroxylation. Vitamin C needed in the body can be made by fruits and vegetables and supplements. Food with alkali neutralizes acid and thus prevents bone degradation. Bone catabolism is decreased by inducing potassium in food. Vitamin C renders depletion in risk of fracture by osteocalcin decarboxylation (Khan et al, 2016).

3.6.1.4 HERBS AND PHYTOCHEMICALS

Alfalfa has an increased osteoclasto genesis effect on bone. The active constituent of alfalfa is ipriflavone and works by enhancing alkaline phosphatase and type 1 collagen. Also shows action of decreased secretion of cathepsin K, L & β glucuronidase. Red clover has key constituents genistein and daidzein. (Arcoraci et al., 2017) It functions by inhibiting tyrosine kinase and topoisomerase II. Inhibits bone resorption by osteoclasts. Horsetail has active compounds silica, quercetin, kaempferol and luteolin that helps in formation of collagen, activates anti- inflammatory pathway and decreases bone marrow mononuclear cell differentiation respectively. Silica leads to bone mineralization, quercetin and kaempferol inhibits osteoclasts differentiation and luteolin decreases bone resorption. Citrus fruit contains high amounts of limonoids and flavonoids which tackles oxidative stress and leads to low bone resorption (Kaleem & Ahmad, 2018; Keservani & Sharma, 2014; Keservani et al., 2016a, 2016b). Green tea effects of activation by caspase-3 by active element epigallocatechin-3-galate and modulates osteoclasts cell death. Also, it suppresses osteoclast formation by metalloproteinase-a matrix suppression or inhibiting IL-6 and down regulation of RANKL tends to osteoclast differentiation suppression. Enhancement of osteoblastogeneis is done by increment of alkaline phosphatase and suppress TNF-α. Berries contains anthocyanidins and plum contains polyphenols, both inhibits RANKL receptors further leads to inhibition. Resveratrol enhances osteoblastogenesis in bone and functionalizes to modulate osteoclast and osteoblast gene expression. Also reduce oxidative stress and inflammatory cytokines and shows antioxidant and anti-inflammatory action on bone (Nieves et al., 2021; Rajput et al., 2018).

3.6.1.5 MACRONUTRIENTS

Proteins help in development and maintenance and growth of bone are carried out by proteins. Nevertheless, it has contributed to negative calcium homeostasis and decrease bone content. It works by enhancement of endogenous acid that leads to movement of calcium in bone to salt forms and therefore regulates acidity. 30% of bone is made up of protein. Metabolism of bone is looked upon by dietary protein. Hampered protein level in bone causes disturbance in bone development and growth. Vegetative protein has no effect on hip fracture cases. Carbohydrates shows minus effect on calcium absorption and reabsorption of estrogen from intestine. In another research carbohydrate successfully denoted reduction in osteoporosis and fracture. Lipids high levels acquaints for frequent fracture development. Depletes bone density of spine and ends of radius. 2 types of lipids monosaturated and polysaturated deals with low BMD in neck bone. Olive oil gives positive response in BMD of radius ends and it is related with Vitamin E concentration. Upsurge of lipids directs for hypercalciuria and magnesium levels elevation in urine and opposite balance of calcium and magnesium. Lipid content level is indirectly proportional to calcium absorption. Increased lipid increases retinol and follows resorption of bone but high fat (Singh et al., 2017b) diet will lack other important nutrient for bone enhancement (Karpouzos et al., 2017).

3.6.1.6 MICRONUTRIENTS

Vitamin A is stored in the body in retinol form. Over consumption of Vitamin A has the opposite effect on bone. Hypersecretion of Vitamin A retinoic acid forms more osteoclast and decreases osteoblast. This process further initiates bone resorption and distorted bone formation and some cases show hypercalcemia too. Research has proved action of Vitamin A by osteoclast and osteoblast for bone resorption, decrease bone formation and pain in skeletal frame. Excessively sodium has a risk for osteoporosis. Bone and cartilage disorder is due to deficiency of copper. Zinc incompetence during growth decrease bone density and magnesium enhance bone quality. Bone metabolism is caused by manganese. Increasing bone density is a function of a high Vitamin C diet. Vitamin B12 deficiency depletes bone formation and management. Vitamin K regulates blood coagulation and treats arthritis, diabetes and wrinkles. It is obtained from cabbage, spinach, beetroot, broccoli and cauliflower in small quantities. Low Vitamin K in plasma results in hip fracture risk so Vitamin K supplements are used as studies showed osteocalcin level increment in bone. This prevented fracture risk and raised bone density. Vitamin B12 is necessary for bone health and deficiency leads to osteoporosis and negative impact on skeletal system and BMD (Karpouzos et al., 2017).

3.6.1.7 MISCELLANEOUS

Prebiotics are components which are present in the intestinal flora of humans in the form of bacteria. They can produce fatty acids and lactate metabolism. Production of Vitamin C, K, D and folate are mechanisms that improve bone health by probiotics. Where nutrients are associated with bone formation, reducing PTH, enhancing absorption of minerals, initializing antioxidants. Bacteria found in bone are lactobacilli and bifidobacteria.

Probiotics are the supporting food substances for growth of bacteria that are fermented by the intestine and generate a positive effect in the host. Bone mineralization is signified by non-digestible oligosaccharides. Studies were done to find out the effect of galacto-oligosaccharide in rats. To conclude the research calcium absorption was eventually increased with maximum magnesium absorption, calcium obtained in femur and magnesium, calcium reuptake and density of tibia and femur.

PUFA Polyunsaturated fatty acid has two classes' n-6 and n-3 fatty acids. N-6 is located in vegetable oil and gained from linoleic acid and n-3 fatty acid like EPA &DHA present in fish/ fish oil. Studies have shown higher omega-3 and omega 6-fatty acids. Long chains of PUFA are associated with calcium ATPase rise and thus management of calcium in the intestine. N-6 fatty acid in cell membrane replaced by n-3 fatty acid.

Melatonin is a hormone produced by pineal gland that is released and made at night. It increases growth hormone and upsurge metabolism of osteoclast and elevates corticosteroid. Also, it shows antioxidant properties and infuses procollagen in bone matrix. Higher osteoprotegerin level stops osteoclast differentiation. The study on Wistar rats shows greater bone volume, also it showed large flexure, bone rigidness (Rajput et al., 2018).

3.6.2 NUTRIENTS FOR BONE RESTORATION BASED ON CHEMICAL CLASSIFICATION

3.6.2.1 PHYTOESTROGENS

They are estrogen like constituents obtained from natural sources. They act like agonists or antagonists of estrogen. Generally, Phytoestrogens are differentiated into 1) isoflavone which are principally present in soy products that are genistein, daidzein and glycitein. 2) Chickpeas, legumes are source from flaxseed, cereal bran, and active compounds for activity are enterolactone and enterodiol. 3) Coumestans or coumestrols is originated from alfalfa and clover and estrogenic action can be established by black cohosh. For formulating high doses of dietary supplement Phytoestrogens are purified and concentrated. In many naturally obtained food products isoflavones are present but as glycoside that has low absorptivity up on oral intake. Treatment with these leguminous plants will enhance absorption and thus affect bioavailability. Activity of isoflavones in case of fracture is activating osteoblast for its effect and inactivating osteoclast in bone. Also, mainly shows agonism of estrogen receptors by means of α and β estrogen binding by isoflavones. Estrogen receptor dependent and independent both effects of isoflavones are seen on the bone property of some isoflavones. Genistein like isoflavones that are dietary supplement have bone resorption defeating activity in post-menopausal women. Isoflavone is a proven supplement for bone calcium retention. In a study for isoflavone action on bone, soy protein with isoflavone stated fundamental decrease in bone resorption and depletion of type 1 procollagen N- peptide. Isoflavonea glycone part used for enhancing bone function and their properties. Bone density is boosted with trimming down fracture threat. Women with intake of regular soy diet less than 13.26 mg soy/ day have reduced risk of fracture than low soy diet women (genistein). The soy protein combined with other nutrients resulted in BMD and resorption elevation. Food containing isoflavones in aglycone form are considered as medical food.

Red clover is a leguminous plant obtain which treats cholesterol and osteoporosis and menopausal effects. Safety and duration of time are inversely proportional in recovering menopause indications. Also, it is used in treatment of breast and hormonal cancers.

Ipriflavoneis an extracted substance of isoflavone of soy. These daidzein derived ipriflavones intake has revealed bone mass development. Also, it had side effects like diarrhea, dizziness, stomach pain and WBC count depletion.

Black cohosh helps in improving skeletal health by developing estrogenic activity. It is isolated from the roots of buttercup. Side effects like headache, GI disturbance, weight issues are to a lesser extent. It is a herbal dietary supplement marketed in liquid, capsule, tablet form with different doses ranging from 120–540 mg. Bone formation is seen in 12 weeks of black cohosh supplements ingestion in menopause. (Rajput et al., 2018).

3.6.2.2 DEHYDROEPIANDROSTERONE

A steroid hormone DHEA is the primary element in synthesis of androgens and estrogens in male and female respectively. Assertion relating to supplement of energy, immunity, fat cutting, muscle and anti-aging. It has acquired potential for BMD in hip, lumbar spine and trochanter region. Also associated with few adverse effects like acne, masculinizing, disturbed liver function (Karpouzos et al., 2017).

3.6.2.3 ANTIOXIDANTS:

Oxidation renders risk diseases and bone loss and decreases mechanical strength. As bone grows older chances of affecting survival of osteocytes, osteoblasts and osteoclast increases. So, antioxidants inhibit disease and develop healthy condition by involving vit E, C, A, flavonoid, omega 3 fatty acid in diet. It provides bone strength by making an alkaline environment decreasing calcinuria excretion condition and generating other active constituents like flavonoids and phenols. Flavonoids are sourced from tea, wine, juice, and vegetables. Flavanones, anthocyanins, flavan-3-ols, polymers, polymers, flavanols, flavones, isoflavones are the different classes of flavonoids. The intake amount of flavonoid is proportional to BMD in spine and hip bone. An antioxidant quercetin improves skeletal condition (Kaleem & Ahmad, 2018). It is obtained from fruits, apples, tea, red wine, parsley, onion, and other supplements. Antioxidants isolated from tea and green tea are related so blended results in bone density and fracture.

Omega 3 fatty acid is provided from eggs, flaxseed, fish and walnuts. They are the polyunsaturated fatty acids with α- linoleic acid and eicosapentanoic and docosahexaenoic acid.

Omega helps in skeleton development by antioxidants, maintenance of calcium absorption by intestine. Omega-6 fatty acid is predominant in oils obtained from vegetables, poultry, seeds and nuts. Low levels of omega-6 when compared with omega 3 may have better bone density. Fish oils, flaxseed and walnuts are some items which are well nourished with omega-3 are mostly preferred. Olive oil, coconut oil and palm oil are less in omega-6 and should be applied in diet rather than oil from omega-3 effect acid in cottonseed, safflower, corn, sunflower. Increased calcium in diet has a direct relation to omega-3 fatty acid on bone health (Nieves, 2021).

Carotenoids are plant derived pigments and phytoconstituents that's known for possessing antioxidant power. Sources of carotenoids are carrots, bell pepper, tomatoes, vegetables, spinach, sweet potato and a lot more.

Different forms of carotenoids are α, β, γ carotene and β cryptoxanthin that could be transformed in Vitamin A. Retinol untransformed carotenoids like lutein, lycopene and zeaxanthin also act as antioxidant lower BMD are evaluated for low lycopene, β carotene and cryptoxanthin concentration. A result of study determined towered BMD level in assistance to high carotenoid contained food. Another study case had low fracture risk with raised intake of lycopene and carotenoids (Pandey et al., 2018).

3.6.2.4 BICARBONATES

A lower pH background for bone formation deals with depressing skeletal conditions. Alkaline substances neutralize acids and urinary calcium excretion is shortened and calcium is penetrated faster and bone resorption decreases eventually. Bicarbonates that are present in vegetables, fruits, potato spinach are potassium based, sodium bicarbonate is also called baking soda used in food. In postmenopausal women it was seen that when serum level of bicarbonates depleted BMD of total body and spine decreased in loss of bone depletes with higher HCO_3 plasma concentration with maintained BMD (Khan et al., 2016; Nieves, 2021).

3.6.2.5 VITAMIN B AND HOMOCYSTEINE

Water dissolving vitamins like pyridoxine, folate and cyanocobalamin tend to give strength to bone. The adult folate daily intake is 400 μg. Folate is the Vitamin B 9 found in dark green vegetables, breads, nuts, cereals and folic acid as fortified food. Vitamin B12 RDA value is about 2.4 μg, while Vitamin B12 is acquired from shellfish, beef, cheese, eggs. Currently bone fragility and hip fracture is related to homocysteine serum level and raised homocysteine level is due to lack of folic acid, Vitamin B6, Vitamin B12. Decreased bone density is linked to low Vitamin B12 in females and male which ultimately lead to fracture and delicate bone structure. A follow up research study showed insignificant risk of bone breaking due to increased Vitamin B6 concentration (Nieves, 2021; Singh, 2017).

3.7 SAFETY AND RISK FACTORS OF NUTRACEUTICALS IN BONE REPAIR

Millions of people consume nutraceuticals daily and they are healthy. Even if not approved by authority, nutrient based products are marketed. This is because the items used in formulating nutraceutical and pharmaceuticals are identical. Some types of nutraceuticals can show potential health risks. Risk factors associated with nutraceutical primarily is raw material that can be adulterated by contaminants by biased practices. A review of risk of dose of Vitamin D supplement resulted in insufficiency to predict profit and damage by Vitamin D and calcium supplements for inhibiting fracture. Risk that aroused later was impurities, A herbal (Rane et al., 2020a; Rane et al., 2020b; Rane et al., 2020c; Sharma et al., 2020) dietary supplement consisted of heavy metal which was related to risk as included mercury content. Effect of nutraceutical can be seen according to the time it is taken. Gravidity, safety dose in children, supplements to be given for chronic illness are different factors that are to be justified for harmless products like multivitamins or minerals when given regularly. To a relief, a clinical study stated the safety of these products to be more than 10 years. (Qureshi & Pisulkar, 2021).

3.8 CONCLUSION AND FUTURISTIC APPROACH

Bone physiology and maintenance of bone health depends mostly on supply of Vitamin D and calcium, scarcity of calcium develops need for supplementary calcium intake. Calcium is primarily consumed through food whereas dietary intake of Vitamin D is necessary for bone formation. In some cases, nutraceuticals are used to regulate skeletal growth and development. The dietary products supplier should have an idea about safety status, efficacy range and data regarding the nutraceuticals. For many circumstances the dose of nutraceutical is equal to the level of pharmacologic response but nutrients from food are not at a level of nutraceuticals. Nevertheless, it is clearly proven that in conditions like osteoporosis or fracture treatment with proper medicated therapy it is must and it cannot be solely replaced by nutraceuticals (Telessy, 2019). The market needs for natural ingredients and organic treatment aids have been catered by providing nutraceutical with utilizing, isolating and extracting various nutrients that can supplement the main therapy when used as an adjuvant. Moreover, nutraceuticals properties are satisfactorily expected for therapeutic with lower side effects. Preventive therapies and methodologies with low adverse effects are preferred for bone health because treatment of low bone density have varying regimen of drug usage for quality life (Ronis et al., 2018). Different drug strategies are available but none of them has a higher level of efficacy with less side effects and cost effectiveness. So naturally derived plant originated products are good substitutes to outdated drug regimens. Showing similar effects as that of pharmacological drug response, they are not accepted by the health specialists as an alternative approach. The biggest drawback in natural agents is that it fails to explain detailed pharmacokinetics, of formulation and non-targeting effects, which reflects in depletion of success ratio in clinical trials (Valsamidou et al., 2021). Parameters that matter for evaluating nutraceuticals intake on a regular basis are its quantity, consumption methods, dosing frequency, safety and efficacy, for skeletal health management (Schmitz et al., 2020). Combinational therapies of different nutraceuticals or nutraceuticals with drug molecules can be used for well-functioning of the skeletal system and to decrease the incidences of bone abnormalities.

KEYWORDS

- **bone**
- **bone marrow density**
- **composition**
- **milk basic proteins**
- **nutraceuticals**
- **structure and anatomy**

REFERENCES

Arcoraci, V., Atteritano, M., Squadrito, F., D'Anna, R., Marini, H., Santoro, D., Minutoli, L., et al., (2017). Antiosteoporotic activity of genistein aglycone in postmenopausal women: Evidence from a post-hoc analysis of a multicenter randomized controlled trial. *Nutrients, 9*(2), 179. https://doi.org/10.3390/nu9020179.

Föger-Samwald, U., Dovjak, P., Azizi-Semrad, U., Kerschan-Schindl, K., & Pietschmann, P., (2020). Osteoporosis: Pathophysiology and therapeutic options. *EXCLI J., 19*, 1017–1037. https://doi.org/10.17179/excli2020-2591.

Jani, R., (2019). Nutraceuticals for bone health in pregnancy. In: Bhatt, P., Miraghajani, M. S., Pathak, S., & Pathak, Y., (eds.), *Nutraceuticals for Prenatal, Maternal and Offspring's Nutritional Health* (pp. 211–234). CRC Press. https://doi.org/10.1201/9780429437632-15.

Kaleem, M., & Ahmad, A., (2018). Flavonoids as nutraceuticals. In: *Therapeutic, Probiotic, and Unconventional Foods* (pp. 137–155). Elsevier. https://doi.org/10.1016/B978-0-12-814625-5.00008-X.

Karpouzos, A., Diamantis, E., Farmaki, P., Savvanis, S., & Troupis, T., (2017). Nutritional aspects of bone health and fracture healing. *Journal of Osteoporosis, 2017*, 1–10. https://doi.org/10.1155/2017/4218472.

Keservani, R. K., & Sharma, A. K., (2014). Flavonoids: Emerging trends and potential health benefits. *Journal of Chinese Pharmaceutical Sciences, 23*(12), 815.

Keservani, R. K., Kesharwani, R. K., Sharma, A. K., Vyas, N., & Chadoker, A., (2010b). Nutritional supplements: An overview. *International Journal of Current Pharmaceutical Review and Research, 1*(1), 59–75.

Keservani, R. K., Kesharwani, R. K., Vyas, N., Jain, S., Raghuvanshi, R., & Sharma, A. K., (2010a). Nutraceutical and functional food as future food: A review. *Der Pharmacia Lettre, 2*(1), 106–116.

Keservani, R. K., Sharma, A. K., & Kesharwani, R. K., (2016a). Nutraceutical and functional foods for cardiovascular health. In: *Food Process Engineering* (pp. 257–278, 291–312). Apple Academic Press, CRC Press. ISBN: 9781771884020.

Keservani, R. K., Sharma, A. K., & Kesharwani, R. K., (2016b). Medicinal effect of nutraceutical fruits for the cognition and brain health. *Scientifica*, 10. Article ID 3109254. doi: 1 0.1155/2016/3109254.

Keservani, R. K., Sharma, A. K., & Kesharwani, R. K., (2017). An overview and therapeutic applications of nutraceutical and functional foods. *Recent Advances in Drug Delivery Technology*, 160–201. ISBN:9781522507543.

Keservani, R. K., Sharma, A. K., & Kesharwani, R. K., (2020). *Nutraceuticals and Dietary Supplements: Applications in Health Improvement and Disease Management*. CRC Press. ISBN: 9781771888738.

Khan, M. U., Elhassan, G. O., Khalilullah, H., & Eid, E. E. M. (2016). Nutraceuticals – an opportunity of healthcare: A review. *World Journal of Pharmaceutical Research, 5*(11), 229–239.

Natarajan, T. D., Ramasamy, J. R., & Palanisamy, K., (2019). Nutraceutical potentials of synergic foods: A systematic review. *J. Ethn. Food, 6*(1), 27. https://doi.org/10.1186/s42779-019-0033-3.

Nieves, J. W., (2021). Nutrients beyond calcium and vitamin D to treat osteoporosis. In: *Marcus and Feldman's Osteoporosis* (pp. 1679–1693). Elsevier. https://doi.org/10.1016/B978-0-12-813073-5.00071-X.

Nwosu, O. K., & Ubaoji, K. I., (2020). Nutraceuticals: History, classification and market demand. In: Egbuna, C., & DableTupas, G., (eds.), *Functional Foods and Nutraceuticals* (pp. 13–22). Springer International Publishing: Cham. https://doi.org/10.1007/978-3-030-42319-3_2.

Pandey, M. K., Gupta, S. C., Karelia, D., Gilhooley, P. J., Shakibaei, M., & Aggarwal, B. B., (2018). Dietary nutraceuticals as backbone for bone health. *Biotechnology Advances, 36*(6), 1633–1648. https://doi.org/10.1016/j.biotechadv.2018.03.014.

Qureshi, A., & Pisulkar, G., (2020). Evaluation of nutritional status in fracture healing of elderly patients. *JEMDS, 9*(52), 3939–3944. https://doi.org/10.14260/jemds/2020/862.

Rajput, R., Wairkar, S., & Gaud, R., (2018). Nutraceuticals for better management of osteoporosis: An overview. *Journal of Functional Foods, 47*, 480–490. https://doi.org/10.1016/j.jff.2018.06.013.

Rane, B. R., Bharath, M. S., Patil, R. R., Keservani, R. K., & Jain, A. S., (2020b). Novel approaches in nutraceuticals; Chapter 11. In: Rajesh, K. K., Raj, K. K., & Anil, K. S., (eds.), *Enhancing the Therapeutic Efficacy of Herbal Formulations Through Novel Drug Delivery*

Systems (pp. 241–266). IGI Global International Publisher, Pennsylvania, USA. ISBN13: 9781799844532.

Rane, B. R., Patil, A. S., Keservani, R. K., & Jain, A. S., (2020a). Novel approaches in herbal formulation; Chapter 2. In: Rajesh, K. K., Raj, K. K., & Anil, K. S., (eds.), *Enhancing the Therapeutic Efficacy of Herbal Formulations Through Novel Drug Delivery Systems* (pp. 43–68). IGI Global International Publisher, Pennsylvania, USA. ISBN13: 9781799844532.

Rane, B., R., Tadavi, S. A., & Keservani, R. K., (2020c). Naturopathy; Chapter 12. In: Anil, K. S., Raj, K. K., & Surya, P. G., (eds.), *Herbal Product Development* (pp. 321–347). Apple Academic Press, CRC Press, Taylor & Francis Group. ISBN: 9781771888776.

Roberts, J. L., & Drissi, H., (2020). Advances and promises of nutritional influences on natural bone repair. *J. Orthop. Res., 38*(4), 695–707. https://doi.org/10.1002/jor.24527.

Ronis, M. J. J., Pedersen, K. B., & Watt, J., (2018). Adverse effects of nutraceuticals and dietary supplements. *Annu. Rev. Pharmacol. Toxicol., 58*, 583–601. https://doi.org/10.1146/annurev-pharmtox-010617-052844.

Sale, C., & Elliott-Sale, K. J., (2019). Nutrition and athlete bone health. *Sports Med., 49*(S2), 139–151. https://doi.org/10.1007/s40279-019-01161-2.

Schmitz, S. M., Lopez, H. L., Mackay, D., Nguyen, H., & Miller, P. E., (2020). Serious adverse events reported with dietary supplement use in the United States: A 2.5 year experience. *Journal of Dietary Supplements, 17*(2), 227–248. https://doi.org/10.1080/19390211.2018.1513109.

Sharma, A. K., Keservani, R. K., & Gautam, S. P., (2020). *Herbal Product Development*. Apple Academic Press, CRC Press, Taylor & Francis Group. ISBN: 9781771888776.

Singh, P., Kesharwani, R. K., & Keservani, R. K., (2017a). Antioxidants and vitamins: Roles in cellular function and metabolism. In: *Sustained Energy for Enhanced Human Functions and Activity* (pp. 385–407). Academic Press. ISBN:9780128054130.

Singh, P., Kesharwani, R. K., & Keservani, R. K., (2017b). Protein, carbohydrates, and fats: Energy metabolism. In: *Sustained Energy for Enhanced Human Functions and Activity* (pp. 103–115). Academic Press. ISBN:9780128054130.

Singh, V., (2017). Medicinal plants and bone healing. *Natl. J. Maxillofac. Surg, 8*(1), 4. https://doi.org/10.4103/0975-5950.208972.

Télessy, I. G., (2019). Nutraceuticals. In: *The Role of Functional Food Security in Global Health* (pp. 409–421). Elsevier. https://doi.org/10.1016/B978-0-12-813148-0.00024-4.

Valsamidou, E., Amerikanou, C., Tzavara, C., Zoumpoulakis, P., Mariolis-Sapsakos, T. D., Skarpas, G., & Kaliora, A. C., (2021). Nutraceuticals for knee osteoarthritis pain relief. Results from a preliminary randomized clinical trial. *Dietetics, 1*(1), 2–14. https://doi.org/10.3390/dietetics1010002.

Whiting, S. J., Kohrt, W. M., Warren, M. P., Kraenzlin, M. I., & Bonjour, J. P., (2016). Food fortification for bone health in adulthood: A scoping review. *Eur. J. Clin. Nutr., 70*(10), 1099–1105. https://doi.org/10.1038/ejcn.2016.42.

CHAPTER 4

THERAPEUTIC POTENTIAL OF NUTRACEUTICALS IN THE MANAGEMENT OF SCOLIOSIS AND RELATED COMPLICATIONS

GUNIKA GULATI, SIDDHI BAGWE PARAB, DHRUV SANJAY GUPTA, and GINPREET KAUR

Shobhaben Pratapbhai Patel School of Pharmacy and Technology Management, SVKM'S NMIMS, Mumbai, Maharashtra, India

ABSTRACT

Scoliosis is a very common spinal deformity which affects 2–3% of the population globally and is characterized by the abnormal curvature of the spine. There key complications associated to with scoliosis are pain, bone loss, digestion issues, abnormal skeletal growth, neurotransmitter imbalances, hormonal imbalances, etc. It is also linked to several other syndromes including Marfan's adn Ehlers-Danlos syndrome. Currently, the treatment options are limited considering our limited knowledge on the pathogenesis of scoliosis and that a majority of cases are idiopathic. The treatment options include physical treatment (bracing, exercise, and surgery), pharmacotherapy (mainly includes symptom management – analgesics, anti-inflammatory, etc.), and nutraceuticals. Alternative therapies are being sought due to high cost of treatment, and in order to reduce dependence on synthetic drugs. Nutraceuticals are effective, natural, comparatively cheaper, and safer for long term use. Research through *in vivo* and *in vitro* studies has shown links between scoliosis and vitamin deficiencies which can be treated by use of nutraceuticals with further trials. Till now, results of several

Nutraceuticals and Bone Health. Deepak Sharma, Madan Mohan Gupta, Anil K. Sharma, Raj K. Keservani, Rajesh K. Kesharwani (Eds.)

studies involving vitamins like Vitamin D3, C, B6, B9, B12, K2; minerals like calcium, magnesium, manganese; hormones like melatonin, estrogen and phytoconstituents like quercetin, genistein, resveratrol, curcumin and cannabidiol have shown promise in assisting the treatment of scoliosis. Nutraceutical therapy, used concomitantly with traditional treatment options may help to manage complications in a better manner compared to using synthetic drugs used currently. Evidence shows that using nutraceuticals as concurrent therapy along with physical therapy shows positive results in reducing curve progression. Nutraceuticals are be highly effective particularly in early stages of scoliosis and not only help reduce our dependence on synthetic drugs but also making the treatment comparatively cheaper. This chapter focuses on the potential nutraceutical therapies that may be used in the management of scoliosis and related complications.

4.1 INTRODUCTION

Scoliosis, first described by Hippocrates, comes from the Greek word scoliosis, which may be translated to crooked or twisted (Talasila, Gorantla, & Thomas, 2017). It is a three-dimensional spinal deformity, characterized by an abnormal curvature of the spine (Fadzan & Bettany-Saltikov, 2018). In India, there has been a significant increase in its incidence, with nearly 10 million people being affected annually, and a global occurrence rate of 2–3% of the population is observed. Over 80% of all cases are idiopathic, while neuromuscular, syndromic, or congenital predispositions make up the remainder. Scoliosis has been seen to affect about 4% of adolescents between the ages of ten and sixteen (Grivas, 2012). According to a recent study, females between the ages of 10 and 18 have a higher risk of developing adolescent idiopathic scoliosis (AIS) than males in the same age group (Fadzan & Bettany-Saltikov, 2018).

Scoliosis may be broadly categorized into two types: structural and non-structural. In structural scoliosis, the curving is characterized by fixed rotational prominence, where the vertebrae are turned towards the convex side of the curve. The curve is subject to worsening with time. Its subtypes include idiopathic, congenital, neuropathic, de novo, traumatic, and iatrogenic scoliosis. In non-structural scoliosis, there is no rotatory component to the curve, neither does it worsen with time. The subtypes of non-structural scoliosis include: postural, sciatic and inflammatory scoliosis. Scoliosis may also be classified based on the age of onset, as infantile (affecting infants under age of 3), juvenile (affecting children between the age 3 and 9), adolescent (age 10–18), and adults (18 and above) (Machida, Weinstein, & Dubousset, 2018; Goldstein & Waugh, 1973; Silva & Lenke, 2010; Lenke & Dobbs, 2007; Arlet, Odent, & Aebi, 2003; Marc et al., 2021; Allam & Schwabe, 2013). Curve advancement, which is linked to the patient's rapid skeletal growth (most typically during the pubertal growth spurt), is the most serious condition associated with scoliosis (Grivas, 2012). The most commonly observed characteristics of scoliosis include uneven shoulders, a protruding shoulder blade in the upper back, unequal hip arrangement, and ribs that bulge out when leaning forward (Marc et al., 2021).

Physical examination tests, such as Adam's forward bend test, may be used for its detection. Scoliotic patients experience lateral bending of the spine, but a noticeable rib hump will provide a confirmatory diagnosis, due to the spinal curvature. A scoliometer is the device used to measure the angle of the curvature. Curves less than 5° are often deemed negligible and do not require further investigation. Curves between 5° and 9° require re-examination every 6 months, whereas curves greater than 10° require a radiological examination (such as an X-ray or MRI) to determine the Cobb angle. A curve of 10°–24° is considered mild, 25°–39° is considered moderate and above 40° is severe (Horne, Flannery, & Usman, 2014; Langensiepen et al., 2013).

There are several theories proposed regarding its etiology. Some of them include genetic factors (providing a hereditary basis), biomechanical theories, neurological theories, as well as explanations based on environmental and metabolic factors, etc. (Rajasekaran, Sukumaran, & Shetty, 2020; Acaroglu et al., 2012; Kikanloo, Parshottam Tarpada, & Cho, 2019).

Possible risk factors for scoliosis include genetic predisposition, certain sports/activities like rhythmic gymnastics, joint hypermobility, growth factors, and delayed puberty, late onset of menarche and low BMI, to name a few (Fadzan & Bettany-Saltikov, 2018). Scoliosis may lead to serious health complications, including bone loss and irregular skeletal growth. In addition, several patients face difficulties such as postural control, neurological impairment, neurotransmitter imbalance, gastrointestinal issues, and pain. Diseases like congenital heart disease have been linked with scoliosis, owing to the disease itself or a consequence of cardiac surgery in immature thoracic cage. A recent theory suggested that DiGeorge (22q11.2 deletion) syndrome may play a role in its pathogenesis (Cohen et al., 2021).

The current treatment strategies include pharmacotherapy, physical therapy, and surgical intervention. The pharmacotherapy includes usage of analgesics, and NSAIDs (such as acetaminophen and ibuprofen respectively). In cases of severe curves where pain persists, opiates maybe administered

(Everett & Patel, 2007; Rehman, 2020). The requirement of physical therapy depends on the severity and gradual worsening of the curve. Conservative treatment is generally applied for curvature below 45° in growing adolescents. This includes physiotherapeutic scoliosis specific rehabilitation (PSSR) like Schroth Therapy, chiropractic care and spinal bracing (using braces like Boston orthosis) intended to reduce the curvature below the degree requiring surgical intervention. For patients with curvatures exceeding 45°, surgery is indicated. However, with reference to Adolescent Idiopathic Scoliosis (AIS), studies show no evidence of long-term complications, with surgical intervention showing little to no benefits (Marc et al., 2021; Everett & Patel, 2007; Kuru et al., 2016; Letts et al., 1992).

However, the current treatment scenario has certain drawbacks. Pharmacotherapeutic intervention offers a short-term solution, which may pose several challenges, as outlined further. The marketed medications available for osteoporosis, which is a major associated complication, are cost intensive. Treatment for patients having severe curves is also expensive, considering cost of surgery and subsequent post-operative care. Physical therapy for mild curves, such as bracing, poses a patient compliance issue considering the physical discomfort of wearing braces. Studies are being undertaken to cultivate a better understanding of the etiology of scoliosis, in order to determine treatment options that are inexpensive, efficacious and provide a greater number of benefits in comparison to generally prescribed pharmacotherapy. Although the scope of pharmacotherapy remains limited, newer concepts are emerging for surgical intervention, like magnetic controlled growth rods, that may be used in the management of early-onset scoliosis, so as to conduct lengthening non-invasively after initial surgery as opposed to traditional growing rods which require regular surgical lengthening (Tsirikos & Roberts, 2020).

The major aim of using various treatment strategies is to help prevent curve progression. It is important to note that three of five predictors for curve progression in AIS patients have nutritional implications, namely- age of onset of menarche, bone mineral density (BMD) and relatively shortened spinal cord (in relation to the vertebral column). Macronutrients and micronutrients have been reported to affect the onset of puberty, BMD and fibrosis of nerve tissues, which in turn may influence the progression of the disease (Marc et al., 2021).

Nutraceuticals have emerged as the best line of therapy for the management of nutritional deficiencies associated with scoliosis (such as Vitamin D and melatonin deficiencies). Furthermore, nutraceuticals also positively influence disease related complications, mainly bone loss and neurotransmitter imbalance. Scoliosis is a complex disorder, and treatment must be tailored to the patient's individual needs, especially in idiopathic cases. Nutraceuticals may be used to manage scoliosis concomitantly with first line therapies for maximizing patient benefit. The highlight of this chapter is to present an overview of nutraceuticals that have the potential to manage scoliosis and related complications. The rationale of treatment includes regulation of bone loss using vitamins, and minerals, reduction of oxidative stress using antioxidants, decreasing hormonal, and neurotransmitter imbalances by dietary supplements chiefly composed of plant extracts and phytoconstituents, which may also aid to the overall maintenance of good health. This treatment strategy will help manage complications using natural agents, in addition to providing a boost to patients receiving physical therapy. This chapter provides a brief insight into the involvement of nutraceuticals in the treatment of scoliosis and their varied actions. Also, this chapter provides a brief highlight on the recent clinical and pre-clinical investigations conducted to establish the safety and efficacy studies of nutraceuticals.

4.2 COMPLICATIONS ASSOCIATED WITH SCOLIOSIS

Scoliosis is a complex disorder, characterized by multiple complications such as bone loss, hormonal, and neurotransmitter imbalances, pain, digestion, etc. All of these may contribute to the worsening of curve progression. All of these are severe, and hence, may interfere with the day-to-day life of the patient and cause other health issues subsequently. Hence, managing these complications will help a patient lead a better life and also prevent possible curve progression. The treatment of scoliosis is very holistic with regards to the management of these complications. The major complications associated to scoliosis are detailed in Table 4.1.

4.3 NUTRACEUTICALS USED IN MANAGEMENT OF SCOLIOSIS

Nutraceuticals have an emerging potential for usage in the management of scoliosis and associated complications, owing to a variety of benefits offered. These include their ready availability in nature, established safety profile for long term administration, effectiveness for treatment and lower cost when compared to synthetic drugs. The nutraceuticals that have been discussed in this chapter for the management of scoliosis can be broadly categorized into the following categories – hormones, vitamins and minerals, nutraceuticals from plant sources (including plant extracts, and phytoconstituents) and nutraceuticals from animal sources. The different categories of nutraceutical

TABLE 4.1 Complications Associated to Scoliosis

Complications	Pathophysiology	Studies	References
1. Bone Loss	Decreased osteoclast levels. Abnormal bone metabolism and Osteopenia in AIS patients.	29.5%–38% of AIS patients suffer from osteopenia.	Ng et al., (2018); Suh, Eun, and Lee, (2010); Batista et al., (2014); Annika et al., (2013); Balaji V., Chin K., Samir F., Kouklinos A., Tucker S. (2014); Jaovisidha et al., (1998); Cheng, Guo, and Sher, (1999); Routh et al., (2005); Marc et al., (2021); Li et al., (2008)
2. Hormonal and Neurotransmitter Imbalances	Decreased levels of estrogen compared to normal. Repression of TRPH1 by action of Vitamin D.	Estrogen prevented bone loss in osteoporosis induced mice.	Morningstar, (2016); Liang et al., (2021); Ng et al., (2018); Walther and Bader, (2003); Berridge, (2017); Menendez et al., (2001); Kim & Choi, (2013)
3. Pain	Facet joint loading leads to hyper excitability of the spinal cord causing chronic pain.	Increased back pain in 90% AIS patients and 124 AIS patients	Teles et al., (2020)
4. Digestion	Impair functioning of peripheral nerves due to irregular bone arrangement causing digestive issues.	Scoliosis can cause acid reflux, heartburn, constipation, stomach pain, irritable bowel syndrome.	Kevin, (2011); Acaroglu et al., (2012); Nalda, (2019); Mompoint, Carlier & Orlikowski, (2010)
5. Diseases associated with Scoliosis	Marfan's Syndrome, osteogenesisimperfecta, Ehlers-Danlos Syndrome and Neurofibromatosis.	1. The incidence of scoliosis in neurofibromatosis varies from 10–60%. 2. The 40–70% of patients with Marfan's Syndrome have scoliosis.	Hresko, (2013)

compounds, their pharmacological properties and broad range of therapeutic effects have been listed in Table 4.2.

TABLE 4.2 Properties of Nutraceuticals in Management of Scoliosis

Category	Nutraceuticals Included	Mechanism of Action	References
Hormone	Melatonin	Anti-oxidant	Mottaghi & Nasri, (2021)
Vitamins	Vitamin D	Bone-building (through increasing Ca absorption and bone mineralization)	Gil, Plaza-Diaz & Mesa, (2018)
	Vitamin C	Bone- Building	Morton & Schneider, (2001)
	Vitamin B (B6, B9 & B12)	Involved in biosynthesis of amino acids and neurotransmitters, Anti-oxidant	Parra, Stahl, & Hellmann, (2018); Khan & Jialal, (2018); Elmer et al., (2013)
	Vitamin K2	Bone-building	Guillemant et al., (2001)
Minerals	Manganese	Bone-building, anti-oxidant, immunity booster	Chen, Bornhorst, & Aschner, (2018); Lüder, (2004)
	Magnesium	Bone-building	Zittermann, (2013)
	Calcium	Bone-building	Cashman, (2002)
	Copper	Anti-oxidant, bone-building	Rondanelli et al., (2021)
Plant Extracts	Coconut Oil	Digestive aid	Wang et al., (2015)
	Ananascomosus	Balances neurotransmitters and has analgesic and anti-inflammatory activities.	Kalhaa, Damgude & Dabade, (2021); Pavan et al., (2012)
	Rhodiolarosea	For over physiological health and maintenance of hormonal levels.	Khanum, Farhath, Amarinder Singh Bawa, (2006)
Phytoconstituents	Curcumin	Anti-oxidant, analgesic, anti-inflammatory	Chin, (2016); Megan Strauchman, (2015)
	CBD	Analgesic	Megan Strauchman, (2015)
	Resveratrol	Phytoestrogen, anti-oxidant and anti-inflammatory	Wong et al., (2020)
	Genistein	Phytoestrogen, anti-oxidant and bone-building	Polkowski & Mazurek, (2000)
	Tryptophan	Balancing neurotransmitters	Richard et al., (2009)
	Quercetin	Anti-oxidant, anti-inflammatory & balancing neurotransmitter levels	Mlcek et al., (2016) Haggag et al., (2003)
Animal Source	Gelatin	Bone-building	Koyama et al., (2001)
	Cod liver oil	Digestive aid and analgesic & anti-inflammatory activities.	Huang, Fan & Zhang, (2011).

The various pharmacological potentials detailed in Table 4.2 aid the management the various complications associated with the disease, as detailed in Table 4.1, through a variety of mechanisms and biochemical pathways. A combination of these therapeutic agents may help in the management complications in an effective manner.

Figure 4.1 demonstrates an overview of the potential role of these nutraceuticals in the management of scoliosis and related complications.

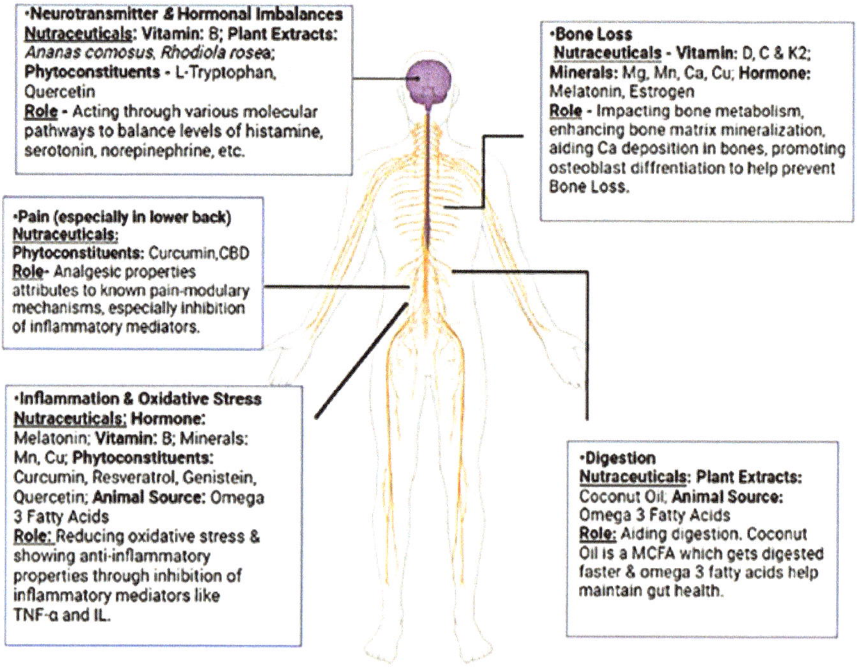

FIGURE 4.1 The role of nutraceuticals in management of scoliosis and related complications.

4.4 HORMONES

Hormones perform a role as chemical messengers and have a massive impact on the growth and development of the body. Some theories supporting the etiology of scoliosis focus on disruption of hormonal homeostasis as a primary causative factor. This is based on the data obtained from pre-clinical and clinical studies, which point to the abnormal levels of hormones that have been observed in scoliotic subjects. This may be attributed to a complex interaction involving other hormones, or malfunctioning of a signaling pathway. The outcomes of imbalance of agents thought to have a role in the progression of scoliosis have been discussed below.

4.4.1 *MELATONIN*

Melatonin (N-acetyl-5-methoxytryptamine) has been long suspected to play a role in the etiology of scoliosis, which may be explained through its effect on bone metabolism. Melatonin influences bone metabolism by stimulating osteoblast proliferation and osteoprotegerin production, leading to a suppression of osteoclast differentiation (Cardinali et al., 2003). A pre-clinical study conducted by Kono et al. on pinealectomy induced chickens showed that administration of intraperitoneal injections of melatonin (8mg/kg) helped prevent the manifestation of scoliosis (Kono et al., 2011). Similarly, experimentation in humans by Sadat-Ali M et al. has revealed that AIS patients have found to have relatively lower melatonin levels, as compared to the control group (Sadat-Ali, Ai-Habdan, & Al-Othman, 2000). Recent research suggests that dysfunction in the melatonin signal transduction pathway may be a pre-disposing factor for the pathogenesis of scoliosis. It is important to note that this dysfunction would show consequences similar to that of melatonin deficiency. The cAMP pathway is responsible for regulating osteoblastic function, and metabolism. Melatonin reduces cAMP levels and suppresses forskolin-stimulated adenylyl cyclase activity. In normal individuals, melatonin is responsible for the inhibition of forskolin-stimulated adenylyl cyclase activity. This was determined by the observation of a 60–70% reduction in cAMP levels. In comparison, osteoblasts from AIS patients revealed a lack of inhibition of forskolin-stimulated adenylyl cyclase activity by melatonin, which may indicate that the melatonin signaling pathway was impaired (Moreau et al., 2004). In a prospective study, a total of 16 patients with low melatonin levels were treated with oral melatonin supplements (in doses of 3 mg, 1.5–2.0 hours before sleep time) for 3–6 years. 12 of the patients developed a stability in scoliosis progression, whereas 4 had progressive scoliosis. Based on the experimental data obtained, it was concluded that melatonin supplements may be effective in preventing curve progression in mild cases (<35°) (MacHida et al., 2009).

4.5 VITAMINS AND MINERALS

Vitamins and Minerals contribute to the proper functioning of the body via their roles in various biochemical pathways. Studies indicate prevalent deficiencies of several vitamins and minerals in scoliotic patients. Administration of these compounds as nutraceutical formulations may be undertaken, for the wide range of beneficial effects offered. These include bone-forming minerals like calcium, which are co-factors in enzymatic reactions. Thus, these compounds have a role in the prevention and potential treatment of malfunctions in molecular pathways that maybe associated with the pathogenesis of scoliosis.

4.5.1 VITAMIN D

Vitamin D plays a vital role in the maintenance of bone health by regulation of calcium absorption in the intestine. The active form of Vitamin D, referred to as calcitriol (1,25-dihydroxycholecalciferol), regulates serum calcium levels and aids in bone formation and growth (Gil, Plaza-Diaz, and Mesa, 2018). Several clinical studies have reported that a significant number of AIS patients suffer from Vitamin D deficiency (Mayes, Anadio, and Sturm, 2017; Annika et al., 2013; Batista et al., 2014). The deficiency may result in improper bone mineralization, and over a prolonged duration, bone mineralization may not occur at all owing to the calcium and phosphate depletion (Gil, Plaza-Diaz, and Mesa, 2018). A randomized, double-blind, controlled trial examined the effects of Vitamin D, and calcium supplements 257 elderly (mean age 71 ± 4 years) subjects. They all received Calcium citrate (1000 mg/day) and Vitamin D (3750 IU/day or 600 IU/day) based on group allotted for the duration of 1 year. BMD increase in lumbar spine was observed in both the study groups, whereas the subtotal BMD increased in the high dose group only.

It was concluded that Vitamin D and calcium are beneficial for bone health, however, the high dosage provides little additional benefit when compared to the usually recommended dose in the elderly (Rahme et al., 2017). Vitamin D also interacts with several hormones like leptin, estrogen and melatonin which play a role in maintenance of BMD. The consequent deficiency in these hormones could lead to a reduction in BMD. The resulting osteopenia could lead the vertebrae to become prone to wedge deformities in early stages of AIS and worsen curve progression (Ng et al., 2018). Figure 4.2 shows how Vitamin D interacts with various hormones and their molecular pathways.

4.5.2 VITAMIN C

Vitamin C is an essential vitamin and anti-oxidant, and its properties may be attributed to its electron donating action. Vitamin C acts as an electron

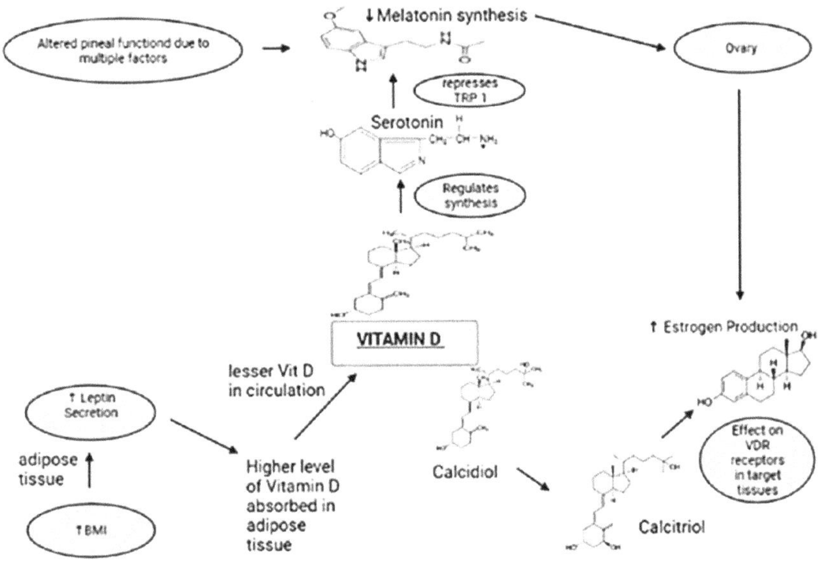

FIGURE 4.2 Interaction of vitamin D with hormones.

Source: Adapted from Berridge, 2017; Kim & Choi, 2013; Menendez et al., 2001; Ng et al., 2018; Walther & Bader, 2003).

donor for a total of 8 enzymes, 3 of which are involved in collagen hydroxylation. These reactions lead to increased stability of collagen molecule structure by addition of hydroxyl group proline or lysine (amino acids) in the collagen molecule (Padayatty et al., 2003). By enhancing collagen synthesis (precursor of bone matrix mineralization), it stimulates the overall process of osteoblast differentiation, and increases the rate of osteoclast formation (Morton & Schneider, 2001). Other enzymes regulated by vitamin C play a role in dopamine synthesis and tyrosine metabolism. Therefore, Vitamin C, via these biochemical pathways influences bone health and synthesis and modulation of neurotransmitters, etc. (Padayatty et al., 2003). A study conducted by Morton et al. demonstrated the effect of Vitamin C on BMD of post-menopausal women. In a sample size of 994 subjects, 277 regularly consumed vitamin C. The mean daily dose was 745mg and 85% of the subjects had taken Vitamin C supplements for more than 3 years. Based on the results obtained from the trial, it was observed that Vitamin C had beneficial effects on BMD (Morton & Schneider, 2001).

4.5.3 VITAMIN B

Vitamin B6, B9 & B12 are of clinical importance in the management of scoliosis. Vitamin B6 plays a role in the biosynthesis of neurotransmitters. It also exerts antioxidant properties (Parra, Stahl, and Hellmann, 2018). Vitamin B9 brings about the production of numerous coenzymes, among other roles (Gaby, 2010). Vitamin B12 is essential for the metabolism of amino acids, such as glycine and serine, which are important neurotransmitters necessary for cognition and limbic function (Elmer et al., 2013). A clinical case report analyzed the case of a 36-year-old female who had severe scoliosis, osteopenia and homocystinuria. High homocysteine levels may lead to impairment in bone cell functions, that regulate bone remodeling and bone material properties. Hence, low BMD may occur as a result of imbalance between bone resorption and formation. It was concluded that these vitamins (B6, B9, and B12) supplements must be considered as an adjuvant therapy option in such cases of low BMD or related conditions (Al-Allaf, Paul, and Awadh, 2020).

4.5.4 VITAMIN K2

Vitamin K2, also referred to as menaquinone, improves bone health, reduces vascular calcification and cardiovascular risk. It helps in the formation of γ-carboxyglutamate (Gla). One of the Gla proteins is osteocalcin, the synthesis of which hence becomes Vitamin K dependent (Beulens et al., 2013). Vitamin K2 aids calcium deposition in the bones by carboxylating osteocalcin, the synthesis of which is triggered by Vitamin D3 (Guillemant et al., 2001). Osteocalcin is one of the most abundant proteins in bone, and recent discoveries imply a role for it in whole body metabolism, reproduction and cognition (Zoch, Clemens& Riddle, 2016). Vitamin K acts as a cofactor to promote the gamma-carboxylation of osteocalcin for bone formation. This process is hampered in case of Vitamin K deficiency in which case the concentration of under carboxylated osteocalcin increases (Je et al., 2011). In a study conducted by Chen et al. including 99 patients with varying severity of curves, the serum osteocalcin levels were found to be negatively associated with the Cobb angle (Chen et al., 2019). A clinical trial conducted by Hyeon Je et al. demonstrated the effects of Vitamin K on BMD and under carboxylated osteocalcin. 38 women were treated with 15 mg of Vitamin K2 (3 times/day), 400 IU Vitamin D3 (1 time/day) and 315 mg Calcium (2 times/day) for a duration of 6 months. It was concluded that Vitamin K given along with Vitamin D and Calcium improved BMD and reduced under carboxylated osteocalcin levels (Je et al., 2011).

4.5.5 MAGNESIUM

Magnesium impacts both matrix and mineral metabolism in bone. Its deficiency results in hypomagnesemia, causing reduced osteoblast, and osteoclast activity. A large percentage of magnesium in the bone is present in apatite crystals. In case of deficiency, larger crystals form and bone stiffness decreases. Thus, bone fragility increases as a result of hypomagnesemia. Indirect effects of hypomagnesemia include impact on PTH levels resulting in lowering of Vitamin D levels (Castiglioni et al., 2013; Zittermann, 2013). Therefore, magnesium acts on bone metabolism by influencing these biochemical pathways. Hypomagnesemia, measured by an increased retention of an intravenous magnesium level, is apparent in osteoporotic patients (Cohen, 1988; Cohen, and Kitzes, 1982). A study conducted by Marrero-Barrera et al. on AIS patients who had undergone posterior spinal fusion surgery reported that 53% of the subjects had hypomagnesemia (Marrero-Barrera et al., 2021). Magnesium can also be used post-operatively to lessen morphine consumption as demonstrated by a prospective randomized double-blind study. 50 patients undergoing scoliosis surgery were evaluated. The treatment group, after induction, was administered IV bolus ketamine 0.2 mg/kg and magnesium 50 mg/kg followed by continuous infusion of ketamine 0.15 mg/kg/hr and magnesium 8mg/kg/hr until extubation. All patients were administered IV morphine via patient-controlled analgesia pump. The post-operative morphine consumption was lowered in magnesium treated group by 29.5% (Jabbour et al., 2014).

4.5.6 MANGANESE

Manganese is an essential nutrient, having a vital role in immune responses, neuronal regulation, digestion, development, and antioxidant defense (Chen, Bornhorst, & Aschner, 2018). Manganese acts as a cofactor for several enzymes, some of which are involved in bone formation. It stimulates the synthesis of bone matrix by acting as a cofactor for an enzyme called glycosyltransferase. This enzyme is required for the synthesis of proteoglycans that are further required for bone formation (Chen, Bornhorst, & Aschner, 2018; Peroni et al., 2021). Pre-clinical studies conducted on chickens show that manganese deficiency resulted in worsening of scoliosis in chickens (Rucker et al., 1986; Greve et al., 1987). A pre-clinical trial conducted by Rico et al. on Wistar rats to observe the effects of manganese supplements on ovariectomy inducted osteopenia in the rats. 15 rats in the treatment group received 40 mg/kg of manganese per kilogram of feed for 30 days. It was concluded that

Manganese supplements are an effective inhibitor of bone loss (Rico et al., 2000). A placebo-controlled double-blind clinical trial was conducted by Strause et al. on post-menopausal women to determine the effect of Calcium given with, and without trace minerals like manganese, copper and zinc. The treatment group received Calcium (1000 mg/day), Manganese (5 mg/day), Zinc (15 mg/day) and Copper (2.5 mg/day) for 2 years. The changes in BMD in the placebo group was -3.53 ± 1.24% whereas in the treatment group it was 1.48 ± 1.40%. This suggests that trace minerals given along with calcium supplements results in further increase in BMD (Strause et al., 1994).

4.5.7 CALCIUM

Calcium is a vital mineral for bone formation. 99% of the total body calcium is present as calcium phosphate and calcium carbonate in mineralized tissues (mainly bones). It helps with maintaining the structure, integrity and rigidity of the skeleton. The plasma calcium levels are regulated via a complex system involving the interaction of Vitamin D, PTH and calcitonin (Cashman, 2002). Calcium intake directly influences BMD in AIS (W. T. K. et al., 2018). A randomized double-blinded placebo-controlled trial on AIS patients was conducted by Lam et al. to determine the therapeutic effect of Vitamin D and Calcium supplements on management of scoliosis. The patients were provided varying doses of supplements with the highest being 600 mg Calcium and 800 IU Vitamin D3/day for a period of 2 years. It was concluded that the use of these supplements improved bone strength and prevented curve progression in scoliosis trials (Lam et al., 2017).

4.5.8 COPPER

Copper is an essential mineral required for energy metabolism at the cellular level because it is required for the formation of enzymes that transfer electrons (oxidases) to achieve the reduction of molecular oxygen. Among these enzymes is lysyl oxidase, which employs the amino acids lysine and hydroxylysine (found in collagen and elastin) as substrates to form the cross-links required for connective tissue development, including bone formation (Rondanelli et al., 2021). A pre-clinical study was conducted by Greve et al. to test the influence of diet in the expression of scoliosis in chickens. The chickens were fed skim-milk based diets varying in copper levels (0–10μg per gm of diet). The expression of scoliosis decreased from 75% to 50% in chickens fed diets containing more than 4μg (per gm of diet) copper (Carl

Greve et al., 1987). A randomized double-blinded study conducted by Eaton-Evans et al. investigated the effects of copper supplementation on vertebral trabecular BMD (VTBMD) of 73 women aged between 45–56 years. The patients were provided with copper supplements (3 mg/day) or placebo pills for a duration of 2 years. There was no difference observed in the initial and final VTBMD of the group taking copper supplementation, however, a significant reduction in VTBMD of the women taking a placebo pill was observed. This suggests that copper supplementation prevented the loss of VTBMD in middle-aged women (Eaton-Evans et al., 1996) (Table 4.3).

TABLE 4.3 Vitamins and Minerals – Sources and Optimal Levels for Intake

Vitamins and Minerals	Optimal Levels	Plant-based Sources	Animal-Based Sources	References
Vitamin D	15–20 µg/d	Mushroom, yeast, dairy products (milk, cheese, yogurt), butter and margarine	Fish, shellfish, organ meats, eggs,	Lee et al., (2008); Vaes et al., (2017)
Vitamin B6	1.3–1.7 mg/d	White potato, watermelon, banana, sweet potato	Calf, chicken, pork liver & kidney, tuna and trout	Field & Stover, (2015)
Vitamin B9/ Folate	300–400 µg/d	Legumes (beans, nuts, peas), leafy green vegetables (spinach), citrus, broccoli, liver	Dairy products like milk, yogurt, etc.	Dietrich, Brown& Block, (2005)
Vitamin B12	1.8–2.4 mg/d	Dried green and purple lavers, fortified breakfast cereal	Meat, milk, egg, fish, shellfish	Watanabe (2007)
Vitamin C	75–90 mg/d	Oranges, mango, orange, water melon, asparagus, broccoli, brussels sprouts, cabbage, cauliflower, kale, pepper, etc.	Lamb liver, calf adrenals, lamb heart, lamb tongue, human milk, goat milk, cow milk	Padayatty et al., (2003); Raveendran & Devaki, (2017)
Vitamin K	75–120 mg/d	Spinach, broccoli, kale, brussels sprouts	Dairy products like cheese, curd	Beulens et al., (2013)
Magnesium	240–420 mg/d	Almonds, cashew, spinach, cereal, oatmeal, peanuts, wheat bran, avocado	Milk	Guerrera, Volpe, & Mao (2009)
Calcium	1,000–1,330 mg/d	Legumes, nuts, seeds, grain and grain products	Dairy products like milk, cheese, fish, meat, etc.	Fleming & Heimbach (1994)
Manganese	1.9–2.3 mg/d	Coffee, black pepper, soybean, whole grain, mussel, nuts, soybeans, rice, leafy vegetables, tea,	Milk	Chen, Bornhorst, & Aschner (2018)

*Dietary intake levels as per food and nutrition board.

4.6 PLANT EXTRACTS FOR MANAGEMENT OF SCOLIOSIS

Plant extracts are bioactive compounds sourced from medicinal plants (Srivastava et al., 2020). Research indicates that plant extracts offer variety of therapeutic properties, including the ability to restore neurotransmitter levels, act as digestive aid and boost the overall metabolism of the body.

4.6.1 COCONUT OIL

Coconut oil is commercially sourced from *Cocos Nucifera*, which belongs to the family Arecaceae. It primarily acts as a digestive aid. It is a highly saturated oil (approx. 90%) and medium chain fatty acids (MCFA) make up 60% of its total composition. MCFAs have a chain length of 6–12 carbon atoms and are absorbed directly into the portal circulation, unlike long chain fatty acids (LCFA). They have been observed to reduce fat deposition, and enhance serum lipid profiles because of their faster metabolism (as compared to LCFA) and reduced storage in adipocytes. Therefore, the quick absorption and metabolism of coconut oil will put lesser strain on the digestive system and is beneficial for scoliosis patients.

4.6.2 ANANAS COMOSUS

Ananas comosus, commonly known as pineapple, belongs to the family *Bromeliaceae* and has been used as a medicinal plant in several native cultures. The medicinal qualities of this agent are attributed to a phytochemical called Bromelain. It exhibits a variety of pharmacological activities like anti-inflammatory, fibrinolytic, anti-thrombotic and anticancer effects. It is used in the treatment of osteoarthritis, angina pectoris, bronchitis, sinusitis, surgical trauma, thrombophlebitis and various cardiovascular disorders (Pavan et al., 2012). In addition to this, bromelain helps in balancing neurotransmitters levels. A recent *in vitro* study was conducted by Kalhaa et al. to observe the effect of bromelain in reduction of histamine levels in white wine. In a sample treated by 200 mg/L bromelain, the histamine concentration decreased 62.34% which the highest of reduction when compared to white wine treated with 50 mg/L quercetin and 400 mg /L ascorbic acid showing 61.64% and 46.41% reduction respectively, thereby indicating the promising potential of bromelain in the regulation of inflammatory mediators such as histamine (Kalhaa, Damgude, & Dabade, 2021).

Bromelain offers potent anti-inflammatory and analgesic activities. Research suggests that it acts via regulating levels of prostaglandins, decreasing fibrinogen levels in plasma and reducing amount of bradykinin, which leads to reduced vascular permeability, pain and edema (Kalhaa, Damgude, & Dabade, 2021). A prospective randomized double-blinded trial was conducted by Akhtar et al. to observe the effect of bromelain combined with rutin and trypsin, in comparison with diclofenac in patients with osteoarthritis of the knee. A total of 103 patients were treated for 6 weeks with either bromelain, rutin and trypsin or diclofenac. It was concluded that this combination of phytochemicals is a safe and efficacious alternative to NSAIDs for use in osteoarthritis pain (Akhtar et al., 2004). Bromelain is absorbed well into the body when taken orally and has shown no major side effects, even after prolonged use (Pavan et al., 2012).

4.6.3 RHODIOLA ROSEA

Rhodiola rosea (rose root), belonging to the Crassulaceae family, is a popular medicinal plant. It regulates a range of physiological processes, including modulation of neurotransmitter levels, CNS activity, and cardiovascular function, which have been reported to be attributed to the clinical significance of extracts obtained from its roots. Constituents including salidroside (rhodioloside), rosavins, and p-tyrosol have been credited for a majority of these effects. It has been established to be a powerful antioxidant and anti-carcinogenic. R. rosea primarily exerts its effects by regulating the following proposed mechanisms- firstly, it stimulates neurotransmitters like epinephrine, dopamine, serotonin, and brings about nicotinic cholinergic effects in the CNS, and secondly, it enhances the effects of these neurotransmitters on the brain by increasing the permeability of precursors of DA and 5-HT across the blood brain barrier. (Khanum, Farhath, & Amarinder Singh Bawa, 2006) Norepinephrine has been reported to be one of the neurotransmitters that is imbalanced in scoliosis patients (Morningstar, 2016). In a pre-clinical study conducted by Verpeut et al, adult male Sprague Dawley rats were treated with *Rhodiola rosea* (20 mg/kg) and/or *Citrus aurantium* (5.6 mg/kg) for 10 days. This resulted in an elevation of hypothalamic norepinephrine by 15% (Verpeut, Walters& Bello, 2013). Supplementation of resveratrol as an adaptogenic herbal supplement aids in modulating the adrenal output of norepinephrine.

4.7 PHYTOCONSTITUENTS FOR MANAGEMENT OF SCOLIOSIS

Numerous phytoconstituents have been screened for their therapeutic potentials in the management and alleviation of complications associated with scoliosis as well as related disorders. The experimental evidence obtained demonstrates their ability to combat oxidative stress by the exertion of potent anti-oxidant properties, thereby arresting the formation of free radicals. In addition to this, these compounds offer anti-inflammatory and analgesic properties, which are brought about by various mechanisms, most prominently the inhibition of inflammatory mediators.

4.7.1 CURCUMIN

Curcumin is a polyphenol derived from turmeric, which is the grounded root and rhizome of the plant *Curcuma longa* and has a long history of use in traditional Asian medicine (Chin, 2016; Sharma, Gescher, & Steward, 2005). It is known to have well established anti-nociceptive and antioxidative effects (Chin, 2016; Megan Strauchman, 2015). It is mainly indicated for pain management, and as an anti-inflammatory therapy for adult scoliotic patients. When compared to adolescents, adults suffering from are relatively more likely to seek therapeutic intervention for scoliosis, owing to a lowered pain threshold. In most cases, there is an inflammation-driven pain response as well as an asymmetrical mechanical strain on the supportive soft tissues in cases of adult scoliosis, which is usually a chronic condition (Megan Strauchman, 2015). Curcumin has been reported to inhibit IKB activation and phosphorylation of PKB, as well as the IL-1β triggered association of the two signaling molecules. This results in the inhibition of phosphorylation and degradation IKBα (NF-κB blocker). As a result, curcumin inhibits the translocation of NFκ-B into the nucleus, inhibiting the cellular inflammatory response. (Chin, 2016).

The extract of *C. longa* has been shown to be an effective treatment for osteoarthritis on its own. A clinical study was conducted by Madhu et al. on patients suffering from osteoarthritis. Over 42 days of the study, the subjects of one experimental group were supplemented solely with *C. longa* extract, while the other groups were given glucosamine, a combination of *C. longa* extract and glucosamine, and a placebo respectively. In all parameters, including WOMAC subscales and total score, clinician global impression of change, joint discomfort, crepitation, effusion, and limitation of movement, *C. longa* extract alone outperformed the combination and placebo groups. In subjects taking C. longa extract alone, subjects who needed paracetamol administrations and the likelihood of occurrence of adverse responses were the lowest across all groups (Madhu, Chanda, & Saji, 2013).

Curcumin also offers analgesic properties, based on evidence obtained from experimental studies. A clinical study was conducted by Morningstar et al. to observe effects of Curcumin supplements on pain of patients suffering from adult degenerative scoliosis. The supplementation of curcumin was carried out alongside a scoliosis-specific rehabilitation program in which all the subjects of the study took part. Only half of the subjects were given Curcumin supplements (CurcuPlex CR – containing 1.4g of C. longa extract/ twice a day) following the program, for 6 months. The average pain score in the treatment group went down to a significantly greater extent in the experimental group, as compared to the control group (Megan Strauchman, 2015).

4.7.2 CANNABIDIOL

Cannabidiol, abbreviated as CBD, is a phytocannabinoid and a non-psychoactive component of Cannabis sativa L. It is being considered an emerging treatment option for the management of chronic pain in a variety of illnesses (Argueta et al., 2020). Cannabinoids are known to be effective for alleviating pain arising from musculoskeletal disorders (like scoliosis), central pain, neuropathic pain and fibromyalgia, to name a few (Hegazy and Platnick, 2019). In addition to its analgesic property, CBD shows anti-convulsive, sedative, anxiolytic, hypnotic, anti-inflammatory and anti-oxidative effects. (Hegazy and Platnick, 2019; Mechoulam, Parker and Gallily, 2002). A clinical case report by Hegazy et al. mentioned a woman suffering from Neurofibromatosis type 1, who had complained of chronic pain among other complications. As a part of the study design, CBD oil was administered sublingually for a duration of three months. On completion of the experimental period, the patient reported a significant reduction in pain, fewer episodes of migraines, better mood, and an overall improvement in the quality of life. CBD exerts antagonist activity on CB2 receptors, which leads to inhibition of immune cell migration. (Hegazy and Platnick, 2019).

4.7.3 RESVERATROL

Resveratrol (3,4,' 5 trihydroxystilbene) is a well-known polyphenolic phytoalexin which is found in many natural sources, including grapes, peanuts, and berries (Berman et al., 2017). Polygonurn cuspidatum ("Kojokon" in Japanese) roots, which have long been utilized in traditional oriental medicine, are used to synthesize resveratrol (Frémont, 2000). It exerts anti-cancer, anti-oxidant, anti-inflammatory, neuroprotective, and cardioprotective properties. It may bring about modulation of lipid metabolism, alongside blocking the aggregation of platelets, and oxidation of low-density lipoproteins (Frémont, 2000; Tan et al., 2015). It is a phytoestrogen and also plays a role in bone protection according to some evidence gathered from *in vitro* and *in vivo* studies (Wong et al., 2020).

A randomized, placebo-controlled, double-blind clinical trial examined the effects of resveratrol in improving BMD of post-menopausal women. Resveratrol (75 mg/twice a day) was given for a duration of 12 months. An improvement in the lumbar BMD was observed. Additionally, a 7.24% reduction was also observed in a bone resorption marker in comparison to placebo (Wong et al., 2020).

4.7.4 GENISTEIN

Genistein (4′,5,7-trihydroxyisoflavone) is a phytoestrogen and isoflavone derived from soy. Its concentration in soy food material ranges from 0.2–1 mg/g, largely as glycosidic conjugates which form genistein on undergoing hydrolysis. It has found to have a wide range of pharmacological effects like inhibition of tyrosine kinase, topoisomerase II and phosphatidylinositol turnover, chemotherapeutic, antioxidant, anti-angiogenic and immuno-suppressive effects, to enlist a few. It is used in treatment of menopause, osteoporosis, and cardiovascular diseases (Polkowski, and Mazurek, 2000). Being a phytoestrogen, bears structural similarity to 17β-estradiol. It exhibits estrogen-like activity and can bind to estrogen receptors. It has been reported to cause an increased expression of estrogen-responsive genes on binding (Polkowski, and Mazurek, 2000).

Estrogen promotes bone metabolism by promoting osteoblast differentiation, bringing about osteoclast apoptosis and inhibiting osteoblast apoptosis. The downregulation of NK-κB (RANK) and upregulation of osteoprotegerin (OPG) is the mechanism of action (Ng et al., 2018; Walther, and Bader, 2003; Berridge, 2017; Menendez et al., 2001; Kim & Choi, 2013). Hence, a phytoestrogen on binding to its receptor may also impact bone metabolism. A clinical trial was conducted by Marini et al. to assess the effects of genistein on BMD in 389 postmenopausal women suffering from osteopenia. The subjects given Genistein (56 mg), in combination with vitamin D (400 IU) and calcium carbonate (500 mg) showed a significant improvement in BMD of femoral neck. By the end of the study, an increase in BMD from 3.5% to 9.3% was observed (Marini et al., 2007). The rise in osteoprotegerin, responsible for the differentiation of osteoclasts, was credited for the increase in BMD (Marini et al., 2008).

4.7.5 L -TRYPTOPHAN (TRP)

Tryptophan is an α-amino acid, and is a key component of the protein (Keservani et al., 2015; Singh et al., 2017b) biosynthesis process. It is a precursor of many compounds like serotonin, melatonin, quinolinic acid, kynurenic acid and tryptamine. Some common sources include oats, banana, cheese, bread, turkey, peanuts and milk. Tryptophan is the sole precursor for serotonin, which is in turn a precursor for melatonin. Hence, the biosynthesis of both these compounds becomes tryptophan dependent. (Richard et al., 2009). L-tryptophan supplements may be used to increase serotonin, and melatonin levels (Morningstar, 2016). A pre-clinical study was conducted by Akiyama et al. to examine the role of 5-Hydroxy-L-Tryptophan (5-HTP) in preventing TRP induced scoliosis. Chum salmon fry *Oncorhynchus keta* was fed diets low in TRP (TRP 50 mg/100g diet) and were given different doses of 5-HTP supplement (5-HTP; 0.1, 1, 10 & 100mg/100g diet). On administration of dose less than 10mg, 62–65% fish showed development of scoliosis within 4 weeks of rearing, and at 100 mg it reduced to 1.7%. The levels of norepinephrine also increased with increased supplementation. (Akiyaina et al., 1989).

4.7.6 QUERCETIN

Quercetin (3,3′,4′,5,7-pentahydroxyflavone) is a polyphenol, which may be further classified as a flavonol. Its natural sources include fruits like apples, wine, herb, peppers, onion, tea, broccoli, grapes and berries, to name a few. It possesses antioxidant, anti-inflammatory and anti-allergic properties. The implications of these activities include boosting of immunity, anti-allergic activity elicited by suppressing histamine release, suppressing of pro-inflammatory cytokines, leukotriene generation, and inhibition of IL-4 synthesis. It may also help in the improvement of Th1/Th2 balance and prevent the formation of antigen-specific IgE antibodies. It also exerts its actions by inhibiting enzymes including lipoxygenase, eosinophil, and peroxidase, as well as certain inflammatory mediators. These mechanisms largely contribute to its pharmacological effects (Mlcek et al., 2016).

Research shows patients suffering from scoliosis tend to have increased histamine levels. Quercetin supplements have been observed to aid the inhibition of histamine release. A pre-clinical study by Haggag et al. investigated quercetin's ability to reduce histamine release in rats. It was found that quercetin inhibited the histamine released by immunoglobin E and anti-immunoglobin E in peritoneal mast cells by 95% and 97% respectively (Haggag et al., 2003). In addition to this, a study conducted by Zhang et al. analyzed effects of quercetin on a rat model of osteoarthritis induced by papain injection. Over the course of the study, the serum IL-1β, TNF-α & NF-κB concentrations were seen to proportionately decrease as doses of quercetin were increased. Hence, quercetin was observed to exert anti-inflammatory effects (Zhang et al., 2019).

4.8 NUTRACEUTICALS OF ANIMAL SOURCE

Nutraceuticals of animal origin have been widely studied for their diverse pharmacological effects. They have been examined for their potential role in improving bone health, digestion, and cardiovascular heath, and have shown promising results in the experimental studies conducted to establish their quality, safety and efficacy. The use of these nutraceuticals is growing and consequently fulfilling the gaps in nutrition requirement of patients, in addition to therapeutic agents obtained from other sources. (Raghunandan, 2009).

4.8.1 GELATIN

Gelatin is one of the richest natural sources of collagen, which is a major constituent of bones. Several studies have reported that gelatin has the potential to influence BMD. A pre-clinical study was conducted by Nomura et al. to examine the effects of shark skin gelatin on ovariectomized rats. The rats were administered increasing doses of gelatin (1 mg/kg, 2 mg/kg, and 4 mg/kg). The control group was given ovalbumin 2 mg/kg. Results indicated that the rats administered a dose of 2 mg/kg had the highest BMD among all the groups assessed. Elevated levels of Type-I collagen were detected, particularly apparent in the rats belonging to this experimental group. This suggests that the mechanism followed by gelatin in increasing

BMD involved promoting the synthesis of type-I collagen (Nomura et al., 2005). Data obtained from experimental studies supports the benefits offered by oral administration of gelatin in increasing BMD (Koyama et al., 2001).

4.8.2 COD LIVER OIL

Dietary supplements containing cod liver oil are extensively utilized, owing to the wide range of therapeutically beneficial compounds they offer. These formulations contain Vitamin D, Vitamin A and essential omega-3 fatty acids [Eicosapentaenoic acid (EPA) and Docosahexaenoic acid (DHA)]. The anti-inflammatory properties of cod liver oil are attributed primarily to the presence of EPA and DHA. These supplements reduce inflammation by altering cell membrane function, suppressing the release of inflammatory mediators like IL-6, and lowering the formation of arachidonic acid derivatives and reactive oxygen species (Calder, 2017). This makes them useful in the treatment and management of spinal conditions, such as rheumatoid arthritis and scoliosis. This is especially helpful in cases where scoliosis patients concurrently suffer from inflammation and pain in joints. Sarcopenia is a complication linked to degenerative lumbar scoliosis (DLS) and it becomes critical to maintain muscle mass in this condition. Omega 3 fatty acid supplements are suggested to be taken concomitantly with physical therapy for DLS patients and have been observed to exert positive effects in regulating muscle mass levels in the body (Ng, 2019).

Table 4.4 summarizes the pre-clinical studies of nutraceuticals (Keservani et al., 2010a, 2020) and gives an overview of their promising therapeutic potentials in the management of scoliosis and related disorders. Table 4.5 summarizes the clinical trials of nutraceuticals in the chapter. Table 4.6 reviews the marketed formulations of nutraceuticals in this chapter.

CONCLUSION

Scoliosis is a spinal deformity involving multiple complications, with 80% of the cases categorized as idiopathic. Currently, the treatment options include surgical intervention, physical therapy (such as scoliosis-specific exercises), and pharmacotherapy (analgesics, anti-inflammatory agents, etc.). However, nutraceuticals may be employed in the management of scoliosis concomitantly, with the first line therapies. This chapter highlights the potential use of nutraceuticals that may be used in the management of scoliosis and associated complications. Vitamins, minerals, plant extracts, phytoconstituents,

TABLE 4.4 Pre-clinical Trials of Nutraceuticals in Management of Scoliosis

Nutraceutical	Animal Model	Method	Results	References
Hormones				
Melatonin	Melatonin deficient female mice	Oral administration of melatonin (8mg/kg/d)	The rate of scoliosis occurrence in bipedal mice was – 80% (control) & 22% (treated). In quadrupedal mice – it was 30% (control) & 5% (treated).	Liu et al., (2019)
Vitamins and Minerals				
Vitamin D and Calcium	Female Wistar Rats	1. Vitamin D rich diet (5120 IU/kg/day), 2. Calcium rich diet (1.2%/kg/day)	Vertebral BMD increased in subjects with vitamin D enriched diet by 76%, calcium diet by 70%	Schapira et al., (1995)
Calcium, Vitamin D3 and Vitamin K2	Ovariectomized adult female Sprague Dawley rats	1. Calcium (0.4 mg/kg) 2. Vitamin D3 (0.2 μg/kg) 3. Vitamin K2 (0.4 μg/kg) combinationfor 8 weeks.	1. Enhanced femoral BMD, BMC 2. Higher relative bone volume 3. Increased periosteal bone apposition	Omelka & Svik, (2020)
Vitamin C	Heterobranchuslongifilis fingerlings	Vitamin C upto 50 mg/kg/day for 12 weeks.	Increased specific growth rate, protein efficiency ratio and survival rate	Ibiyo, Madu& Eze, (2006)
Vitamin K	Ovariectomized mice	Vitamin K2 (30mg/kg/day) and Alendronate (56μg/kg/week s.c.) for 4 weeks.	Improvement in bone strength.	Sasaki & Miyakoshi, (2010)
Calcium	Ovariectomized Sprague Dawley rats	Dietary calcium (0.2%/kg-0.4%/kg)	1. Increase in BMD levels 2. Decrease in oxidative stress	Tian et al., (2003)
Magnesium	Wistar male rats	150mg mg/kg for 3 months.	1. Decrease oxidative stress and inflammatory markers. 2. Normalized bone metabolism.	Markiewicz-Górka et al., (2019)

TABLE 4.4 *(Continued)*

Nutraceutical	Animal Model	Method	Results	References
Manganese	Sprague-Dawley rats	0.001% for 12 weeks.	Increase of lumbar vertebrae.	Bae & Kim, (2008)
Phytoconstituents				
Curcumin	Ovalbumin sensitized rats	0.15 mg/mL, 0.30 mg/mL & 0.60 mg/mL	Improved levels of total WBC, PLA2, TP, IgE, IL-4, IFN-γ, IFN-γ/IL-4 ratio, SOD, thiol, NO_2, and NO_3	Shakeri & Boskabady, (2017)
Resveratrol	Ovariectomized female Sprague-Dawley rats	5mg/kg/d, 25 mg/kg/d and 45 mg/kg/d for 8 weeks.	Dose dependent increase in BMD levels	Feng et al., (2014)
Genistein	Ovariectomized female Sprague-Dawley rats	4.5mg/kg/d, 9mg/kg/d and 18mg/kg/d for 12 weeks.	Dose dependent increase in BMD levels	Miao et al., (2012)
Animal Source				
Gelatin	Ovariectomized female Wistar rats	Shark skin gelatin (1–4 mg/kg/day for 2 weeks.	10x increase in Type I collagen	Nomura et al., (2005)

TABLE 4.5 Clinical Trials of Nutraceuticals in Management of Scoliosis

Nutraceuticals	Study Design	Method	Results	References
Hormones				
Melatonin	Prospective Study	16 patients with low melatonin levels were treated with oral melatonin supplements (3 mg, 1.5–2 hours before sleep time). The patients were screened annually for serum melatonin levels and curve progression for 3–6 years.	Of those 16 patients – 12 had stable scoliosis & 4 had progressive scoliosis.	MacHida et al., (2009)
Vitamins & Minerals				
Vitamin D and Calcium	Randomized, Double-blinded, Placebo-controlled	1. Treatment 1: 600 mg Calcium + 400 IU Vitamin D3/ day 2. Treatment 2: 600mg Ca+ 800 IU Vitamin D3/ day for 2 years.	1. Improved bone strength. 2. Prevention of curve progression	Lam et al., (2017)
Vitamin K, Vitamin D and Calcium	Randomized	1. Group 1: Vitamin K2 (15mg/thrice a day) + Vitamin D (400 IU/day) + Calcium (315 mg/ twice a day) 2. Group 2: Vitamin D (400 IU/day) + Calcium (315 mg/twice a day) for 6 months.	L3 (lumbar) BMD was significantly increased in Group 1 when compared to the Group 2	Je et al., (2011)
Calcium, Zinc, Manganese and Copper	Double-Blind, Placebo-controlled	Treatment group: Calcium (1000 mg/day) + Zinc (15 mg/day) + Manganese (5mg/day) + Copper (2.5 mg/day) for 2 years.	Changes observed in bone density of treatment group was significant compared to the other groups	Strause et al., (1994)
Copper	Randomized, Double-blinded, Placebo-controlled	Copper supplement (3mg/day) for 2 years.	Significant decrease in the bone loss	Eaton-Evans et al., (1996)

TABLE 4.5 *(Continued)*

Nutraceuticals	Study Design	Method	Results	References
Phytoconstituents				
Curcumin	Retrospective, Open-Labeled, Case-Controlled	Curcumin–1.4g/twice a day	Pain score was reduced by 22 compared to 12 in the control.	Megan Strauchman, (2015)
Resveratrol	Randomized, Double-blind, Placebo-Controlled	75 mg/twice daily for 12 months.	Significant increase in the BMD of the lumbar spine	Wong et al., (2020)
Genistein	Randomized, Double-blind, Placebo-Controlled	54 mg/day) for 24 months.	Increase in lumbar BMD	Marini et al., (2007)
5-HTP, L-theanine, pyridoxal-5-phosphate, quercetin	Controlled Comparative Chart Review	5-HTP (100 mg thrice a day) + L-theanine (100 mg/day) + pyridoxal-5-phosphate (50 mg/day) + quercetin (100–400 mg/day) for 6 months.	The difference in the total mean Cobb angle post-treatment and pre-treatment for Group 1 was 12.29° as compared to placebo (4.25°)	Aatif Siddiqui, (2014)

Table 4.6 Marketed Nutraceuticals for Treatment and Management of Scoliosis

Nutraceuticals	Marketed Formulations	Company	Dose	Mechanism of Action	References
Vitamins & Minerals					
Vitamin D	HK Vitals Vitamin D3 softgels	HealthKart	50mcg/ 2000IU	1. Calcium homeostasis, by increasing the calcium absorption from the intestine and enhancing osteoclast differentiation 2. Matrix calcification and calcium resorption of the bone after binding to VDR	Gil, Plaza-Diaz & Mesa, (2018)
Vitamin K2	Vitamin K2 Capsule	WOW Life Sciences	100 mg	1. Aids in calcium deposition in the bones by carboxylatingosteocalcin, whose synthesis is triggered by Vitamin D3. 2. Osteocalcin formation is Vitamin K2 dependent	Beulens et al., (2013); Guillemant et al., (2001)
Vitamin C	Plant based Vitamin C Tablets	Himalayan Organics	1000mg	1. Electron donor for enzymes involved in collagen hydrolysis 2. Promotes bone matrix mineralization, 3. Increase osteoblast differentiation & osteoclast formation.	Padayatty et al., (2003); Morton & Schneider, (2001).
Vitamin B Complex	Naturewell B complex with glycine Capsules	Naturewell Organics	1000mg	1. Aids in neurological function 2. Plays a role in amino acid metabolism	Parra, Stahl & Hellmann, (2018); Khan & Jialal, (2018); Elmer et al., (2013)
Calcium	Pure encapsulations Calcium citrate Capsules	Pure encapsulations	300 mg	1. Essential bone forming mineral 2. Promotes bone mineralization.	Lam et al., (2017); (2015).

Table 4.6 *(Continued)*

Nutraceuticals	Marketed Formulations	Company	Dose	Mechanism of Action	References
Copper	Solaray Copper Capsules	Nutraceutical Corp.	2mg	Formation of enzymes including lysyloxidase which have a role in bone formation.	Rondanelli et al., (2021).
Magnesium	GNC Magnesium Capsules	Guardian Healthcare	310 mg	Reduction in the bone fragility.	Zittermann, (2013).
Manganese	Nature Plus Manganese Tablets	Natural Organics Laboratory	50 mg	Biosynthesis of mucopolysaccharides in bone matrix.	Chen, Bornhorst & Aschner, (2018)
Phytoconstituents					
Curcumin	CurcuPlex CR Tablets	Xymogen	1.4 mg	1. Antioxidant & anti-inflammatory properties. 2. Immune cell modulation 3. Inhibits inflammatory mediators like IL, TNF-α, COX-2, NF-κb and, iNOSenzyme.	Chin, (2016)
Quercetin	HealthVitQuercetin Capsules	DF Pharmacy Ltd	100 mg	Anti-inflammatory, antioxidant, anti-allergic and anti-viral properties	Mlcek et al., (2016).
Resveratrol	Healthy Hey Resveratrol Extract (98%) Capsules	Healthy Hey Foods LLP	260 mg	Systemic & cerebral circulator functions by activating endothelial estrogen receptors and also has a bone protective effect.	Wong et al., (2020).
Genistein	Isoflavone with Genistein Capsules	Douglas Laboratories	250 mg	Bone protective effect	Polkowski & Mazurek, (2000).
Animal Source					
Gelatin	Solgar Gelatin Capsules	Solgar Inc.	1,680 mg	It has an impact on circulating calcium levels in the blood and helps increase BMD.	Phillips & Williams, (2011).

Table 4.6 *(Continued)*

Nutraceuticals	Marketed Formulations	Company	Dose	Mechanism of Action	References
Omega 3 Fatty Acids	True Basics Omega 3 Fatty Acids Capsules	HealthKart	960 mg	Reduces inflammation by altering cell membrane function, suppressing IL-6 and other pro-inflammatory cytokines and lowering formation of arachidonic acid derivative and reactive oxygen species.	Huang, Fan & Zhang, (2011); Calder, (2017)

hormones, and nutraceuticals from animal sources are being studied for their use as a combinatorial therapy in the management of scoliosis. The mechanisms by which these nutraceuticals (Keservani et al., 2010b, 2017) primarily act are exertion of antioxidant and anti-inflammatory effects, bone-forming, digestive-aids and correcting hormonal, and neurotransmitter imbalances. Pre-clinical and clinical studies have helped identify the different properties and potential of nutraceuticals. Hormones, such as melatonin act by stimulation osteoblast differentiation, and influence cAMP levels. Vitamins (Singh et al., 2017a), particularly Vitamin D play a major role in bone-building by influencing intestinal absorption of calcium and through its interactions with other hormones like serotonin, estrogen, and leptin. Minerals such as calcium and magnesium are of vital importance, since they form an integral part of the bone formation process and have direct effects on bone metabolism. Phytoconstituents such as Curcumin and Cannabidiol have analgesic properties and are used for the treatment of chronic pain related to conditions like scoliosis. Other such as Resveratrol and Genisteinare phytoestrogens and act by binding on estrogen receptors and influencing bone metabolism. One of the major targeted biochemical pathway of phytoconstituent is anti-inflammatory property by inhibition of inflammatory mediators such as IL-4, TNF-α, COX-2, NF-κb and iNOS enzyme. Plant extracts, like *Rhodiola rosea* can be used to correct neurotransmitter levels. It not only stimulates neurotransmitter levels, but also enhances their effect by increasing the permeability of blood-brain barrier to precursors of 5-HT. Nutraceuticals from animal sources, like cod liver oil are a rich source of omega 3 fatty acids. These have a range of properties, like anti-inflammatory action via altering cell membrane function, suppressing inflammatory mediators, etc. Therefore, these nutraceuticals can be used in the management of scoliosis related complications like inflammation, pain, bone loss, etc. Nutraceuticals offer a huge potential in this therapeutic area since they are economical, easy to administer, and have less side effects. However, the exact dose of these nutraceuticals and their treatment duration are yet to be determined. Thus, further research and clinical trials are required to re-instigate the therapeutic properties of these nutraceuticals so they can be utilized for patient benefit. These nutraceuticals can be formulated in a patient complaint dosage form so that they can improve the overall quality of life of the patients. Nutraceuticals have emerging potential for use in scoliosis treatment concomitantly with traditional treatment, and offer an opportunity for further, more focused research in this domain to gain a deeper insight into the broad range of therapeutic benefits offered by these agents.

Conflict of Interest: The authors declare that the study was conducted in the absence of any commercial or financial relationships that could be construed as a potential conflict of interest.

KEYWORDS

- **calcium**
- **nutraceuticals**
- **osteopenia**
- **scoliosis**
- **spinal curvature**
- **spinal health**
- **vitamin D**

REFERENCES

Aatif, S., & Mark, W. M., (2014). Can neurotransmitter status affect the results of exercise-based scoliosis treatment? Results of a controlled comparative chart review. *Alternative & Integrative Medicine, 3*(4). https://doi.org/10.4172/2327-5162.1000177.

Acaroglu, E., Regis, B., Jocelyn, E., Ralph, M., Florina, M., & Alain, M., (2012). The metabolic basis of adolescent idiopathic scoliosis: 2011 report of the "metabolic" workgroup of the Fondation Yves Cotrel. *European Spine Journal, 21*(6), 1033–1042. https://doi.org/10.1007/s00586-012-2245-8.

Akhtar, N. M., Rizwan, N., Abid, Z. F., Wajahat, A., & Mussadeq, N., (2004). Oral enzyme combination versus diclofenac in the treatment of osteoarthritis of the knee – a double-blind prospective randomized study. *Clinical Rheumatology, 23*(5), 410–415. https://doi.org/10.1007/s10067-004-0902-y.

Akiyaina, T., Takeshi, M., Hideaki, K., Katsuyoshi, M., & Midori, H., (1989). Effect of dietary, 5-hydroxy-L-tryptophan for prevention of scoliosis in tryptophan-deficient chum salmon fry. *Nippon Suisan Gakkaishi (Japanese Edition), 55*(1), 99–104. https://doi.org/10.2331/suisan.55.99.

Al-Allaf Abdul-Wahab, Theresa, P., & Basem, A., (2020). P44 homocystinuria is a rare cause of fracture in young patients: A case report. *Rheumatology, 59* (Supplement_2), 111–043. https://doi.org/10.1093/rheumatology/keaa111.043.

Allam, A. M., & Aloysia, L. S., (2013). Neuromuscular scoliosis. *PM and R, 5*(11), 957–963. https://doi.org/10.1016/j.pmrj.2013.05.015.

Annika, A., Kouklinos, A., Julies, P., Shaw, M., & Jacobs, B., (2013). Vitamin D levels and signs of metabolic bone disease in adolescents with idiopathic scoliosis. *Bone Abstracts, 2*, 171. https://doi.org/10.1530/boneabs.2.p171.

Argueta, D. A., Christopher, M. V., Stacy, K., Varun, S., & Kalpna, G., (2020). A balanced approach for cannabidiol use in chronic pain. *Frontiers in Pharmacology*. https://doi.org/10.3389/fphar.2020.00561.

Arlet, V., Odent, T., & Aebi, A., (2003). Congenital scoliosis. *European Spine Journal, 12*(5), 456–463. https://doi.org/10.1007/s00586-003-0555-6.

Bae, Yun-Jung., & Mi-Hyun, K., (2008). Manganese supplementation improves mineral density of the spine and femur and serum osteocalcin in rats. *Biological Trace Element Research, 124*(1), 28–34. https://doi.org/10.1007/s12011-008-8119-6.

Balaji, V., Chin, K., Samir, F., Kouklinos, A., Tucker, S., & Shaw, M., (2014). Vitamin D deficiency in patients with adolescent idiopathic scoliosis (AIS): A cause for concern? *Orthopaedic Proceedings, 96*-B (Supplement, 6), 2.

Batista, R., Delio, E. M., Lilian, F. H., Lazaretti-Castro, M., Eduardo, B. P., & Marcelo, W., (2014). Association between vitamin D serum levels and adolescent idiopathic scoliosis. *Scoliosis, 9*(S1), 45. https://doi.org/10.1186/1748-7161-9-s1-o45.

Berman, A. Y., Rachel, A. M., Maia, Y. W., & Marina, K. H., (2017). The therapeutic potential of resveratrol: A review of clinical trials. *NPJ Precision Oncology, 1*(1), 35. https://doi.org/10.1038/s41698-017-0038-6.

Berridge, M., (2017). Vitamin D and depression: Cellular and regulatory mechanisms. *Pharmacological Reviews, 69*(2), 80–92. http://dx.doi.org/10.1124/pr.116.013227.

Beulens, J. W. J., Sarah, L. B., Ellen, G. H. M., Van, D. H., Elisabeth, S., Athanasia, B., & Cees, V., (2013). The role of menaquinones (vitamin K2) in human health. *British Journal of Nutrition*. https://doi.org/10.1017/S0007114513001013.

Calder, P. C., (2017). Omega-3 fatty acids and inflammatory processes: From molecules to man. *Biochemical Society Transactions*. https://doi.org/10.1042/BST20160474.

Cardinali, D. P., Marta, G. L., Verónica, B., Rodolfo, A. C., & Carlos, M., (2003). Melatonin effects on bone: Experimental facts and clinical perspectives. *Journal of Pineal Research*. https://doi.org/10.1034/j.1600-079X.2003.00028.x.

Cashman, K. D., (2002). Calcium intake, calcium bioavailability and bone health. *British Journal of Nutrition, 87*(S2). https://doi.org/10.1079/bjn/2002534.

Castiglioni, S., Alessandra, C., Walter, A., & Jeanette, A. M. M., (2013). Magnesium and osteoporosis: Current state of knowledge and future research directions. *Nutrients, 5*(8), 3022–3033. https://doi.org/10.3390/NU5083022.

Chen, H., Jiajun, Z., Yujia, W., Ka Yee, C., Alec, L. H. H., Tsz, P. L., et al., (2019). Abnormal lacuno-canalicular network and negative correlation between serum osteocalcin and cobb angle indicate abnormal osteocyte function in adolescent idiopathic scoliosis. *FASEB Journal: Official Publication of the Federation of American Societies for Experimental Biology, 33*(12), 13882–13892. https://doi.org/10.1096/fj.201901227R.

Chen, P., Julia, B., & Michael, A., (2018). Manganese metabolism in humans. *Frontiers in Bioscience – Landmark, 23*(9). https://doi.org/10.2741/4665.

Cheng, J. C. Y., Xia, G., & Andy, H. L. S., (1999). Persistent osteopenia in adolescent idiopathic scoliosis: A longitudinal follow-up study. *Spine, 24*(12), 1218–1222. https://doi.org/10.1097/00007632-199906150-00008.

Chin, K. Y., (2016). The spice for joint inflammation: Anti-inflammatory role of curcumin in treating osteoarthritis. *Drug Design, Development and Therapy, 10*, 3029–3042. https://doi.org/10.2147/DDDT.S117432.

Cohen, L. L., Robert, P., Audrey, C. M., John, B. E., & Daniel, J. H., (2021). Surgical correction of scoliosis in children with severe congenital heart disease and palliated single ventricle physiology. *Spine, 46*(14), 791–796.

Cohen, L., (1988). Recent data on magnesium and osteoporosis. *Magnesium Research: Official Organ of the International Society for the Development of Research on Magnesium, 1*(1, 2), 85–87.

Cohen, L., & Kitzes, R., (1982). Infrared spectroscopy and magnesium content of bone mineral in osteoporotic women. *Israel Journal of Medical Sciences, 17*(12), 1123–1125.

Dietrich, M., Coralie, B., & Gladys, B., (2005). The effect of folate fortification of cereal-grain products on blood folate status, dietary folate intake, and dietary folate sources among adult non-supplement users in the United States. *Journal of American College of Nutrition, 24*(4), 266–274. https://doi.org/10.1080/07315724.2005.10719474.

Eaton-Evans, J., Mcllrath, E., Jackson, W., McCartney, H., & Strain, J., (1996). Copper supplementation and the maintenance of bone mineral density in middle-aged women. *The Journal of Trace Minerals in Experimental Medicine, 9*(3), 87–94. https://doi.org/10.1002/(SICI)1520-670X(1996)9:3%3C87::AID-JTRA1%3E3.0.CO;2-E.

Elmer, R., Mark, M., Clayton, S., Brian, D., & Aatif, S., (2013). Tissue-Specific Characteristics and Requirements for Long-Term Scoliosis Rehabilitation: An ICSB Position Paper. *J. Scoliosis Rehabilitation.* 2013.

Everett, C. R., & Rajeev, K. P., (2007). A systematic literature review of nonsurgical treatment in adult scoliosis. *Spine, 32*(Supplement, 19), 130–134. https://doi.org/10.1097/BRS.0b013e318134ea88.

Fadzan, M., & Bettany-Saltikov, J., (2018). Etiological theories of adolescent idiopathic scoliosis: Past and present. *The Open Orthopaedics Journal, 11*(1), 1466–1489. https://doi.org/10.2174/1874325001711011466.

Feng, J., Shuai, L., Sai, M., Jian, Z., Wei, Z., Wei, Q., Pengchong, C., Zheng, W., & Wei, L., (2014). Protective effects of resveratrol on postmenopausal osteoporosis: Regulation of SIRT1-NF-KB signaling pathway. *Acta Biochimica et Biophysica Sinica, 46*(12), 1024–1033. https://doi.org/10.1093/abbs/gmu103.

Field, M., & Patrick, S., (2015). Vitamin B6. *Advances in Nutrition, 6*(1), 132–133. https://doi.org/10.3945/an.113.005207.

Fleming, K. H., & Heimbach, J. T., (1994). Consumption of calcium in the U.S.: Food sources and intake levels. In: *Journal of Nutrition* (Vol. 124). https://doi.org/10.1093/jn/124.suppl_8.1426s.

Frémont, L., (2000). Minireview: Biological effects of resveratrol. *Life Sciences*. https://doi.org/10.1016/S0024-3205(99)00410-5.

Gaby, A., (2010). *Nutritional Medicine*. Fritz Perlberg Publishing, Concord, New Hampshire.

Gil, Á., Plaza-Diaz, J., & María, D. M., (2018). Vitamin D: Classic and novel actions. *Annals of Nutrition and Metabolism, 72*(2), 87–95. https://doi.org/10.1159/000486536.

Goldstein, L. A., & Waugh, T. R., (1973). Classification and terminology of scoliosis. *Clinical Orthopedics and Related Research, 93*, 10–22. https://doi.org/10.1097/00003086-197306000-00003.

Greve, C., Trachtenberg, E., Opsahl, W., Abbott, U., & Rucker, R., (1987). Diet as an external factor in the expression of scoliosis in a line of susceptible chickens. *Journal of Nutrition, 117*(1), 189–193. https://doi.org/10.1093/jn/117.1.189.

Grivas, T. B., (2012). *Recent Advances in Scoliosis*. https://doi.org/10.5772/2497.

Guerrera, M. P., Stella, L. V., & Jun, J. M., (2009). Therapeutic uses of magnesium. *American Family Physician*. https://doi.org/10.1201/b13470-137.

Guillemant, J., Le, H. T., Maria, A., Allemandou, A., Peres, G., & Guillemant, S., (2001). Wintertime vitamin D deficiency in male adolescents: Effect on parathyroid function and response to vitamin D3 supplements. *Osteoporosis International, 12*(10), 875–879. https://doi.org/10.1007/s001980170040.

Haggag, E. G., Abou-Moustafa, M. A., William, B., & Theoharis, C. T., (2003). The effect of a herbal water-extract on histamine release from mast cells and on allergic asthma. *Journal of Herbal Pharmacotherapy, 3*(4), 41–54. https://doi.org/10.1300/J157v03n04_03.

Hegazy, O., & Howard, P., (2019). Cannabidiol (CBD) for treatment of neurofibromatosis-related pain and concomitant mood disorder: A case report. *Cureus, 11*(12). https://doi.org/10.7759/cureus.6312.

Horne, J. P., Robert, F., & Saif, U., (2014). Adolescent idiopathic scoliosis: Diagnosis and management. *American Family Physician, 89*(3), 193–198.

Hresko, T., (2013). Idiopathic scoliosis in adolescents. *The New England Journal of Medicine, 368*(9), 834–841. https://doi.org/10.1056/NEJMcp1209063.

Huang, W. B., Qian, F., & Xiu, L. Z., (2011). Cod liver oil: A potential protective supplement for human glaucoma. *International Journal of Ophthalmology*. https://doi.org/10.3980/j.issn.2222-3959.2011.06.15.

Ibiyo, L., Chuka, M., & Stanley, E., (2006). Effects of vitamin C supplementation on the growth of *Heterobranchus longifilis* fingerlings. *Archives of Animal Nutrition, 60*(4), 325–332.

Jabbour, H. J., Naccache, N. M., Jawish, R. J., Abou, Z. H. A., Jabbour, K. B., Rabbaa-Khabbaz, L. G., Ghanem, I. B., & Yazbeck, P. H., (2014). Ketamine and magnesium association reduces morphine consumption after scoliosis surgery: Prospective randomized double-blind study. *Acta Anaesthesiologica Scandinavica, 58*(5), 572–579. https://doi.org/10.1111/aas.12304.

Jaovisidha, S., Jong, K. K., David, J. S., Enrique, B., Sharon, E., Barrett-Connor, E., & Parichart, R., (1998). Scoliosis in elderly and age-related bone loss: A population-based study. *Journal of Clinical Densitometry, 1*(3), 227–233. https://doi.org/10.1385/JCD:1:3:227.

Je, S. H., Nam, S. J., Beom, H. C., Kwang, M. K., Bom, T. K., Sat, B. P., et al., (2011). Vitamin K supplement along with vitamin D and calcium reduced serum concentration of undercarboxylated osteocalcin while increasing bone mineral density in Korean postmenopausal women over sixty-years-old. *Journal of Korean Medical Science, 26*(8), 1093–1098. https://doi.org/10.3346/jkms.2011.26.8.1093.

Kalhaa, R. K., Gauri, D., & Ashish, D., (2021). Reduction of food histamine in white wine by bromelain, quercetin and ascorbic acid. *Food Science and Applied Biotechnology, 4*(2), 130–137. https://doi.org/10.30721/fsab2021.v4.i2.143.

Keservani, R. K., Kesharwani, R. K., Sharma, A. K., Vyas, N., & Chadoker, A., (2010b). Nutritional supplements: An overview. *International Journal of Current Pharmaceutical Review and Research, 1*(1), 59–75.

Keservani, R. K., Kesharwani, R. K., Vyas, N., Jain, S., Raghuvanshi, R., & Sharma, A. K., (2010a). Nutraceutical and functional food as future food: A review. *Der Pharmacia Lettre, 2*(1), 106–116.

Keservani, R. K., Sharma, A. K., & Jarouliya, U., (2015). Protein and peptide in drug targeting and its therapeutic approach. *ArsPharmaceutica (Internet), 56*(3), 165–177.

Keservani, R. K., Sharma, A. K., & Kesharwani, R. K., (2017). An overview and therapeutic applications of nutraceutical and functional foods. *Recent Advances in Drug Delivery Technology*, 160–201. ISBN:9781522507543.

Keservani, R. K., Sharma, A. K., & Kesharwani, R. K., (2020). *Nutraceuticals and Dietary Supplements: Applications in Health Improvement and Disease Management*. CRC Press. ISBN: 9781771888738.

Kevin, L., (2011). *Your Plan for Natural Scoliosis Prevention and Treatment: Health in Your Hands*. Createspace Independent Pub, California.

Khan, K. M., & Jialal, I., (2022). *Folic Acid Deficiency*. In: *StatPearls* [Internet]. Treasure Island (FL): StatPearls Publishing; 2023 Jan-. Available from: https://www.ncbi.nlm.nih.gov/books/NBK535377/.

Khanum, F., Amarinder, S. B., & Brahm, S., (2006). *Rhodiola rosea*: A versatile adaptogen. *Comprehensive Reviews in Food Science and Food Safety, 4*(3), 55–62. https://doi.org/10.1111/j.1541-4337.2005.tb00073.x.

Kikanloo, S. R., Sandip, P. T., & Woojin, C., (2019). Etiology of adolescent idiopathic scoliosis: A literature review. *Asian Spine Journal, 13*(3), 1976–1902. https://doi.org/10.31616/asj.2018.0096.

Kim, J. H., & Jin-Ho, C., (2013). Pathophysiology and clinical characteristics of hypothalamic obesity in children and adolescents. *Ann. Pediatr. Endocrinol. Metab., 18*(4), 161–167. https://doi.org/10.6065/apem.2013.18.4.161.

Kono, H., Masafumi, M., Masashi, S., Yuji, N., Hiroyuki, K., Naobumi, H., Kazuhiro, C., et al., (2011). Mechanism of osteoporosis in adolescent idiopathic scoliosis: Experimental scoliosis in pinealectomized chickens: Osteoporosis in idiopathic scoliosis. *Journal of Pineal Research, 51*(4), 387–393. https://doi.org/10.1111/j.1600-079X.2011.00901.x.

Koyama, Y. I., Hirota, A., Mori, H., Takahara, H., Kuwaba, K., Kusubata, M., Matsubara, Y., et al., (2001). Ingestion of gelatin has differential effect on bone mineral density and body weight in protein undernutrition. *Journal of Nutritional Science and Vitaminology, 47*(1), 84–86. https://doi.org/10.3177/jnsv.47.84.

Kuru, T., İpek, Y., Elçin, D. E., Fatih, D., İlker, Ç., & Arzu, R. Ö., (2016). The efficacy of three-dimensional Schroth exercises in adolescent idiopathic scoliosis: A randomized controlled clinical trial. *Clin. Rehabil., 30*(2), 181–190. https://doi.org/10.1177%2F0269215515575745.

Lam, T. P., Benjamin, H. K. Y., Fiona, W. P. Y., Echo, K. L. T., Kenneth, K. W. T., et al., (2015). Improving low bone mass in girls with adolescent idiopathic scoliosis (AIS) using calcium and vitamin D supplementation – a randomized controlled trial. *Bone Abstracts*. https://doi.org/10.1530/boneabs.4.p93.

Lam, T. P., Benjamin, H. K. Y., Gene, C. W. M., Wayne, Y. W. L., Elisa, M. S. T., et al., (2017). Effective therapeutic control of curve progression using calcium and vitamin D supplementation for adolescent idiopathic scoliosis – a randomized double-blinded placebo-controlled trial. *Bone Abstracts*. https://doi.org/10.1530/boneabs.6.oc8.

Langensiepen, S., Semler, O., Sobottke, R., Fricke, O., Franklin, J., Schönau, E., & Eysel, P., (2013). Measuring procedures to determine the cobb angle in idiopathic scoliosis: A systematic review. *European Spine Journal, 22*(11), 2360–2371. https://doi.org/10.1007/s00586-013-2693-9.

Lee, Meei-Shyuan, Hsin-Lu, L., Tsui-Hsia, H., Hsing-Yi, C., Feili, L. Y., & Mark, W., (2008). Vitamin D intake and its food sources in Taiwanese. *Asia Pac. J. Clin. Nutr., 17*(3), 397–407.

Lee, W. T. K., Tse, Y. K., Cheung, C. S. K., Chau, W. W., Qin, L., & Cheng, J. C. Y., (2018). Determinants of osteopenia in adolescent idiopathic scoliosis–abnormal growth, escalated bone turnover, poor nutrition or inadequate physical activity. *Orthopedic Proceedings, 90-B*(S-III), 432.

Lenke, L. G., & Matthew, B. D., (2007). Management of juvenile idiopathic scoliosis. *Journal of Bone and Joint Surgery, 89*(Supplement, 1), 55–63. https://doi.org/10.2106/jbjs.f.00644.

Letts, M., Rathbone, D., Yamashita, T., Nichol, B., & Keeler, A., (1992). Soft Boston orthosis in management of neuromuscular scoliosis: A preliminary report. *Journal of Pediatric Orthopedics, 12*(4), 470–474. https://doi.org/10.1097/01241398-199207000-00010.

Li, Xin-Feng, Hai, A. E., Li, A., Zu-De, L., Li, A. E., & Yang, D., (2008). Low bone mineral status in adolescent idiopathic scoliosis. *European Spine Journal, 17*, 1431–1440. https://doi.org/10.1007/s00586-008-0757-z.

Liang, Zhou-Tao, Chao-Feng, G., Jiong, L., & Hong-Qi, Z., (2021). The role of endocrine hormones in the pathogenesis of adolescent idiopathic scoliosis. *The Journal of the Federation of American Societies for Experimental Biology, 35*(9), e21839. https://doi.org/10.1096/fj.202100759R.

Liu, H., Zhen, L., Chi-Wai, M., Jing, G., Xiao, H., Zongshan, H., Tzi, B. N., et al., (2019). The effect of exogenous melatonin on reducing scoliotic curvature and improving bone quality in melatonin-deficient C57BL/6J mice. *Sci Rep*. https://doi.org/10.1038/s41598-019-42467-5.

Lüder, (2004). *Krause's Food, Nutrition, & Diet Therapy* (Vol. 11). Philadelphia: Saunders. https://doi.org/10.1002/food.19970410330.

MacHida, M., Jean, D., Thoru, Y., & Jun, K., (2009). Serum melatonin levels in adolescent idiopathic scoliosis prediction and prevention for curve progression – a prospective study. *Journal of Pineal Research, 46*(3), 344–348. https://doi.org/10.1111/j.1600-079X.2009.00669.x.

Machida, M., Stuart, L. W., & Jean, D., (2018). *Pathogenesis of Idiopathic Scoliosis*. https://doi.org/10.1007/978-4-431-56541-3.

Madhu, K., Chanda, K., & Saji, M. J., (2013). Safety and efficacy of curcuma longa extract in the treatment of painful knee osteoarthritis: A randomized placebo-controlled trial. *Inflammopharmacology, 21*(2), 129–136. https://doi.org/10.1007/s10787-012-0163-3.

Marini, H., Letteria, M., Francesca, P., Alessandra, B., Domenica, A., Marco, A., Agostino, G., et al., (2007). Effects of the phytoestrogen genistein on bone metabolism in osteopenic postmenopausal women: A randomized trial. *Annals of Internal Medicine, 146*(12), 839–847. https://doi.org/10.7326/0003-4819-146-12-200706190-00005.

Marini, H., Minutoli, L., Polito, F., Bitto, A., Altavilla, D., Atteritano, M., Gaudio, A., Mazzaferro, S., Frisina, A., Frisina, N., et al. (2008). OPG and SRANKL serum concentrations in osteopenic, postmenopausal women after, 2-year genistein administration. *Journal of Bone and Mineral Research, 23*(5), 715–720. https://doi.org/10.1359/jbmr.080201.

Marc, M., Borysov, M., Ng. S. Y., (2021). *Schroth's Textbook of Scoliosis and Other Spinal Deformities*. Cambridge Scholars Publishing, Newcastle upon Tyne.

Markiewicz-Górka, I., Krystyna, P., Aleksandra, J., Lidia, J., Paweł, P., & Natalia, P., (2019). Alleviating effect of α-lipoic acid and magnesium on cadmium-induced inflammatory processes, oxidative stress and bone metabolism disorders in Wistar rats. *International Journal of Environmental Research and Public Health, 16*(22). https://doi.org/10.3390/ijerph16224483.

Marrero-Barrera, P., Norman, R., Gerardo, O., Juan, C., Lenny, R., & Marrero-Ortiz, P., (2021). Hypomagnesemia following posterior spinal fusion in adolescent idiopathic scoliosis.

Journal of Pediatric Orthopaedics Part B, 30(3), 225–229. https://doi.org/10.1097/BPB.0000000000000741.

Mayes, T., Jennifer, M. A., & Peter, F. S., (2017). Prevalence of vitamin D deficiency in pediatric patients with scoliosis preparing for spinal surgery. *Spine Deformity, 5*(6), 369–373. https://doi.org/10.1016/J.JSPD.2017.03.008.

Mechoulam, R., Linda, A. P., & Ruth, G., (2002). Cannabidiol: An overview of some pharmacological aspects. *Journal of Clinical Pharmacology*. https://doi.org/10.1002/j.1552-4604.2002.tb05998.x.

Megan, S., & Morningstar, M. W., (2015). Controlled-release curcumin for the treatment of pain related to adult degenerative scoliosis: A retrospective, open-label, case-controlled series. *Journal of Pain & Relief, 4*(4). https://doi.org/10.4172/2167-0846.1000192.

Menendez, C., Lage, M., Peino, R., Conchiero, P., Diéguez, C., Casanueva, F., & Baldelli, R., (2001). Retinoic acid and vitamin D3 powerfully inhibit *in vitro* leptin secretion by human adipose tissue. *Journal of Endocrinology, 170*(2), 425–431.

Miao, Q., Jing-Ge, L., Shan, M., Nan, H., Jin, Z., Song, Z., Yan-Hua, X., et al., (2012). The bone-protective effect of genistein in the animal model of bilateral ovariectomy: Roles of phytoestrogens and PTH/PTHR1 against post-menopausal osteoporosis. *Int. J. Mol. Sci, 13*(1), 56–70. https://doi.org/10.3390/ijms13010056.

Mlcek, J., Tunde, J., Sona, S., & Jiri, S., (2016). Quercetin and its anti-allergic immune response. *Molecules*. https://doi.org/10.3390/molecules21050623.

Mompoint, D., Carlier, R. Y., & Orlikowski, D., (2010). Pathologies neuromusculaires: Imagerie des complications thoraciques, digestives et urinaires. *Journal de Radiologie, 91*(12), 1419–1436. https://doi.org/10.1016/S0221-0363(10)70221-3.

Moreau, A., Da Shen, W., Steve, F., Bouziane, A., Debora, A., Franco, F., Hubert, L., et al., (2004). Melatonin signaling dysfunction in adolescent idiopathic scoliosis. *Spine, 29*(16), 1772–1781. https://doi.org/10.1097/01.BRS.0000134567.52303.1A.

Morningstar, M., (2016). Neurotransmitter status and idiopathic scoliosis: A commentary on pathways, testing, clinical utility, and treatment. *Current Pediatric Research, 20*(1, 2).

Morton, D. J., Barrett-Connor, E. L., & Schneider, D. L., (2001). Vitamin C supplement use and bone mineral density in postmenopausal women. *Journal of Bone and Mineral Research, 16*(1), 135–140. https://doi.org/10.1359/jbmr.2001.16.1.135.

Mottaghi, P., & Parto, N., (2021). Antioxidant and bone; protect your future: A brief review. *Iran J. Public Health, 50*(9), 1783–1788.

Nalda, T., (2019). *Scoliosis Hope*: How New Approaches to Treatment Are Transforming Lives. Celebration, Florida.

Ng, S. Y., (2019). Conservative treatment of degenerative lumbar scoliosis. Spinal deformities in adolescents, adults and older adults. In: *Spinal Deformities in Adolescents, Adults and Older Adults*.

Ng, S. Y., Bettany-Saltikov, J., Irene, Y. K. C., & Karen, K. Y. C., (2018). The role of vitamin D in the pathogenesis of adolescent idiopathic scoliosis. *Asian Spine Journal, 12*(6), 1127–1145. https://doi.org/10.31616/asj.2018.12.6.1127.

Nomura, Y., Keigo, O., Mutsuto, W., & Shohei, K., (2005). Increase in bone mineral density through oral administration of shark gelatin to ovariectomized rats. *Nutrition, 21*(11, 12), 1120–1126. https://doi.org/10.1016/J.NUT.2005.03.007.

Omelka, R., & Svik, K., (2020). The effects of eggshell calcium (Biomin H®) and its combinations with alfacalcidol (1α-Hydroxyvitamin D3) and menaquinone-7 (vitamin

K2) on ovariectomy-induced bone loss in a rat model of osteoporosis. *Journal of Animal Physiology and Animal Nutrition, 105*(2), 336–344. https://doi.org/10.1111/jpn.13458.

Padayatty, S. J., Arie, K. Y. W., Peter, E., Oran, K., Je Hyuk, L., Shenglin, C., et al., (2003). Vitamin C as an antioxidant: Evaluation of its role in disease prevention. *Journal of the American College of Nutrition, 22*(1), 18–35. https://doi.org/10.1080/07315724.2003.10719272.

Parra, M., Seth, S., & Hanjo, H., (2018). Vitamin B6 and its role in cell metabolism and physiology. *Cells, 7*(7), 84. https://doi.org/10.3390/cells7070084.

Pavan, R., Sapna, J., Shraddha, & Ajay, K., (2012). Properties and therapeutic application of bromelain: A review. *Biotechnology Research International, 2012*. https://doi.org/10.1155/2012/976203.

Peroni, G., Mariangela, R., Milena, A. F., Vittoria, I., Clara, G., Giancarlo, I., Simone, P., et al., (2021). Essentiality of manganese for bone health: An overview and update. *Natural Product Communications, 16*.

Phillips, G. O., & Peter, A. W., (2011). *Handbook of Food Proteins. Handbook of Food Proteins*. https://doi.org/10.1533/9780857093639.

Polkowski, K., & Aleksander, P. M., (2000). Biological properties of genistein. A review of *in vitro* and *in vivo* data. *Acta Poloniae Pharmaceutica – Drug Research*.

Raghunandan, Y., (2009). Nutraceuticals with animal origin. In: Pathak, Y., Schapiro, S.J. (eds.). *Handbook of Nutraceutical Volume I. Ingredients, Formulations, and Applications*. CRC Press, Boca Raton, Florida.

Rahme, M., Sima, L. S., Rafic, B., Robert, H. H., Georges, H., Asma, A., Ravinder, J. S., et al., (2017). Impact of calcium and two doses of vitamin D on bone metabolism in the elderly: A randomized controlled trial. *Journal of Bone and Mineral Research, 32*(7), 1486–1495. https://doi.org/10.1002/jbmr.3122.

Rajasekaran, S., Vijay, S., & Ajoy, S., (2020). Etiology and natural history of adolescent idiopathic scoliosis: A review. *Indian Spine Journal, 3*(2), 131–142.

Raveendran, R., & Sudha, D., (2017). Vitamin C: Sources, functions, sensing and analysis. In: *Vitamin C*. https://doi.org/10.5772/intechopen.70162.

Rehman, A., (2020). Scoliosis treatments and medications. *Single Care, 2020*.

Richard, D. M., Michael, A. D., Charles, W. M., Ashley, A., Hill-Kapturczak, N., & Donald, M. D., (2009). L-tryptophan: Basic metabolic functions, behavioral research and therapeutic indications. *International Journal of Tryptophan Research*. https://doi.org/10.4137/ijtr.s2129.

Rico, H., Gómez-Raso, N., Revilla, M., Hernández, E. R., Seco, C., Páez, E., & Crespo, E., (2000). Effects on bone loss of manganese alone or with copper supplement in ovariectomized rats a morphometric and densitomeric study. *European Journal of Obstetrics and Gynecology and Reproductive Biology, 90*(1), 97–101. https://doi.org/10.1016/S0301-2115(99)00223-7.

Rondanelli, M., Milena, A. F., Vittoria, I., Clara, G., Giancarlo, I., Simone, P., Antonella, R., et al., (2021). Copper as dietary supplement for bone metabolism: A review. *Nutrients, 13*(7), 2246. https://doi.org/10.3390/nu13072246.

Routh, R. H., Sara, R., Ram, D. P., Alan, L. B., & Eric, A. N., (2005). The relationship between bone mineral density and biomechanics in patients with osteoporosis and scoliosis. *Osteoporosis International, 16*(12), 1857–1863. https://doi.org/10.1007/s00198-005-1951-z.

Rucker, R., Opsahl, W., Abbott, U., Greve, C., Kenny, C., & Stern, R., (1986). Animal model of human disease. Scoliosis in chickens. A model for the inherited form of adolescent scoliosis. *American Journal of Pathology, 123*(3), 585–588.

Sadat-Ali, M., Ai-Habdan, I., & Al-Othman, A., (2000). Adolescent idiopathic scoliosis. Is low melatonin a cause? *Joint Bone Spine, 67*(1), 62–64.

Sasaki, H., & Naohisa, M., (2010). Effects of combination treatment with alendronate and vitamin K2 on bone mineral density and strength in ovariectomized mice. *Journal of Bone and Mineral Metabolism, 28*(4), 403–409. https://doi.org/10.1007/s00774-009-0148-5.

Schapira, D., Linn, S., Sarid, M., Mokadi, S., Kabala, A., & Silbermann, M., (1995). Calcium and vitamin D enriched diets increase and preserve vertebral mineral content in aging laboratory rats. *Bone, 16*(5), 575–582. https://doi.org/10.1016/8756-3282(95)00088-U.

Shakeri, F., & M, B., (2017). Anti-inflammatory, antioxidant, and immunomodulatory effects of curcumin in ovalbumin-sensitized rat. *BioFactors, 43*(4), 567–576. https://doi.org/10.1002/biof.1364.

Sharma, R. A., Gescher, A. J., & Steward, W. P., (2005). Curcumin: The story so far. *European Journal of Cancer, 41*(13), 1955–1968. https://doi.org/10.1016/j.ejca.2005.05.009.

Silva, F. E., & Lawrence, G. L., (2010). Adult degenerative scoliosis: Evaluation and management. *Neurosurgical Focus, 28*(3), E1. https://doi.org/10.3171/2010.1.FOCUS09271.

Singh, P., Kesharwani, R. K., & Keservani, R. K., (2017a). Antioxidants and vitamins: Roles in cellular function and metabolism. In: *Sustained Energy for Enhanced Human Functions and Activity* (pp. 385–407). Academic Press. ISBN:9780128054130.

Singh, P., Kesharwani, R. K., & Keservani, R. K., (2017b). Protein, carbohydrates, and fats: Energy metabolism. In: *Sustained Energy for Enhanced Human Functions and Activity* (pp. 103–115). Academic Press. ISBN:9780128054130.

Srivastava, P., Singh, M., Chaturvedi, R. (2020). Herbal medicine and biotechnology for the benefit of human health. In: Verma, A. S & Singh, A. (eds.). Animal Biotechnology (Second Edition). Amsterdam: Academic Press, Elsevier, pp. 613–629.

Strause, L., Saltman, P., Smith, K. T., Bracker, M., & Andon, M. B., (1994). Spinal bone loss in postmenopausal women supplemented with calcium and trace minerals. *Journal of Nutrition, 124*(7), 1060–1064. https://doi.org/10.1093/jn/124.7.1060.

Suh, K. T., Il, S. E., & Jung, S. L., (2010). Polymorphism in vitamin D receptor is associated with bone mineral density in patients with adolescent idiopathic scoliosis. *European Spine Journal, 19*(9), 1545–1550. https://doi.org/10.1007/s00586-010-1385-y.

Talasila, S. S. A., Misha, G., & Vimala, T., (2017). A study on screening for scoliosis among school children in the age group of, 10–14 using a cost effective and an innovative technique. *International Journal of Community Medicine and Public Health, 4*(6), 2118–2123. https://doi.org/10.18203/2394-6040.ijcmph20172187.

Tan, R. R., Shi, J. Z., Bun, T., Wen, S. H., Xiao, J. Z., Xiao, Y. C., Nan, Y., et al., (2015). A natural product, resveratrol, protects against high-glucose-induced developmental damage in chicken embryo. *Journal of Asian Natural Products Research, 17*(5), 586–594. https://doi.org/10.1080/10286020.2015.1043901.

Teles, A., St-Georges, M., Fahad, A., Leonardo, S., Fan, J., & Neil, S., (2020). Back pain in adolescents with idiopathic scoliosis: The contribution of morphological and psychological factors. *European Spine Journal, 29*(8), 1959–1971. https://doi.org/10.1007/s00586-020-06489-2.

Tian, Y., Wanli, L., Guangfan, H., & Yongxi, Z., (2003). Effect of calcium supplement on superoxide dismutase and malonaldehyde of disuse osteoporosis in young rats. *Wei Sheng Yan Jiu = Journal of Hygiene Research, 32*(1), 49, 50.

Tsirikos, A. I., & Simon, B. R., (2020). *Magnetic Controlled Growth Rods in the Treatment of Scoliosis: Safety, Efficacy and Patient Selection*. https://doi.org/10.2147/MDER.S198176.

Vaes, A. M. M., Brouwer-Brolsma, E. M., Van, D. Z. N. L., Van, W. J. P., Berendsen, A. A. M., N. Van, S. N., Van, D. V., et al. (2017). Food sources of vitamin D and their association with, 25-hydroxyvitamin D status in dutch older adults. *The Journal of Steroid Biochemistry and Molecular Biology, 173*, 228–234. https://doi.org/10.1016/J.JSBMB.2016.10.004.

Verpeut, J. L., Amy, L. W., & Nicholas, T. B., (2013). Citrus aurantium and *Rhodiola rosea* in combination reduce visceral white adipose tissue and increase hypothalamic norepinephrine in a rat model of diet-induced obesity. *Nutrition Research, 33*(6), 503–512. https://doi.org/10.1016/j.nutres.2013.04.001.

Walther, D. J., & Michael, B., (2003). A unique central tryptophan hydroxylase isoform. *Biochemical Pharmacology, 66*(9), 1673–1680. https://doi.org/10.1016/S0006-2952(03)00556-2.

Wang, J., Xiaoxiao, W., Juntao, L., Yiqiang, C., Wenjun, Y., & Liying, Z., (2015). Effects of dietary coconut oil as a medium-chain fatty acid source on performance, carcass composition and serum lipids in male broilers. *Asian-Australasian Journal of Animal Sciences, 28*(2), 223–230. https://doi.org/10.5713/ajas.14.0328.

Watanabe, F., (2007). Vitamin B12 sources and bioavailability. *Experimental Biology and Medicine*. https://doi.org/10.3181/0703-MR-67.

Wong, R. H. X., Jay, J. T. Z., Cory, J. X., & Peter, R. C. H., (2020). Regular supplementation with resveratrol improves bone mineral density in postmenopausal women: A randomized, placebo-controlled trial. *Journal of Bone and Mineral Research, 35*(11), 2121–2131. https://doi.org/10.1002/jbmr.4115.

Zhang, J., Jing, Y., Daohong, Z., Chaoran, W., Yuhao, Z., Yingsong, W., & Tao, L., (2019). Therapeutic effect and mechanism of action of quercetin in a rat model of osteoarthritis. *Journal of International Medical Research, 48*(3). https://doi.org/10.1177/0300060519873461.

Zittermann, A., (2013). Magnesium deficit – overlooked cause of low vitamin D status? *BMC Medicine*. https://doi.org/10.1186/1741-7015-11-229.

Zoch, M. L., Thomas, L. C., & Ryan, C. R., (2016). New insights into the biology of osteocalcin. *Bone, 82*, 42–49. https://doi.org/10.1016/j.bone.2015.05.046.

CHAPTER 5

ROLE OF NUTRACEUTICALS IN THE TREATMENT OF RHEUMATOID ARTHRITIS

DEVESH U. KAPOOR[1] and MANISH JAIMINI[2]

[1]*Dr. Dayaram Patel Pharmacy College, Bardoli, Gujarat, India*

[2]*Department of Pharmacy, Maharishi Arvind University, Jaipur, Rajasthan, India*

ABSTRACT

In the 21st century rheumatoid arthritis (RA) is very common autoimmune disorder having no exact cure and on priority for biological investigation for scientific community. Due to limitation in safety and efficacy in existing treatment options, there is need to search for novel and more safe and effective alternatives for this disorder. The current experimental investigations shed light on the effectiveness of nutraceuticals for RA management. There are imperative biochemical findings established by researchers for various nutraceuticals against RA. The several phytochemicals have the potential to treat RA due to their immunomodulatory, anti-inflammatory, anti-oxidative and enzyme inhibition properties. The nutraceuticals have mammoth potential as safe and effectual treatment option against RA disorder.

5.1 INTRODUCTION

In our body, naturally the Immune system is the dynamic system which plays a vital role in protecting the body from foreign invaders or foreign

Nutraceuticals and Bone Health. Deepak Sharma, Madan Mohan Gupta, Anil K. Sharma, Raj K. Keservani, Rajesh K. Kesharwani (Eds.)

body (which is due other individuals or animals or environmental pollution or dust or smoke or other heavy industries pollutions etc.) from different types of infections or diseases. Being an auto immune and inflammatory disease it attacks the healthy cells in body due to oversight or pseudo identification thereby causing inflammation, Both smaller and large joints of hands and feet are subject to painful swellings. In other words in Rheumatoid arthritis, or RA, the body's own immune system attacks its own tissues, including joints (Rindfleisch & Muller, 2005). In extreme severe cases the immune system even attacks and affects the Internal organs including the vital organs including lungs, heart and eyes, where the disease is major. The keys areas which are easily attacked are the inner joints linings ultimately causes severe swellings, and over the period of time it leads to bone erosion, brittle nature of the bones, tendency of twisting of the bones and thereby causing to bone deformation. The lining of the joint becomes inflamed, causing damage to joint tissue. Such tissue damage is long-lasting or with severe chronic pain, causing unsteadiness at maximum area leading to lack of balance, and deformity in short and long bones and around joints resulting in mis shapenness. As the main cause of RA is the immune system attacking healthy body tissue but it's not yet known what triggers this. In fact the immune system normally makes antibodies that attack bacteria and viruses, helping to fight infection. The major symptoms are Inflammation which includes swollen tender, swelling in joints with pain, regular stiffness in joints especially in knees (most commonly seen during morning hours, also called as morning stiffness) and also after long continuous inactivity and also Fatigue, fever and loss of appetite. The condition is more painful during walking. The disease affect smaller joints of the fingers to hands and toes to feet, rather the defined area are more prone to disease with moderate to severe pain and swelling. Due to progressive nature of the disease it is likely to spread to the wrists, knees, ankles, elbows, hips and shoulders (Sanderson, Calnan, Morris, Richards & Hewlett, 2011). The condition is more worst as the symptoms occur in the same joints on both sides of the body, that means the symptoms spreads the entire joint. Rheumatoid arthritis signs and symptoms are variable in severity and may even come and go. In fact it is seen that the Periods of increased disease activity called flares, alternate with periods of relative remission. It also can cause joints to deform and shift out of place (Karami, Aslani, Jamshidi, Garshasbi & Mahmoudi, 2019).

5.2 PATHOGENESIS OF RA

Rheumatoid arthritis (RA) is an autoimmune systemic disease that primarily affects joints. In fact the etiology and the pathogenesis of RA are complex as it involves many types of cells, and others macrophages, T and B cells, fibro- blasts, chondrocytes and dendritic cells. It is seen that the predisposing factors are not very clear although role of genes and epigenetic modifications, development and evolution are of the disease are well studied. Moreover environmental factors like smoke from cigarette, pollutants from industries, and disturbances of intestinal, lungs are key factors for RA pathogenesis.

At initial stage both quantitative and qualitative disturbances of peptide citrullination, protein modifications, antigen presenting cell and fibroblast like synoviocytes activation are key factors. Most microbes foster the process by antigen presenting cell and fibroblast like synoviocytes direct and indirect activation. In next stage or second stage the antigen presenting cell elicit specific humoral B cell response resulting in antibodies production and T cell reactivity. It is also seen that inherited and acquired defects in T and B Cell response are caused by repetition of activation of innate immunity as well as loss of tolerance, elicit autoimmune inflammation of synovial membranes and development of cellular pannus (Karami et al., 2019; McInnes & Schett, 2007) (Figure 5.1).

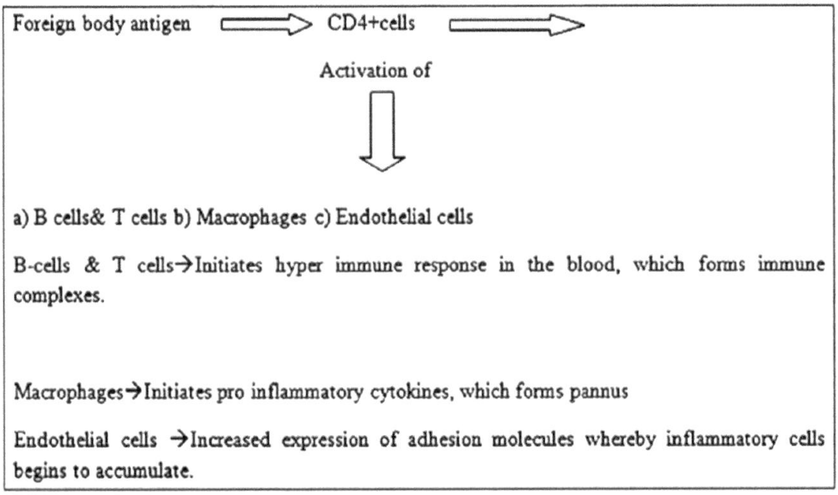

FIGURE 5.1 Pathogenesis of RA.

5.2.1 *PATHOPHYSIOLOGY OF ARTHRITIS*

In RA, the body's immune system attacks the lining of the joint capsule which is a tough membrane enclosing the joint parts. This lining (synovial membrane) is attacked by the body's own defense immune system, becomes inflamed and swollen. The disease process which ultimately and eventually destroy cartilage and bone within the joint. There are as many as twelve early warning signs of RA which include, Fatigue. Before experiencing any other symptoms, a person with RA may feel extremely tired and lack energy, Slight fever. Inflammation associated with RA may cause people to feel unwell and feverish, Weight loss, Stiffness, Joint tenderness, Joint pain, Joint swelling, Joint redness, Joint warmth, Numbness and tingling, Decrease in range of motion, Joints affected on both sides. RA (late stage)-Includes Boutonnire deformity of thumb, Ulnar deviation of metacarpophalangeal joints, Swan-neck deformity of fingers (Cooles & Isaacs, 2011).

5.2.2 *DIAGNOSIS*

Clinically RA can be easily diagnosed with inflammatory arthritis involving at least three or more joints. Positive RF and or anti citrullinated peptide/protein antibody, Disease duration of more than six-eight weeks, Raised or elevated CRP or ESR but without evidences of disease with similar clinical features. Rheumatoid arthritis diagnosis is directly proportional to increases with the number of small joints involved. More the joints severe are the pain. In fact rheumatoid factor or anti-citrullinated protein antibody, or elevated C-reactive protein level or erythrocyte sedimentation rate suggests a diagnosis of rheumatoid arthritis (Majithia & Geraci, 2007). CBC (complete blood count), differential and assessment of renal and hepatic function are the key initial laboratory evaluation. If taking biologic agents the individual must be tested for hepatitis B, hepatitis C, and tuberculosis. Importantly earlier diagnosis of rheumatoid arthritis offers better understanding of the progression of the disease and proper line of earlier treatment with disease-modifying antirheumatic agents, so that the disease further progression can be controlled thereby preventing further deterioration (Aletaha & Smolen, 2018).

5.3 CONVENTIONAL TREATMENT FOR MANAGEMENT OF RA

First line Management include, self-esteem management strategies and medications can help for effective treatment. The line of treatment includes medications for slowing the progression of the disease, preventing the deformity of the joints, which is also called as disease-modifying antirheumatic drugs (DMARDs). The initial choice of the line of treatment of the drug is Methotrexate which is dynamic as the first-line drug for rheumatoid arthritis, although the drug is toxic in nature especially who are prone or sensitive to hepatic disease (Bullock et al., 2018).

Second-Line management is based upon biological response modifiers (biologicals) are medications. These so called Biologic agents are tumor necrosis factor inhibitors, and are better option for the second-line agents or are highly effective for dual therapy. Important is to minimization the joint pain and swelling, prevention of radiographic damage and visible deformity, and continuation of work and personal activities and boosting individual moral high. The simple techniques to control RA, help to provide both best opportunity by boosting moral, develops skills and confidence to effectively manage their condition on their own without much of the support, Interactive workshops teach self-management skills, Regularly attending the programs will help to learn ways to manage pain, exercise safely, and gain control of arthritis and build self-esteem, Physical activity like exercise decreases pain, improves function and delays disability for adults with arthritis. Moderate physical activity each week help substantial health benefits, Personal exercise programs like Fitness club and Walk with ease help adults with arthritis increase their physical activity safely and comfortably, Maintaining optimum healthy weight can limit arthritis progression and activity limitation, Need is to controlled overload on priority (Bullock et al., 2018).

Optimum reduction in the load exerted on the knee, ultimately reduce arthritis-related pain and disability, Sports related or work related injuries to joints can increase the likelihood of developing osteoarthritis, repeated knee bending, place individuals at higher risk for this type of arthritis, Consulting early and diagnosis early helps to maintaining a good quality of life, Essential disease modifying drugs are beneficial in rheumatoid arthritis and other inflammatory arthritis conditions. To manage RA with self-management strategies proven to reduce pain and disability, and allow to pursue the activities important to them instead of compelling them. While there's no cure for rheumatoid arthritis, physiotherapy and medication can help slow the disease's progression. Most cases can be managed with a class of medications called anti-rheumatic drugs (DMARDS) (Monti et al., 2017).

5.3.1 OMEGA 3 FATTY ACIDS

Marine sources having long chain omega-3 fatty acids like algae are novel nutraceutical due to their effective and proven therapeutic effects properties like hypotriglyceridemic, antithrombotic and anti-inflammatory properties. This is due to the presence of eiocosapentaenoic acid (EPA) and docosahexaenoic acid (DHA) in omega-3 fatty acids offers effective therapeutic effects except in alpha linolenic acid which has least therapeutic effect. The effectiveness of therapeutic agents are well clinically proven and documented in patients with diseases like RA, Crohn's disease, colitis, psoriasis, asthma, burns etc., and acute respiratory distress syndrome. In fact Omega-3 fatty acids (EPA and DHIA) provide therapeutics benefits in combination but alone DHIA results are yet to be confirmed. Thus, omega 3-fatty acid (EPA & DHA) nutraceuticals are significant adjunct therapeutic. Omega-3 fatty acid (EPA and DHA) nutraceuticals offer significant adjunct therapeutic benefits in patients with for treatment of inflammatory disease (Rajaei et al., 2016).

5.3.2 VITAMIN E

Alpha tocopherol (also called as Vitamin E) is lipid soluble, chain breaking antioxidant is available biologically membranes. Lipid is core constituents of normal synovial fluid and any changes or alterations in synovial lipids causes RA. Hence any decrease in the levels of alpha tocopherol exhibits detrimental effects in RA. OA (Osteoarthritis) a disease of the joint involves cartilage degeneration and chondrocytes apoptosis. The main cause is the oxidative stress mechanisms under pinning joint degeneration in OA. The therapy includes management of pain symptomatically but do not alter the biological processes underlying the cartilage degeneration.

Vitamin E prevents or treats OA due to antioxidant and anti-inflammatory effects. Vitamin E mitigated stress in cartilage explants or chondrocyte by mechanical stress or free radicals (Xiong et al., 2014).

5.3.3 CAROTENOIDS AND PHENOLIC COMPOUNDS

Phenolics compounds are essential compounds for the suppression of inflammation among phytochemicals, and they revealed their potent anti-inflammatory capacity. In fact the nutritional components direct the cellular metabolism and also control pathological inflammatory process, play vital role as coadjunct in the effective treatment of chronic diseases like RA. The dietary phenolic compounds are sensitive to modulate pro-oxidant and pro inflammatory pathways so as to reduce the onset of RA progression. Dietary phenolic compounds may influence the course of RA, ameliorating the RA symptoms and down regulating the inflammation at the molecular level (Chisté & Mercadante, 2012).

5.3.4 PHYTOSTEROLS

Naturally occurring compounds found in plant cell membranes also called as plant sterol and stanol esters) As phytosterols are structurally similar to the body's cholesterol, when they are consumed they compete with cholesterol for absorption in the digestive system. In fact Phytosterols are safe, easy and effective nutritional supplement which control Total and LDL cholesterol, Triglycerides, risk factor for CVD. If fact Phytosterols foods leads to lifestyle modifications, dyslipidemic individuals achieves lowering of lipid values (Vilahur, Ben-Aicha, Diaz-Riera, Badimon & Padró, 2019).

5.3.5 GAMMA LINOLENIC ACID (GLA)

Plant seeds such as Evening Primrose, Borage oil rich in Fatty acid. Oil of Evening primrose oil contain Gamma linolenic acid is the preferred choice as medicine for skin disease of the skin like systemic sclerosis, psoriasis, and eczema. Arachidonate 5-lipoxygenase produces no leukotrienes and the conversion by the enzyme of arachidonic acid to leukotrienes is inhibited, when acting on GLA. Being an omega-6 fatty acid, GLA in the body can convert to substances that reduce inflammation and cell growth. Polyps in the mouth, high cholesterol and other blood fats, heart disease, metabolic syndrome or syndrome-X, diabetic nerve pain, attention deficit-hyperactivity disorder (ADHD), depression, depression after childbirth, chronic fatigue syndrome (CFS), and hay fever or allergic rhinitis, also to prevent cancer and to help breast cancer patients respond faster to treatment with the drug tamoxifen the choice is Gamma linolenic acid or GLA. Gamma linolenic acid is present in traces in green leafy vegetables and in nuts. The key source of GLA is breast milk for infants. GLA is metabolized to dihomogamma lineic acid which undergoes oxidative metabolism by cycloxegenases and lipoxygenase to produce anti-inflammatory eicosanoids (Sergeant, Rahbar & Chilton, 2016).

5.3.6 AVOCADO-SOYBEAN UNSAPONIFIABLES (ASU)

ASU is a chemical found in avocado and soybean oils. The chemicals are mixed together to make avocado soy unsaponifiable products. ASU products are being used for pain, inflammation, OA, back pain, hot flashes and presently their use are based on traditional response and not scientifically as on date. ASU products includes chemicals like Vitamin E, plant fats etc whose mixtures actually reduces the swelling in joints and other parts of the body. As ASU is a combination of vegetable extract of one third avocado oil and two third soybean oil, it blocks inflammatory chemicals and prevent deterioration of the cells that line the joints. ASU act as an adjuvant treatment for osteoarthritic pain mixture of. The slow-acting symptomatic approach for the treatment of hip and knee osteoarthritis; effectively controls autoimmune diseases; diffuse scleroderma and scleroderma-like states (e.g., morphea, sclerodactyly, scleroderma in bands). ASE mixture can improve the mood and quality of life of postmenopausal women in reducing menopause-related symptoms. Moreover the mixture of ASU provides support by increase in the collagen content in tissues, Increase in tissue lipids levels, There is also increase in the proportion of extractable constituents in relation to insoluble ones, also increase in tissue proteases and collagenases activity of serum leucine-peptidase. Unsaponifiable extracts contains substance characteristic of soybean and avocado seed extracts, which modify the metabolism of connective tissue (Salehi et al., 2020). Importantly, Avocado extract stimulates stroma-related enzymes, whereas soybean extract alone sensitively stimulates lysosomal enzymes with an acidic pH and, to a lesser extent, some neutral lysosomal proteases. It is the granulomas composition that defines the action of unsaponifiables. Any change in the values or upward increase in the ratio of the macromolecules (collagen, glycoproteins) in both the soluble and insoluble fractions of granuloma extracts are being taken or analyzed as a sign of the increased degradation of these tissue constituents. In fact after ASU administration favorable results are seen which are due are due to a collagenolytic effect. The positive benefits of ASU extract are not only in osteoarthritic disorders and autoimmune diseases in fact it also cover other disorders.

5.4 NUTRACEUTICALS USED TO TREAT RA

Nutraceuticals are presently gaining appreciation as they have advantageous result against RA. They exhibited different action such as analgesic, immunomodulatory, activation of antioxidant and anti-inflammatory action (Al-Okbi, 2014). These properties offer critical role in protection against pathologies of diverse chronic and age related disease such as RA. Theses therapeutic efficiency of nutraceuticals is due to the physiologically active secondary metabolites like flavonoid, tannin, polyphenol, anthocyanine, glycoside etc. (Nayak, Jena, Sahu, Tiwari & Chaudhury, 2021).

5.4.1 FLAVONOIDS

These are plant secondary metabolites, mostly found in vegetables, tea, bear, wine, fruits, seeds and nuts. The flavonoids (Keservani and Sharma, 2014; Keservani et al., 2016a, 2016b) are very beneficial due to anti-inflammatory, anti-oxidant antiradical and immunomodulatory properties. The flavonoid immunomodulatory effect in RA is due to proinflammatory cytokines inhibition (Interleukin-6, IL-1β, tumor necrosis factor-α), inhibition of chemokines such as CXCL5, CXCL9, CXCL1, MCP1, apoptosis cell stimulation (neutrophils, macrophages), proliferation of synoviocytes (MMP13, MMP9, MMP3, MMP1) (Leyva-López, Gutierrez-Grijalva, Ambriz-Perez & Heredia, 2016).

5.4.1.1 CITRUS FLAVONOIDS

Citrus plants are well known source for flavonoids. The various flavonoids which are isolated from these citrus plants include naringenin, nobelitin, narirutin, naringin and hesperidin. The chemical structures of these flavonoids are responsible for anti-inflammatory activity. There are some noteworthy features related to structure of flavonoids responsible for anti-inflammatory action include existence of planar ring in flavonoid molecule, keto group presence at C4 position in C ring, existence of unsaturated C rings at C2-C3 position and hydroxyl group present in A ring (C5/C7 position) and B ring (C31/C41 position) (Mohanty, Pal & Si, 2020).

5.4.1.1.1 Hesperidin

It is a flavone glycoside made of rutinose and hespertein (aglycon part). Qi et al. investigated the Hesperidin (HPD) anti-inflammatory activity on fibroblast like synovial cells (FLSC) which is induced by synovial macrophages and lipopolysaccharides (LP) against antigen induced arthritis (AIA). The researchers observed that HPD can significantly inhibit the FLSC inflammatory factors and metalloproteinase which is induced by LPS. HPD also diminish the polarization of M1 macrophages and decrease the inflammatory factors. The investigators observed that HPD delayed arthritis growth by regulating M1 macrophage and FLS cells. The anti-inflammatory effect exerted by HPD by hindering pathway of PI3K/AKT. The researchers concluded that HPD has the ability to hinder the inflammation of synovial cell and polarization of macrophage by PI3K/AKT pathway suppression (Qi et al., 2019).

Xiao et al. studied the HPD effect on human mesenchymal stem cells (MCSC) to repair the cartilage tissue. The researchers were performed the proliferation and colony formation assay to characterize the MCSC self-renewal ability. The MCSC chondrogenesis was evaluated by measurement of Alcian blue staining and Sox9 expression. The assessment of pro inflammatory cytokines IFN-γ, IL-10, IL-4, IL-2 secretion and nuclear factor kappa B (NF-kB) subunit P65 expression were also done by the investigators. The researchers observed that HPD enhanced the self-renewal ability and MCSC chondrogenesis, diminished pro inflammatory cytokines IFN-γ, IL-10, IL-4, IL-2 secretion and decline the P65 expression. The P65 overexpression has the ability to reverse the HPD diminished pro inflammatory cytokines secretions and also obliterate the surging effect of HPD on MCSC chondrogenesis. The researchers delineated that HPD have the potential to increase the MCSC chondrogenesis by decreasing the inflammation and help in tissue repair of cartilage (Xiao et al., 2018).

5.4.1.1.2 Naringenin (NAG)

It is a bitter and colorless flavanone which is found in different types of natural sources like citrus fruit (Keservani et al., 2016b) and grape fruits. The NAG has various different mechanisms to treat RA by hindering several inflammatory processes. The NAG diminished the inflammatory mediators production by hindering proinflammatory cytokines mainly COX-2, as a result decreasing the inflammation and joint pain in RA. Due to inhibition of enhanced COX-2 expression, synthesis of thromboxane, prostacyclin and prostaglandins hindered as well. The COX-2 and PEG2 are stimulatory factors for IL-6 enhancing synthesis from macrophages (Lee & Kim, 2010).

The NAG has the potential to hinder the TNF production by T-lymphocytes, B-lymphocytes, monocytes, macrophages and fibroblasts in cartilage pannus junction as well as synovial tissue, which are very crucial in RA development and progression (Postal & Appenzeller, 2011; Xu et al., 2013).

The synthesis of nitric oxide (NO) occurs due to L-arginine with inducible NO synthetases (I-NOS). This NO is responsible for mitochondrial dysfunction, inflammation and apoptosis in RA. The NAG is capable to modulate the NO synthesis due to murine macrophages cells which are stimulated by LPS (Korhonen, Lahti, Kankaanranta & Moilanen, 2005). The NF-Kb signaling pathway activation plays a vital role in RA pathogenesis. The patients suffering from RA exhibited NF-kB binding and superior amount activated NF-kB proteins mainly in synovial tissues. The activation of this pathway facilitates different process of inflammation like catabolism, destruction of extracellular matrix. The pro inflammatory cytokines such as IL-1β and TNF-facilitated cartilage destruction (Hämäläinen, Nieminen, Vuorela, Heinonen & Moilanen, 2007). The dendritic cells also known as antigen presenting cells extracted out from bone marrow responsible for activation of resting T cells. After maturity these cells reach to lymphoid organs and alter the immune response. The mature dendritic cells plays crucial role for pathogenesis of RA. Mature dendritic cells cause proliferative synovitis and also damage the joints and articular cartilage (Rodríguez-Fernández, 2013). The NAG has the potential to inhibit the dendritic cells maturation, exhibits its anti-rheumatism potential (Li et al., 2015).

Wang et al. formulated nanomicelles which are based on Rebaudiside A (Rb A) to deliver naringenin (NAG) by oral administration. These fabricated nano micelles are able to deliver the NAR with enhanced bioavailability and better therapeutic efficacy. The researcher observed that nano micelles of RbA-NAG formulated by using simple ethanol dissolution evaporation method. The researchers found by mouse model that self nanomicelles enhanced the NAG oral bioavailability and significantly protected from small intestine injuries initiated by indomethacin. The mechanism behind this therapeutic effect was inhibition of pro inflammatory cytokines and oxidative stress. The investigators concluded that formulated nano drug delivery system has noteworthy potential to deliver the NAG orally (Wang, He, Hou, Geng & Wu, 2020).

Zhu et al. investigated the anti-arthritis potential of NAG against the rats suffering from Freund's adjuvant arthritis (FAA). The researchers also made the combination of NAG with dexamethasone (DMS) and evaluated its effect against FAA. The researchers observed that paw swelling was considerably decreased with NAG+DMS. Furthermore, the production of IL-1β, IL-6 and TNF-α significantly plunged with NAG+DMS treatment. The pathological changes regarding tissues of joints considerably suppressed by NAG+DMS revealed by HE staining study. The NAG+DMS induced FAA apoptosis by regulating Bcl-2 and Bax protein expression confirmed by TUNNEL assay. So the investigators concluded that NAG has the noteworthy potential to treat the RA (Wang et al., 2015).

Fan et al. studied the NAG as anti-arthritic nutraceutical against complete Freund's adjuvant arthritis (CFAA) and inflammation induced by formaldehyde. The investigators observed that rats treated with NAG exhibited significant improvement in up regulation of catalase, levels of superoxide dismutase and malondialdehyde and glutathione down regulation. The inflammatory marker TNF-α level was also diminished significantly by NAG. There is enhancement in Nrf-2/HO-1 levels and decline in NF-Kb mRNA levels observed at CFAA rat joints. In addition, the researchers found that expression of metalloproteinase (MMP3 and MMP-9) which are extracellular matrix degrading enzymes declined in CFAA animal model due to NAG. The researchers concluded that NAG have the potential to down-regulate the NF-kB pathway and activate the pathway of Nrf-2/HO-1(Fan, Pan, Zhu & Zhang, 2017).

5.4.1.1.3 Nobiletin (NOB)

It is citrus fruit derivative exhibited attenuation of RA by inhibiting the pro inflammatory cytokines expression. Liu et al. investigated the NOB impact through interleukin-21 (IL-21) mediating effect against MH7A fibroblast like synoviocytes (FLS). The investigators found that IL-21 receptor expression reduced significantly in FLS by NOB. Furthermore, NOB enhanced production of reactive oxygen species (ROS) induced by IL-21 and 4- hydroxynonenal. The NOB also enhanced the expression of TNF-α and IL-6, diminished the potential of mitochondrial membrane. The MMP3 and MMP13, main degradative enzymes responsible for cartilage destruction in RA, attenuated by NOB significantly. The phosphorylation of JAK1 and STAT3 were considerably diminished by NOB. The researchers concluded that NOB offers an effective and safe treatment against RA development and progression (Liu, Guo & Dong, 2020).

Yang et al. investigated the effect of NOB on rat model of CIA and RA, established by bovine type II collagen (BIIC) by employing hematoxylin-eosin staining (H&E), immunosorbent assay (ELISA) linked with serum enzyme and western blotting. The researchers found that there was decline in pro inflammatory cytokines such as interleukin-1β (IL-1β), angiogenesis and synovial tissue inflammatory infiltration. The levels of TNF-α, IL-6 and monocyte chemotactic protein-1 also considerably diminished when treated with NOB. There was downregulation of phosphorylated p38, p65 protein expression levels in CIA rats due to NOB. The investigators concluded that NOB has the potential to inhibit the RA development by declining the inflammatory infiltration and angiogenesis due to down regulation of p38 or signaling pathway of NF-kB in the rat synovium in CIA (Yang, Li, Yuan, Yang & Pan, 2017).

5.4.2 PROBIOTICS

The gut microbiota (Gm) plays a crucial role in RA pathogenesis, is well known fact. There are evidences which suggest that supplementation of probiotics have significant effect against gut dysbiosis responsible for inflammatory response. In RA, the supplementation of probiotics exhibited no noteworthy side effects. The particular bacteria of probiotics modify the immune response by affecting various cells having involvement in acquired and innate immunity like lymphocytes, macrophages, natural killer cells (NKC), epithelial cells and dendritic cells (DCs) (Mohajeri, La Fata, Steinert & Weber, 2018). The maximum studies have been done on pattern recognition receptors (PPRs), able to activate the signaling pathways that affect the cytokine production and proliferation of cell to alter the immune system. The probiotics have the potential to downregulate the expression of TLR, diminishing inflammation (Reid, 2016).

Fan et al. investigated the *Bifidobaceterium adolescentis* (*B. adolescentis*) treatment on rat suffering from collagen induced arthritis (CIA). The researchers have compared the outcome of doses of *B. adolescentis* given as preventive and for therapeutics treatment. The preventive doses of *B. adolescentis* exhibited better results as compared to therapeutic treatment. Early administration of *B. adolescentis* showed excellence in production of short chain fatty acids (SCFAc), noteworthy superior frequency of Trges cells and diminished tumor necrosis factor (TNF) level in comparison with therapeutic doses of *B. adolescentis*. Furthermore, with preventive doses of *B. adolescentis,* Trges were positively correlated with SCFAc, in contrast there is negative correlation of pro inflammatory cytokines with SCFAc. The researchers concluded that preventive doses of probiotics can give positive impact on arthritis inflammation site and probiotics will be beneficial for patient if given an early stage of disease (Fan et al., 2020) (Figure 5.2).

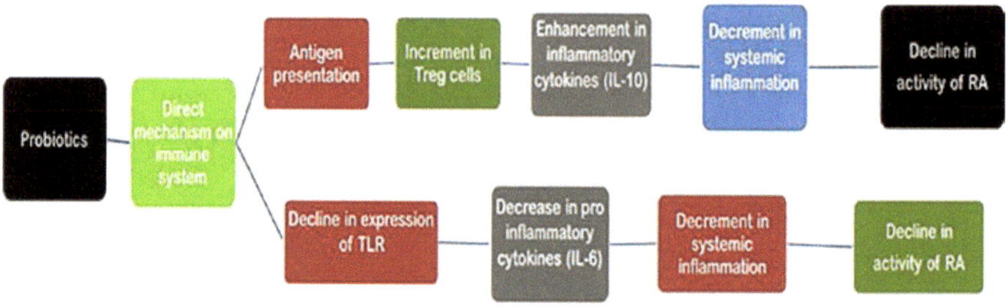

FIGURE 5.2 Direct mechanism of probiotics to decline the RA activity.

Pan et al. examined the *L. casei* potential to treat the adjuvant induced arthritis (AIA). The researchers observed that noteworthy alteration in microbial species was seen in gut. The investigators also observed some clinical alterations at the phases of progression like decline in swelling of joint, bone destruction prevention and decrease in arthritis scores. After thirty days of *L. casei* preventing treatment, noteworthy decline was seen in pro inflammatory levels of cytokines. Furthermore there was also an increment in the different lactobacillus strains like lactobacillus vaginalis, *L. reuteri, L. hominis* and *L. acidophilus* demonstrating that *L. casei* have the ability to improve the arthritis by rebalancing of lactobacillus strains (Pan et al., 2019).

Achi et al. evaluated the prophylactic activity of three different strains of *Bifidobacteria* (BiFi)such as *B. bifidum* (NCIM 5697), *B. berve* (NCIM 5671) and *B. Longum* (NCIM 5672) and against AIA model. The researchers observed that Bifi have the potential to diminish the progression and severity of RA. The administration of Bifi probiotic is very significant to downregulate the arthritis markers. Moreover, *B. berve* exhibited excellent results in Wistar rat model of AIA as compared to other Bifi species. The investigators concluded that anti-arthritic effect is dependent on strain and *B. berve* has significant potential to treat RA (Achi, Talahalli & Halami, 2019). Abhari et al. examined the performance of prebiotic Inulin (PRE) and probiotic (PRO) *Bacillus coagulans* (Independently and in combination) against AIA model of rats. The researchers observed noteworthy clinical development in AIA suffering rats with Inulin and/or *B. coagulans*. The serum amyloid A (Saa) and fibrinogen (FN) production was diminished in AIA model of rats with pretreatment of PRE, PRO and symbiotic (SYM). Furthermore, there was a decline in pro inflammatory parameters such as TNF-α and fibrinogen and suppression of paw swelling was also seen. The researchers concluded that prebiotic Inulin (PRE) and probiotic (PRO*) B. coagulans* have the potential to treat the RA in safe and effective manner (Abhari et al., 2016).

5.4.3 GLUCOSAMINE

It is excellent nutraceutical intended to use for RA, hinders the interleukin-1, COX-2, matrix metalloproteinases and proteoglycan (Fenton, Chlebek-Brown, Caron & Orth, 2002). It is a natural amino sugar which is produced by our body. Nakamua et al. investigated the anti-arthritic effect of glucosamine in placebo controlled study. The researchers have given a daily dose of 1500 mg of glucosamine to 51 patients suffering from RA and 26 patients given placebo. The researchers observed that there was a decline in MMP3 levels but there was no change in CRP and ESR levels. The researchers concluded that there was noteworthy improvement in the symptoms of patients with glucosamine treatment. The glucosamine exhibited only symptomatic effects but no anti-arthritic effects were observed (Nakamura et al., 2007).

5.4.4 KRILL OIL

The krill oil is mainly obtained from Antartic krill, having good amount of phospholipids also known as *Ephausia superba*. The docosahexaenoic acid (DHA) content in krill oil is same as fatty fish but other eicosapentaenoic acid (EPA) content is much higher than fatty fish. Preclinical studies exhibited that paw and joint swelling of AIA suffering mice reduced significantly and decline in arthritic scores with krill oil (Saidane, Semerano & Sellam, 2019). The studies suggested that krill oil has the potential to decrease the pain, C-reactive protein levels and arthritic scores considerably. There is no side effects reported using krill oil (Deutsch, 2007).

5.5 CONCLUSION

Nutraceuticals have enormous potential to treat RA in effective way, but more clinical trials should be conducted to get more evidence on its safety and efficacy. The utilization of nutraceuticals (Keservani et al., 2010a, 2010b, 2017, 2020) have enhanced in current times due to its significant therapeutic effect. The nutraceuticals which include phytosterols, phenolic compounds, tocopherols, carotenoids and polyunsaturated fatty acids have noteworthy and advantageous impact on inflammatory disorder like RA. There are some imperative parameters which are assessing by researchers for novel anti-inflammatory dietary phytochemicals such as colonic microflora, Zn & Cu level and status of antioxidant & oxidative stress. There is prospect for researchers to find out the exact impact of nutraceuticals against some pathophysiological condition in RA such as activation of phagocyte, activation of β-lymphocyte and activation of T-lymphocyte.

KEYWORDS

- **flavonoids**
- **immunosuppression**
- **nutraceuticals**
- **pathogenesis**
- **probiotics**
- **rheumatoid arthritis**

REFERENCES

Abhari, K., Shekarforoush, S. S., Hosseinzadeh, S., Nazifi, S., Sajedianfard, J., & Eskandari, M. H., (2016). The effects of orally administered *Bacillus coagulans* and inulin on prevention and progression of rheumatoid arthritis in rats. *Food & Nutrition Research, 60*(1), 30876.

Achi, S. C., Talahalli, R. R., & Halami, P. M., (2019). Prophylactic effects of probiotic *Bifidobacterium* spp. in the resolution of inflammation in arthritic rats. *Applied Microbiology and Biotechnology, 103*(15), 6287–6296.

Aletaha, D., & Smolen, J. S., (2018). Diagnosis and management of rheumatoid arthritis: A review. *JAMA, 320*(13), 1360–1372.

Al-Okbi, S. Y., (2014). Nutraceuticals of anti-inflammatory activity as complementary therapy for rheumatoid arthritis. *Toxicology and Industrial Health, 30*(8), 738–749.

Bullock, J., Rizvi, S. A., Saleh, A. M., Ahmed, S. S., Do, D. P., Ansari, R. A., & Ahmed, J., (2018). Rheumatoid arthritis: A brief overview of the treatment. *Medical Principles and Practice, 27*(6), 501–507.

Chisté, R. C., & Mercadante, A. Z., (2012). Identification and quantification, by HPLC-DAD-MS/MS, of carotenoids and phenolic compounds from the Amazonian fruit *Caryocar villosum*. *Journal of Agricultural and Food Chemistry, 60*(23), 5884–5892.

Cooles, F. A., & Isaacs, J. D., (2011). Pathophysiology of rheumatoid arthritis. *Current Opinion in Rheumatology, 23*(3), 233–240.

Deutsch, L., (2007). Evaluation of the effect of Neptune krill oil on chronic inflammation and arthritic symptoms. *Journal of the American College of Nutrition, 26*(1), 39–48.

Fan, R., Pan, T., Zhu, A. L., & Zhang, M. H., (2017). Anti-inflammatory and anti-arthritic properties of naringenin via attenuation of NF-κB and activation of the heme oxygenase (HO)-1/related factor 2 pathway. *Pharmacological Reports, 69*(5), 1021–1029.

Fan, Z., Yang, B., Ross, R. P., Stanton, C., Shi, G., Zhao, J., & Chen, W., (2020). Protective effects of *Bifidobacterium adolescentis* on collagen-induced arthritis in rats depend on timing of administration. *Food & Function, 11*(5), 4499–4511.

Fenton, J., Chlebek-Brown, K., Caron, J., & Orth, M., (2002). Effect of glucosamine on interleukin-1-conditioned articular cartilage. *Equine Veterinary Journal, 34*(S34), 219–223.

Hämäläinen, M., Nieminen, R., Vuorela, P., Heinonen, M., & Moilanen, E., (2007). Anti-inflammatory effects of flavonoids: Genistein, kaempferol, quercetin, and daidzein inhibit STAT-1 and NF-κB activations, whereas flavone, isorhamnetin, naringenin, and pelargonidin inhibit only NF-κB activation along with their inhibitory effect on iNOS expression and NO production in activated macrophages. *Mediators of Inflammation, 2007*.

Karami, J., Aslani, S., Jamshidi, A., Garshasbi, M., & Mahmoudi, M., (2019). Genetic implications in the pathogenesis of rheumatoid arthritis; an updated review. *Gene, 702*, 8–16.

Keservani, R. K., & Sharma, A. K., (2014). Flavonoids: Emerging trends and potential health benefits. *Journal of Chinese Pharmaceutical Sciences, 23*(12), 815.

Keservani, R. K., Kesharwani, R. K., Sharma, A. K., Vyas, N., & Chadoker, A., (2010b). Nutritional supplements: An overview. *International Journal of Current Pharmaceutical Review and Research, 1*(1), 59–75.

Keservani, R. K., Kesharwani, R. K., Vyas, N., Jain, S., Raghuvanshi, R., & Sharma, A. K., (2010a). Nutraceutical and functional food as future food: A review. *Der Pharmacia Lettre, 2*(1), 106–116.

Keservani, R. K., Sharma, A. K., & Kesharwani, R. K., (2016a). Nutraceutical and functional foods for cardiovascular health. In: *Food Process Engineering* (pp. 257–278, 291–312). Apple Academic Press, CRC Press. ISBN: 9781771884020.

Keservani, R. K., Sharma, A. K., & Kesharwani, R. K., (2016b). Medicinal effect of nutraceutical fruits for the cognition and brain health. *Scientifica*, 10. Article ID 3109254. doi: 1 0.1155/2016/3109254.

Keservani, R. K., Sharma, A. K., & Kesharwani, R. K., (2017). An overview and therapeutic applications of nutraceutical and functional foods. *Recent Advances in Drug Delivery Technology*, 160–201. ISBN:9781522507543.

Keservani, R. K., Sharma, A. K., & Kesharwani, R. K., (2020). *Nutraceuticals and Dietary Supplements: Applications in Health Improvement and Disease Management.* CRC Press. ISBN: 9781771888738.

Korhonen, R., Lahti, A., Kankaanranta, H., & Moilanen, E., (2005). Nitric oxide production and signaling in inflammation. *Current Drug Targets-Inflammation & Allergy, 4*(4), 471–479.

Lee, J. H., & Kim, G. H., (2010). Evaluation of antioxidant and inhibitory activities for different subclasses flavonoids on enzymes for rheumatoid arthritis. *Journal of Food Science, 75*(7), H212–H217.

Leyva-López, N., Gutierrez-Grijalva, E. P., Ambriz-Perez, D. L., & Heredia, J. B., (2016). Flavonoids as cytokine modulators: A possible therapy for inflammation-related diseases. *International Journal of Molecular Sciences, 17*(6), 921.

Li, Y. R., Chen, D. Y., Chu, C. L., Li, S., Chen, Y. K., Wu, C. L., & Lin, C. C., (2015). Naringenin inhibits dendritic cell maturation and has therapeutic effects in a murine model of collagen-induced arthritis. *The Journal of Nutritional Biochemistry, 26*(12), 1467–1478.

Liu, Z., Guo, S., & Dong, Q., (2020). Nobiletin suppresses IL-21/IL-21 receptor-mediated inflammatory response in MH7A fibroblast-like synoviocytes (FLS): An implication in rheumatoid arthritis. *European Journal of Pharmacology, 875*, 172939.

Majithia, V., & Geraci, S. A., (2007). Rheumatoid arthritis: Diagnosis and management. *The American Journal of Medicine, 120*(11), 936–939.

McInnes, I. B., & Schett, G., (2007). Cytokines in the pathogenesis of rheumatoid arthritis. *Nature Reviews Immunology, 7*(6), 429–442.

Mohajeri, M. H., La Fata, G., Steinert, R. E., & Weber, P., (2018). Relationship between the gut microbiome and brain function. *Nutrition Reviews, 76*(7), 481–496.

Mohanty, S., Pal, A., & Si, S. C., (2020). Flavonoid as nutraceuticals: A therapeutic approach to rheumatoid arthritis. *Research Journal of Pharmacy and Technology, 13*(2), 991–998.

Monti, S., Klersy, C., Gorla, R., Sarzi-Puttini, P., Atzeni, F., Pellerito, R., & Favalli, E. G., (2017). Factors influencing the choice of first-and second-line biologic therapy for the treatment of rheumatoid arthritis: Real-life data from the Italian Lorhen registry. *Clinical Rheumatology, 36*(4), 753–761.

Nakamura, H., Masuko, K., Yudoh, K., Kato, T., Kamada, T., & Kawahara, T., (2007). Effects of glucosamine administration on patients with rheumatoid arthritis. *Rheumatology International, 27*(3), 213–218.

Nayak, S. K., Jena, S., Sahu, P. K., tiwari, P., & Chaudhury, H. C., (2021). Nutraceuticals as supplements in the management of arthritis: A review. *International Journal of Pharmaceutical Research, 13*(1).

Pan, H., Guo, R., Ju, Y., Wang, Q., Zhu, J., Xie, Y., & Lu, L., (2019). A single bacterium restores the microbiome dysbiosis to protect bones from destruction in a rat model of rheumatoid arthritis. *Microbiome, 7*(1), 1–11.

Postal, M., & Appenzeller, S., (2011). The role of tumor necrosis factor-alpha (TNF-α) in the pathogenesis of systemic lupus erythematosus. *Cytokine, 56*(3), 537–543.

Qi, W., Lin, C., Fan, K., Chen, Z., Liu, L., Feng, X., & Zhao, C., (2019). Hesperidin inhibits synovial cell inflammation and macrophage polarization through suppression of the PI3K/AKT pathway in complete Freund's adjuvant-induced arthritis in mice. *Chemico-Biological Interactions, 306*, 19–28.

Rajaei, E., Mowla, K., Ghorbani, A., Bahadoram, S., Bahadoram, M., & Dargahi-Malamir, M., (2016). The effect of omega-3 fatty acids in patients with active rheumatoid arthritis receiving DMARDs therapy: Double-blind randomized controlled trial. *Global Journal of Health Science, 8*(7), 18.

Reid, G., (2016). Probiotics: Definition, scope and mechanisms of action. *Best Practice & Research Clinical Gastroenterology, 30*(1), 17–25.

Rindfleisch, A. J., & Muller, D., (2005). Diagnosis and management of rheumatoid arthritis. *American Family Physician, 72*(6), 1037–1047.

Rodríguez-Fernández, J. L., (2013). Antigen presentation by dendritic cells in rheumatoid arthritis. *Current Topics in Medicinal Chemistry, 13*(6), 712–719.

Saidane, O., Semerano, L., & Sellam, J., (2019). *Could Omega-3 Fatty Acids Prevent Rheumatoid Arthritis?* Elsevier.

Salehi, B., Rescigno, A., Dettori, T., Calina, D., Docea, A. O., Singh, L., & Dowlati, B. A., (2020). Avocado–soybean unsaponifiables: A panoply of potentialities to be exploited. *Biomolecules, 10*(1), 130.

Sanderson, T., Calnan, M., Morris, M., Richards, P., & Hewlett, S., (2011). Shifting normalities: Interactions of changing conceptions of a normal life and the normalization of symptoms in rheumatoid arthritis. *Sociology of Health & Illness, 33*(4), 618–633.

Sergeant, S., Rahbar, E., & Chilton, F. H., (2016). Gamma-linolenic acid, dihommo-gamma linolenic, eicosanoids and inflammatory processes. *European Journal of Pharmacology, 785*, 77–86.

Vilahur, G., Ben-Aicha, S., Diaz-Riera, E., Badimon, L., & Padró, T., (2019). Phytosterols and inflammation. *Current Medicinal Chemistry, 26*(37), 6724–6734.

Wang, H., He, Y., Hou, Y., Geng, Y., & Wu, X., (2020). Novel self-nanomicellizing formulation based on rebaudioside A: A potential nanoplatform for oral delivery of naringenin. *Materials Science and Engineering: C, 112*, 110926.

Wang, J., Wei, T., Gao, J., He, H., Chang, X., & Yan, T., (2015). Effects of naringenin on inflammation in complete Freund's adjuvant-induced arthritis by regulating Bax/Bcl-2 balance. *Inflammation, 38*(1), 245–251.

Xiao, S., Liu, W., Bi, J., Liu, S., Zhao, H., Gong, N., & Gong, M., (2018). Anti-inflammatory effect of hesperidin enhances chondrogenesis of human mesenchymal stem cells for cartilage tissue repair. *Journal of Inflammation, 15*(1), 1–8.

Xiong, R. B., Li, Q., Wan, W. R., Guo, J. Q., Luo, B. D., & Gan, L., (2014). Effects and mechanisms of vitamin A and vitamin E on the levels of serum leptin and other related cytokines in rats with rheumatoid arthritis. *Experimental and Therapeutic Medicine, 8*(2), 499–504.

Xu, C., Chen, J., Zhang, J., Hu, X., Zhou, X., Lu, Z., & Jiang, H., (2013). Naringenin inhibits angiotensin II-induced vascular smooth muscle cells proliferation and migration and decreases neointimal hyperplasia in balloon injured rat carotid arteries through suppressing oxidative stress. *Biological and Pharmaceutical Bulletin*, b13–00247.

Yang, G., Li, S., Yuan, L., Yang, Y., & Pan, M. H., (2017). Effect of nobiletin on the MAPK/NF-κB signaling pathway in the synovial membrane of rats with arthritis induced by collagen. *Food & Function, 8*(12), 4668–4674.

CHAPTER 6

NUTRACEUTICALS INTERVENTIONS FOR EFFECTIVE MANAGEMENT OF PAGET'S DISEASE

SURAJ BAMANKAR and VAISHALI Y. LONDHE

Shobhaben Pratapbhai Patel School of Pharmacy & Technology Management, SVKM's NMIMS, Vile Parle [W], Mumbai, Maharashtra, India

ABSTRACT

Paget's disease is a chronic, slow-progressing skeletal condition characterized by abnormally rapid bone degeneration and reformation. The new bone may develop in one or more locations across the body and is physically aberrant, thick, and fragile. This aberrant growth pattern can result in bone aches, arthritis, malformations, and fractures. While there are several medications available to treat Paget's disease, they are all expensive and have a variety of side effects, which is why the focus is shifting to natural solutions to prevent the adverse effects associated with the therapeutic interventions. This chapter discusses the range of nutraceuticals such as calcium, vitamin D, phytochemicals, and marine products which can support bone health in patients with Paget's disease. While scientific studies demonstrate the therapeutic potential of nutraceuticals, further clinical evaluation is necessary to ensure their safe intake by the pagetic population. Moreover, global regulatory criteria for the pragmatic and safe usage of nutraceuticals can be proposed, given the nutraceutical market's huge development potential in the future ahead.

Nutraceuticals and Bone Health. Deepak Sharma, Madan Mohan Gupta, Anil K. Sharma, Raj K. Keservani, Rajesh K. Kesharwani (Eds.)

6.1 INTRODUCTION

The first explication of the Paget's disease was offered in 1876 by the English physician, Sir James Paget, when he addressed the Surgical and Medical Society of London, United Kingdom, about a kind of chronic bone inflammation. He published his now-classic article on a condition he called "osteitis deformans" in 1877. In that, he noted his meticulous findings in regards to the clinical features of the disease, history, pathological features and chemical characteristics of the bone. Paget inscribed the disease that now beholds his name as chronic disorder of the skeletal system in which there is a focally increased and disorganized bone remodeling in some regions of the bone. It is a metabolic disorder, which is marked by localized regions of elevated bone resorption and dysregulated bone remodeling. Lesions of Paget's disease of the bone (PDB) are highly vascularized and disordered, with deformations and enlargements that are brittle and susceptible to fracture (Rabjohns et al., 2021). While the exact cause of PDB is unknown, it is suspected that the disorder is triggered by a combination of enviro, hereditary, viral, and inflammatory variables. Osteoclasts, which are produced from immunological progenitors, are the key cell type involved in the PDB (Sabharwal et al., 2014).

Pages' disease of bone may damage a single bone or a group of bones, depending on the severity of the condition. In most instances, it affects the axial skeleton, with the pelvis accounting for most of all cases, followed by the femur, lumbar spine, skull, and the tibia. It is usually discovered by incidental detection and frequently manifests itself late in life, with symptoms that are non-specific and common in nature, putting a physician's wits to the test (Ralston, 2013).

PDB is a late-onset disease, which is uncommon before the age of 55, but it becomes more common after that. By the eighth decade of life, it affects around 5 percent of women and 8 percent of men in some countries (Van Staa et al., 2002). Affecting elderly population the most, PDB stands as 2nd most often encountered bone disease after osteoporosis (Seitz et al., 2008). Paget's disease has been shown to affect both men and women. However, men have been found to be more susceptible to the condition. It is also observed that the condition is more prevalent in people of European ancestry and is uncommon among Africans, Indian subcontinent residents, and Asians (Cooper et al., 2006; Joshi et al., 2006). These disparities in susceptibility are likely to have a genetic foundation, and they are consistent with the notion that the condition arose in northern Europe as a result of one or more founder mutations and spread around the world via emigration. Despite the fact that it was first characterized in 1877, evidence of the disease may be found throughout the archaeological record. Pagetic lesions can be detected on skeletons dating back to the Roman era (1st to 4th century CE) in the Europe, and they are more prevalent in the medieval period (Mays, 2010). Additionally, at least one occurrence of PDB in a skull may be found in the Middle East between the fourth and seventh centuries (Kesterke et al., 2019). Another research revealed that frequency of the disease was 1.7% before 1500 CE and grew to 3.1% thereafter (Rogers et al., 2002). However, subsequent reports refute this claimed rise. Extensive investigation of unearthed medieval bones in Northwest England finds that 15% of skeletons have pathologic alterations comparable to current PDB. Some characteristics of this medieval PDB do not correspond to contemporary standards, such as a high incidence of disease, a young age at death, and a large degree of skeletal damage (Shaw et al., 2019). Also worth noting is the fact that life expectancy was far lower prior to the modern era, infections were more common with less efficient treatments, and nutrition and the way of living was distinct. All of these factors may contribute to some of the differences in disease presentation observed.

This chapter aims to discuss the clinical and pathological characteristics of PDB, current diagnostics; therapeutics, promising nutraceutical interventions that can aid in the condition and regulatory aspects of nutraceuticals.

6.2 EPIDEMIOLOGY

The prevalence of PDB varies significantly by ethnic and geographical group. The United Kingdom (UK) has the largest frequency of the condition. When it comes to the elderly in the United Kingdom, PDB affects 1 percent to 2 percent of the population over the age of 55 and older. The frequency of the disease grows with age, reaching 8 percent when people reach their 80s. When compared to other parts of the world, the Indian subcontinent, Southeast and East Asia, and Scandinavia are relatively unaffected by PDB (Cooper et al., 2006; Corral-Gudino et al., 2013; Langston et al., 2004; Van Staa et al., 2002). Gender and age were investigated in a British study to determine the impact of PDB locations on the occurrence and spread of the condition. Study discovered that patients who are younger, had a greater prevalence of monostotic illness, with a stronger trend in females. Females with monostotic illness exhibited a greater degree of tibial involvement than males (Haddaway et al., 2007). According to various estimates, about 15 to 40% of patients with PDB had a 1st-degree relative who is also affected by the condition. This is referred to as "familial PDB," in contrast to the "sporadic PDB," that develops in individuals with no family history. In the familial PDB, the illness seems more likely to be polyostotic, though penetrance is partial. It is also diagnosed at a younger age (Morales-Piga et al., 1995; Seton et al., 2003). According to studies conducted in the 1970s and 1980s, Lancashire had the greatest frequency in England, with a reported prevalence of 6–8%. However, the frequency of this area declined by 60% between 1970 and 1990. Overall, the frequency PDB seems to be reducing, while the age of onset appears to be growing (Cundy, 2018). These results suggest that, both hereditary and environmental variables have a role in the development of this disease.

6.3 ETIOLOGY

The genesis of PDB are still a mystery to this day. In addition to evidence suggesting paramyxoviruses may participate to development of this syndrome dating back to the early 1970s, there has also been a long-standing understanding of the disorder's propensity to cluster in families. Based on these two lines of evidence, a long-held notion has been advanced: that this condition may be caused by a viral infection in a genetically vulnerable host. More than 20 years have passed since each of these areas has been thoroughly investigated, and contemporary advancements in molecular biology have made great progress, but have not yet provided definitive answers to the question of what is generating this abnormality (Siris, 1998).

The first field of growing study into etiology derives from current genomic studies. PDB has long been known as a condition that affects over one member of a family, and a "autosomal dominant pattern of inheritance" has been found (Grahame, 2000). A significant incidence of familial transmission is associated with PDB, with 40% of patients reporting that they have a first-degree relatives who is afflicted with the disease (Morales-Piga et al., 1995). Aside from that, family aggregation studies have shown a sevenfold increase in the relative chance of acquiring Paget's disease among first-degree relatives of diagnosed patients (Siris et al., 1991). Autosomal dominant inheritance is often seen in the disease, with genetic diversity and inadequate penetrance manifesting itself as a consequence of this inheritance pattern. Furthermore, researchers uncovered a number of mutations in the SQSTM1/p62 protein that impair the highly conserved ubiquitin binding site. These mutations were found in both familial and sporadic cases of Paget's disease (Good et al., 2004; Hocking et al., 2004; Johnson-Pais et al., 2003).

A viral etiology for pagetic osteoclasts was hypothesized after the finding of nucleocapsid-like structures in both their nuclei and cytoplasm, as well as the detection of measles and respiratory syncytial virus nucleocapsid protein antigens in these cells (Siris, 1998). As part of an investigation to detect measles virus RNA, researchers discovered that nucleocapsid protein mRNA sequences were present in up to 90% of osteoclasts in the disease affected tissues (Basle et al., 1986). Additional study indicates that the canine distemper virus may be another potential pathogen (Gordon et al., 1991). In the quest for viral identification, research utilizing RNA extraction and PCR techniques have shown conflicting results, where some of the studies revealed the presence of viral RNA inside of affected bone (Gordon et al., 1992) while others concluded that such evidence does not exist (Birch et al., 1994; Ralston et al., 1991). Several studies have recently found measles virus transcripts in bone marrow cells and a range of blood cells from pagetic individuals (Reddy et al., 1995, 1996).

There have been some reports of elevated interleukin-6 (IL-6) levels in the plasma of bone marrow and peripheral blood of affected individuals, while not all studies have confirmed this. IL-6 has been shown to promote osteoclast formation when given to the normal marrow cells, but further research is needed. The formation of osteoclast precursors is thought to be influenced by cytokines in the marrow microenvironment, restricting lesions to small locations once they arise. The etiology of the initial osseous lesion has yet to be determined (Roodman et al., 1992).

6.4 PATHOPHYSIOLOGY

By ordinary, bones tend to undergo a continuous process of remodeling in which old, brittle bone tissue is reabsorbed by a kind of multinucleated bone cell called an osteoclast, which has between 5 and 20 nuclei per cell. The reabsorbed bone tissue is subsequently replaced with new bone tissue produced by a different kind of bone cell called an osteoblast (Singer et al., 2014).

The process is initiated at the cellular level when an osteoblast creates a substance termed receptor activator of nuclear factor κB ligand (RANKL), which interacts with the RANK receptors on the osteoclasts. When RANKL binds to the RANK receptor, it stimulates osteoclasts to produce lysosomal enzymes such as collagenase and hydrochloric acid. When they work in unison, they degrade collagen and disseminate the minerals that make up the bone matrix. After sufficient demineralization of bone, osteoblasts synthesize another protein called osteoprotegerin (OPG), which binds to RANKL and inhibits it from activating RANK receptors (Nakamura et al., 2003). As a result, osteoclasts cease their demineralization of bone. When this happens, the osteoblasts develop a substance termed osteoid seam, which is mostly composed of collagen and acts as a scaffold for the calcium and phosphate deposits. This is the process through which a new bone is produced (Lawrence G. Raisz, M. D., and Barbara E. Kream, 1975; G. David Roodman, 2010).

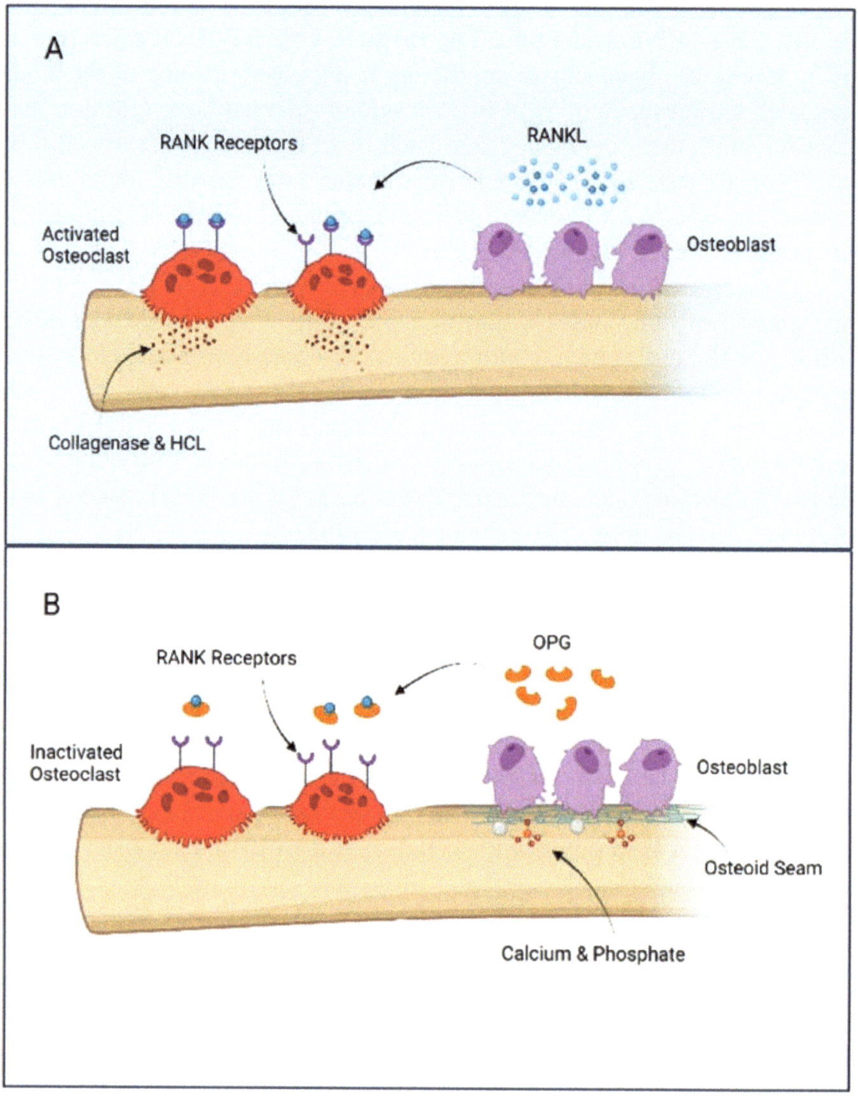

FIGURE 6.1 Bone remodelling process.

Paget's disease of bone manifests itself in three stages:

1. **Lytic Phase:** It is during the lytic phase when osteoclasts with up to 100 nuclei begin to actively demineralize the bone at a rate that is up to 20 times greater than usual. It has also been linked to considerable bone resorption as well as hypervascularization.
2. **Mixed Lytic-Blastic Phase:** It contains both the lytic and blastic stages at the same time. During the blastic phase, a large number of osteoblasts proliferate bone tissue quickly yet randomly, resulting in the production of new bone tissue. As a result, collagen fibers are deposited in an irregular pattern. Repeated cycles of bone removal and formation result in a jigsaw or mosaic pattern comprising numerous little irregularly shaped bone fragments joined by deeply stained hematoxyphilic reversal lines in a jigsaw or mosaic pattern. This special histological characteristic is considered a signature of Paget's disease.
3. **Sclerotic/Burned-out Phase:** With progression of the condition, osteoblastic activity takes over, resulting in an abnormally large quantity of compact and dense bone creation. In this phase, new bone creation outnumbers bone resorption. As a result, bone becomes structurally disordered and hence, weaker than typical healthy bone. The normal trabecular appearance is deformed, with no propensity for the formation of haversian canals or centers around blood vessels. The osteoblastic activity eventually slows down, and the disease enters a latent stage known as the burned-out phase.

Decreased bone results in a hard, thick, and less vascular Pagetic or mosaic bone, which is referred to as the burned-out phase (Ralston, 2008).

6.5 COMPLICATIONS

The skull is one of the most often damaged sections of the body by PDB, and it is also one of the most severe. Skull deformity is a rather common occurrence in the population. A significant variation exists between patients' clinical manifestations and the disease's radiological appearance. The majority of instances show enlargement of the bones, which has largely esthetic ramifications; nevertheless, when the cranial base is weakened, platybasia and sometimes even syringomyelia may develop. In some cases,

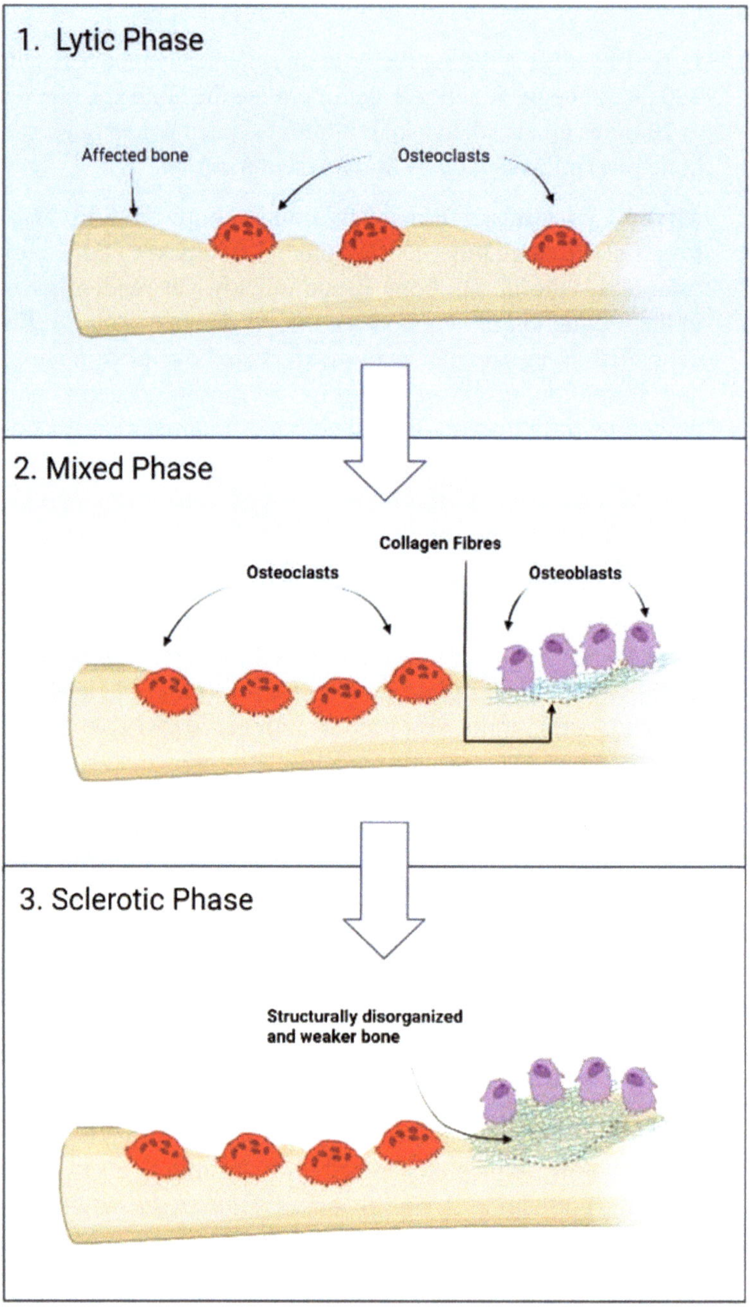

FIGURE 6.2 Stages of Paget's disease of bone.

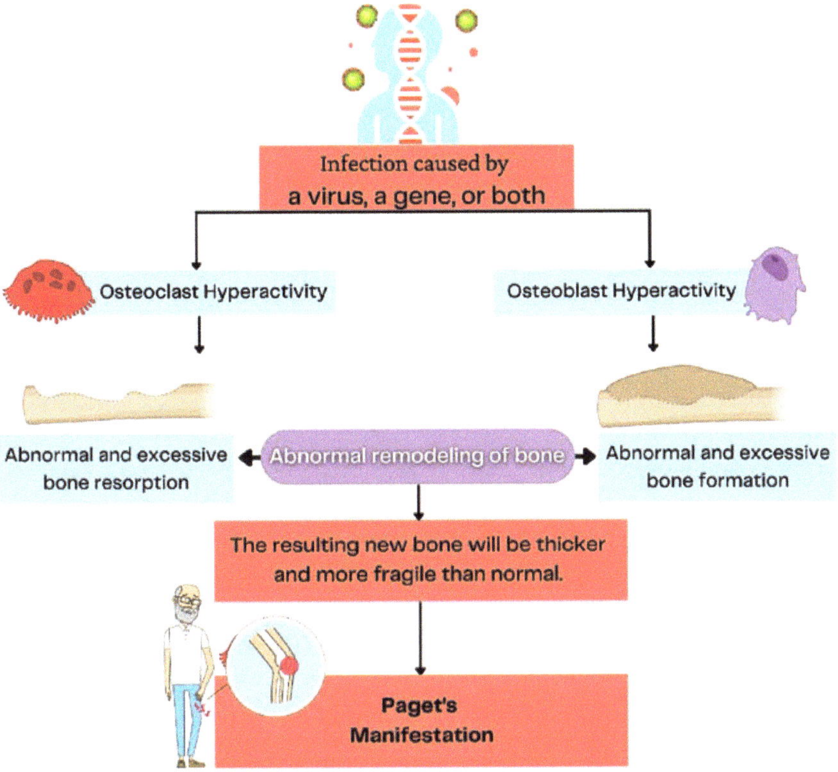

FIGURE 6.3 Paget's disease etiology.

patients experience little or no discomfort and pain, while in others, they experience cephalgia, which may be life-threatening. There is a relationship between PDB and dementia, however it is unclear whether this is due to Paget's disease itself or to another factor (Cushing et al., 2002; McCloskey et al., 2002). The narrowing of the optic foramen has been linked to optic atrophy and papilledema, although increasing intracranial pressure might also be to blame (Kristensen, 1971).

It is possible to develop quadriparesis and perhaps even paraparesis as a consequence of involvement of the vertebral column, which impedes on the spinal cord and reduces blood supply. The degree of the involvement of the vertebral column will determine whether quadriparesis or otherwise paraparesis develops. It is possible that the supply of blood to the affected bone is much larger than the blood suply to normal bone. When this happens in a vertebrae, it is speculated that a vascular "steal" occurs, leading to a temporary ischemia of the spine, which may be treated with medications. This is hypothesized to be responsible for the exceptional reversibility of these major neurologic impairments after Paget's disease therapy (Douglas et al., 1981).

Loss of hearing is a specific and significant complication of PDB. It is rather common in individuals with temporal bone involvement. Despite the fact that Paget's disease of the skull is associated with a greater incidence of hearing impairment, and it might be difficult to differentiate between those who have presbycusis and those who have Paget's disease related loss of hearing. However, both possibilities are undeniably possible in the same person (Monsell, 2002).

Long bones, especially those that bear weight, are more often associated with orthopedic issues. A research observed 236 patient, where 33 fractures were observed throughout the time of study. As a result, one out of every seven patients had a fracture (Melton et al., 2000). Because the apparent incidence of PDB complications is inversely proportionate to the time it takes to diagnose a case, the apparent rate of complications among patients seems to be rather high. When the humerus is heavily involved in the condition, it may cause major degenerative disease of the shoulders and, in rare cases, fracture of the humerus. When the tibia as well as femur are compromised, complications are more probable to appear. These are often susceptible to plastic distortion, which results in the bowing (Glaser et al., 2002).

Over the years, anecdotal evidence has accumulated about the cardiovascular consequences connected with PDB. These were often severe heart failure-related diseases caused by increased cardiac output resulting from increased blood supply. A study discovered that patients with PDB had lower peripheral vascular resistance and a greater cardiac stroke volume than the control subjects. Interestingly, there was a relationship between the estimated rate of osteoclastogenesis and the concentration of hydroxyproline in the urine. However, neither the vascular resistance nor the heart function measures were evaluated in a systematic manner by the investigators in this study (Morales-Piga et al., 2000).

Osteosarcomas associated with PDB are expected to develop at an incidence of 1%. These tumors are more common in patients who have a long history of polyostotic illness, and they strike people in their 7th decade, when osteosarcoma is very unusual in the general population, according to the American Cancer Society. According to epidemiological research, this late rise in osteosarcomas does not occur in locations with a low prevalence of Paget's illness, as previously reported. Uncertainty exists as to what is at the molecular level responsible for the link between osteosarcoma and PDB. Because these osteosarcomas are derived from osteogenic tissue, they often develop in areas of paget's affected bone. New discoveries into the biology of adolescent osteosarcomas, the discovery of VCP (valosin-containing protein) and SQSTM1 (sequestosome 1) mutations in patients with PDB, and the increasing prevalence of stromal lesions in patients with PDB are all changing our understanding of the pathogenicity of Paget's disease and the unusual complications of pagetic osteosarcomas (Hansen et al., 2007).

The rapidity and unexpected aspect of osteogenesis in PDB lesions may interplay with the physiological demands of the damaged bone as well as the functioning of neighboring organs to cause morphological and functional difficulties in the affected bone and surrounding organs (Table 6.1).

6.6 DIAGNOSIS

In most cases, Paget's sickness is found by chance during the testing for another condition. Patients with PDB may also present symptoms that are specific to the disorder, such as pain in the bones, fracture, or deafness, among other things. Despite the fact that rheumatologists see a high number of patients in the UK, there is no proof to say that patients are really benefitted from the treatment which is in the scope of particular medical specialization.

More than 90% of those with untreated PDB have elevated serum alkaline phosphatase (sAP) values that are moderately or considerably higher than normal. As a consequence, when the sAP level is elevated but there is no apparent vitamin D deficiency, a scan of bones is performed to determine if there are any bone abnormalities in the patient. Even in the absence of concurrent liver disease, the bone specific AP has to be measured in order to differentiate between an elevated liver-sAP and an elevated bone-sAP. If there is no history of trauma, radiographic assessment of sore bone regions should be undertaken. Additional imaging is required in parts of the skeleton that elicit concern after the radioisotopic bone scan (Shaker, 2009).

In PDB, conventional X-rays of afflicted skeletal areas reveal a diverse array of radiographic abnormalities. Three distinct phases may be discerned from these observations. The osteolytic phase which is an early phase is marked by a high prevalence of the lytic lesions. Initially developing in the metaphyseal area of the bones, with a prominent distinction from healthy bone (candle-flame or blade of the grass appearance), these lesions eventually migrate to the epiphysis and diaphysis of the bones. Subsequent is

TABLE 6.1 Clinical Characteristics of PDB

CLINICAL FEARURES: • When diagnosed, the majority of individuals are asymptomatic. • Symptoms and indicators vary depending on the affected location, deformity, and bone turnover. • Infection often occurs in the skull, femurs, pelvis, vertebrae and tibia. • Bone or joint discomfort, increased warmth over the bone, fracture, deformation of the bone, and neural compression syndromes are all symptoms. • Development of the osteogenic sarcoma is a very unusual but typically deadly consequence.	
SKULL 	• Frontal bossing, calvarial ridging increase the size of the skull. • Usual: Hearing loss, increased cranial temperature, and headaches • Unusual: Cranial nerve palsy, hydrocephalus caused due to basilar invagination and cerebral spinal fluid flow restriction, and loss of vision.
PELVIS 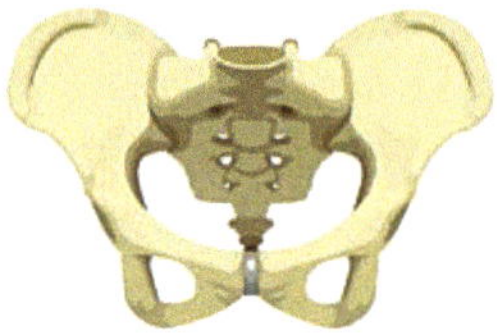	• Deformity of the pelvic bone. • Ache across the sacrum, ischia, or the pelvic ilia. • Acetabular deformity related with tertiary hip osteoarthritis.
VERTEBRAE 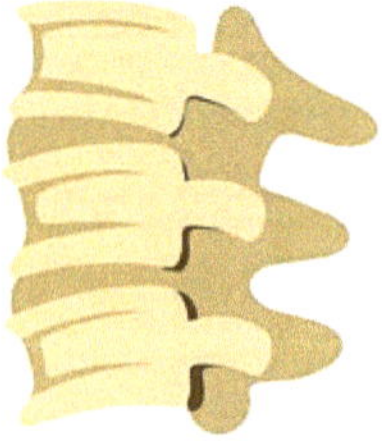	• Ache in the back. • Deformity of the spine (kyphosis, forward tilt, scoliosis) • Radiculopathy of the lumbar spine or sometimes spinal stenosis. • Enlargement of the vertebrae. • Compression of the thoracic spinal cord occurs rarely.
EXTREMITIES 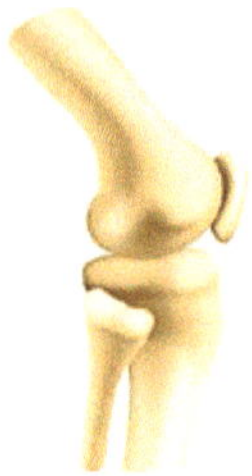	• Ache in the bones. • Bowing deformity related with unequal lengths of leg. • Osteoarthritis secondary to the bent limb joints. • Enlargement of the bones • Warmth around the afflicted tibia • Susceptibility to fracture, particularly when osteolytic illness progresses.

the mixed stage which is characterized by the emergence of osteosclerotic regions, which often results in coarsening of the trabecular patterns.

A "picture-frame" appearance may be created by cortical thickening in the impacted vertebrae, which are enlarged in the spine. A "Rugger–Jersey-vertebra" is a kind of vertebra that may be seen in certain cases of PDB. The combination of lucency and sclerosis in the pelvis leads to anomalies such as acetabular protrusion and hip osteoarthrosis, which are both degenerative conditions. The sclerotic phase is the last stage that leads to a rise in the volume and density of the rebuilt bone. Radiodense foci may be found inside the skull, as well as in other parts of the body. Severe sclerotic alterations may eventually result in nerve tissue compression and associated problems such as amblyacousia.

The majority of the time, PDB is diagnosed based on the appearance of the bone on radiographs; bone biopsy is only used in rare circumstances to further confirm the diagnosis. In individuals with PDB, the histological features of trabecular bone are changed. Additionally, the ordinarily obvious distinction between cortical bone and trabecular bone may be lost. The newly formed bone, on the other hand, does not follow bony trajectory, hence it has less biomechanical stability than the normal bone. (Seitz et al., 2008; Takata et al., 2006).

Bone-scans using radionuclides are far more accurate at identifying the PDB than ordinary X-rays, and many physicians use them to test for the condition and evaluate the degree to which the disease has impacted the skeletal system. The appearance of the bone scan in PDB is often pathognomonic; nonetheless, it has been advised that the diagnosis be validated by diagnostic imaging at least once before starting with the therapy (Langston et al., 2004).

6.7 TREATMENT

Treatment is recommended for people with symptomatic PDB as well as for certain asymptomatic patients. PDB may be treated using a variety of anti-resorptive drugs that target the osteoclasts. Bisphosphonates are the most often utilized of these medications. In the treatment of PDB, the bisphosphonate Zoledronic acid (ZA) is the first-line medication because of its efficacy and ease of administration. (Ralston et al., 2019; Singer et al., 2014). ZA has the benefit of being administered once at a dosage of 5 mg during a 15-minute period by intravenous infusion. Many physicians increase the infusion length to 45 minutes, and pre-and post-infusion acetaminophen may help mitigate the acute phase reaction (Reid et al., 2011).

Is bisphosphonates therapy beneficial in the long run in lives of people with PDB? This is a less certain proposition. According to the results of the Paget's Disease Randomized Trial of Intensive versus Symptomatic Management (PRISM) trial, which examined the findings of aggressive vs symptomatic therapy for individuals with PDB, Between the two groups, there was no statistically significant difference in the rate of fracture, the quality of life, or other disease-related outcomes (Anne L. Langston et al., 2010).

Patients who choose not to receive intravenous medication or who do not have access to ZA might consider a specified oral treatment regimen with alendronate or risedronate as an alternate treatment option. If a patient is unable to tolerate either of these options, then calcitonin is indicated as an alternative treatment. When patients have limited renal function and cannot take bisphosphonates, calcium carbonate is a feasible alternative. Also, it is a good alternative when the objective of treatment is quick pain relief. PDB's first successful drug to reach clinical use was calcitonin; yet, the daily injection required and frequent flushing or nausea restricted patient acceptance, and illness relapsed swiftly upon the termination of therapy (Singer et al., 2014). Patients with PDB may benefit from orthopedic surgery for a variety of reasons, including correction of bone abnormalities, repairing the fracturs, arthroplasty, lumbar decompression, and excision of tumors in bone (Parvizi et al., 2006).

Even though antiresorptive therapies are effective at alleviating pain caused by excessive metabolic activity, patients with PDB frequently experience complications such as deformity of bone and osteoarthritis, which are resistant to antiresorptive therapy. Analgesic and anti-inflammatory medications are commonly required to provide effective symptom management in these situations.

Acupuncture, physiotherapy, hydrotherapy, and TENS (Acupuncture, physiotherapy, hydrotherapy, and transcutaneous electrical nerve stimulation) are examples of non-pharmacological treatments for pain management, and aids like walking sticks and shoe lifts may assist with particular conditions like limb shortening and deformity. Despite the fact that the usefulness of these techniques hasn't been examined, especially in patients with PDB, aids and devices were shown to be associated with considerably improved health quality of life in the participants of PRISM studies (Langston et al., 2007).

6.8 NUTRACEUTICAL INTERVENTIONS IN PDB

Fundamentally, nutraceuticals are meals or dietary components (Keservani et al., 2010a, 2020) that provide substantial contributions to the alteration and maintenance of normal physiological function, which is essential for human health. The present demographic and health trends are the primary drivers of the global nutraceutical industry. Since the beginning of the past decade, the nutraceuticals (Keservani et al., 2010b, 2017) sector has undergone explosive rise in both developed and health-conscious countries throughout the globe. United States of America and Japan have very well-developed markets for nutraceuticals (Pandey et al., 2010). Among others, China, Brazil and India have demonstrated great promise in the sector of nutraceuticals market as compared to the other advanced countries. In recent years, India and China have emerged as important sources of natural raw materials for the nutraceutical sector.

The worldwide market for nutraceuticals was valued at over 250 billion U.S. dollars in 2018. Nutraceutical products were the most valuable market segment in 2007, with a total value of 39.9 billion U.S. dollars. However, by 2013, this figure had climbed to 56.7 billion U.S. dollars (Vandita Kakkar, Nancy Modgill, 2016).

The global market for nutraceuticals, which was estimated to be worth 278.8 billion USD in 2020, is expected to grow by 7.8% per year by 2026, reaching 441.7 billion USD, representing a compound annual growth rate (CAGR) of 7.8% (Global Nutraceuticals Market Report 2021: Market to Reach, n.d.).

The Food and Drug Administration has authorized a number of medications for bone health, but they are all prohibitively costly and come with a variety of negative effects. On the other hand, nutraceuticals are affordable, have no side effects, and yet have anti-inflammatory benefits. Various anomalies have arisen as a result of the last century's neglect of nutrition, a well-balanced diet, and a healthy lifestyle, resulting in fatal consequences all over the world. As a result, nutraceuticals are expected to play a large role as therapeutic agents in the foreseeable future, owing to their suitability for today's lifestyle. There seems to be no question that the amount of accessible information on the subject of nutraceuticals is a big hurdle for nutritionists and clinicians. The use of nutraceuticals as a strong element in the treatment and prevention of nutritionally caused disorders is also being considered by public health authorities. As a result, the use of nutraceuticals in clinical practice is beginning to emerge as an important area of future investigation.

6.8.1 CALCIUM

Approximately 1 kg of calcium is contained in the average adult human body, with bone and teeth accounting for more than 99% of the total. Apatite (hydroxyapatite) is the solid form of calcium that may be found in bone. Because it increases the quantity of bone in the body, calcium has a beneficial effect on bone density. In terms of the development of the skeletal system throughout childhood and the maintenance of bone density in later life, calcium consumption is one of the most important modifiable environmental elements that may be influenced by a person's lifestyle (Zhu et al., 2012).

The accumulation of bone minerals during the developmental phase is a strong predictor of bone density, which is linked to higher risk of developing bone-related conditions as we get older. Bone mass is influenced by several elements such as heredity, gender, endocrine and dietary factors, among others, in order to achieve maximal bone mass.

After being followed for 30 years, a research study comprising boys and girls from Canada, discovered that total body bone mass achieved a pinnacle at the age of 18 in females and at the age of 20 in boys, suggesting that the maximum bone mass is reached around the 20s. Following the attainment of maximal bone mass, it may be anticipated to stay constant for 10 to 20 years without seeing any substantial change (Baxter-Jones et al., 2011). There have only been a few research that have looked at the association of intake of the calcium and bone density in youngsters. According to the findings of two studies, one done with young Japanese women and the other with 125 young female cross-country runners, there is an association among calcium intake and bone density. Requirements of calcium are 20%–30% lower during this era compared to the rapid developing phase, since bone formation has been finished (Nieves et al., 2010).

The majority of calcium or dairy intervention trials were conducted over a period of two years or fewer. One long term research examined the effects of the calcium-supplementation for seven years on American girls who were around ten years old at the start of the study. This research revealed that calcium-supplementation had a statistically significant favorable affect on bone mineral deposition during the pubertal growth period, but that the benefits were lost in late adolescence. Although the research indicated that participants with inadequate calcium consumption might not attain the objective peak bone mass, it also indicated that they were quite unlikely to attain their desired peak bone density by the time they reached adulthood (Matkovic et al., 2005). As long as patient compliance was adequate (75–80%), the majority of randomized-controlled experimentations examining the effects of calcium supplementation alone or calcium supplementation in combination with vitamin D supplementation on the risk of fracture in elderly adults found a reduction in the risk of fracture (Al., 1992; Dawson-Hughes et al., 1997; Prince et al., 2006).

Specific calcium requirements have been established at levels that correspond to the production of maximum peak bone mass during growth and the subsequent reduction in bone loss. These levels are 1300 mg/d for adolescents, 1000 mg/d for adults, and 1200 mg/d for those over the age of 51, respectively (Palacios, 2006).

6.8.2 VITAMIN D

The skin produces vitamin D as a result of exposure to sunlight. In the strictest sense, it is considered as more of a hormone than a vitamin (DeLuca, 2008). Once it is generated in the skin or consumed in the diet. The physiologic purpose of it is to keep calcium concentrations in the extracellular and bloodstream within a healthy range so that cellular processes and neuromuscular function can continue. Vitamin D is necessary not only for the skeletal health of growing children, but also for the maintenance of a healthy skeleton throughout our lives.

Increased recognition of vitamin D in recent years is part of a lengthy history that started in the 1920s with the identification of the antirachitic factor and the subsequent discovery of its production by UV irradiation (DeLuca, 2008). When vitamin D supplementation was suspected to be the cause of idiopathic hypercalcemia (excess calcium in the blood) in the mid-twentieth century, research on the vitamin came to a standstill. Rickets was no longer a severe medical concern in the developed world in the next decade (Pronicka et al., 1997). It wasn't until 1970 that the discovery of vitamin D endocrine system was put forward by researchers, and this rekindled interest in the association between vitamin D and metabolic bone disease (DeLuca, 2014).

When 14 individuals with PDB participated in an observational study, the notion of relative vitamin D deficit in patients with PDB was introduced. When vitamin D levels were in the lower limits of the normal range, areas of rapid turnover revealed localized osteomalacia, which was previously unknown. It was as a consequence of this study's findings that the authors hypothesized that vitamin D deficiency during infancy would be associated with PDB. According to some theories, vitamin D deficiency leads in a reduction in mineralization, which makes the skeleton more prone to the factors that contribute to the onset of PDB (Williams ED, Barr WT, 1981).

The decline in the frequency of PDB in the United Kingdom (Van Staa et al., 2002) coincides with the drop in rickets prevalence in the United Kingdom during the prior century, which was a consequence of efforts to increase children's vitamin D intake (Barker et al., 1974).

During the period of increased osteoblastic activity, it has been claimed that a larger need for vitamin D is present for the mineralization of newly formed osteoid. This increasing demand might result in a full-blown vitamin D deficiency in people who are deficient in the vitamin, which could lead to PDB in the long run (Kannan et al., 2008). Furthermore, PDB is most typically identified in the older population (Cooper et al., 2006), who have a heightened incidence of vitamin D deficiency (Lanham-New et al., 2011).

6.8.3 NATURAL AGENTS

For healthy bone, it is critical to maintain a good equilibrium between the osteoblasts and osteoclasts. Thus, a significant amount of interest would be generated by the natural agents that might modify the expression of key players involved in bone metabolism. The next section discusses many of these natural agents that may have a role in the stated condition.

6.8.3.1 CURCUMIN

Curcumin, the key component in turmeric, has been used for therapeutic purposes for thousands of years. Multiple bodies of data based on *in vitro* and *in vivo* research suggest that curcumin may be utilized to prevent bone loss in certain circumstances. Curcumin suppresses the activation of the transcription factors NF-B, Wnt/-catenin, RANKL, and TNF, all of which are associated with the loss of bone mass. (Aggarwal et al., 2006; J. H. Kim et al., 2011).

The cytokines RANKL and M-CSF, as well as the newly identified chromatin protein high mobility group box 1 (HMGB1), all contribute to osteoclast differentiation. Curcumin inhibits osteoclastogenesis via a P38-MAPK-dependent mechanism that inhibits HMGB1 release (Sakai et al., 2013). According to another study, RANKL causes osteoclastogenesis by activating NF-B, and curcumin supplementation prevents both NF-B activation and RANKL-induced osteoclastogenesis (Bharti et al., 2004). Aside from that, turmeric has been demonstrated to have the ability to prevent bone loss. (J. H. Kim et al., 2011).

6.8.3.2 RESVERATROL

It wasn't until 1992 that resveratrol garnered significant attention when it was postulated that it was responsible for several of red wine's cardioprotective qualities. From then, large numbers of investigations have documented that resveratrol can actually mitigate or delay the development of a broad range of illnesses, including cancer, cardiovascular disease, and ischemic injuries, as well as increase stress resistance and lengthen the lifespans of a variety of organisms. (Baur et al., 2006).

Evidence suggests that resveratrol may have a role in reducing the risk and management of bone loss; for instance, a research found that resveratrol regulates RUNX2 expression and thereby stimulates osteoblast formation (Zainabadi et al., 2017). According to the findings of another study, analogs of resveratrol suppresses osteoclastogenesis in a mouse model via deacetylating FoxOs (H. N. Kim et al., 2015). Manna et al. reported that resveratrol inhibits NF-B activity, a critical factor in bone loss (Manna et al., 2000). Moreover, several animal models suggest that resveratrol administration impedes bone resorption. Nonetheless, more clinical investigations are required to fully evaluate the effectiveness of this stilbene in the prevention of bone loss (Dosier et al., 2012; Tou, 2014; Zhao et al., 2015).

6.8.3.3 QUERCETIN

Quercetin is one of approximately 4000 naturally occurring phenolic compounds found in plants. It is a member of the flavonoids (Keservani and Sharma, 2014; Keservani et al., 2016a, 2016b) family and is one of the most ubiquitously used nutritional antioxidants. It may be found in a broad range of foods, such as vegetables, fruits, tea, and wine, as well as a variety of nutritional items, and provides continual protection against a variety of maladies, including osteoporosis, some forms of melanoma, lung and heart disease, and aging. This flavonoid is thought to regulate bone metabolism by either activating osteoblasts or inhibiting osteoclast differentiation (Casado-díaz et al., 2016). Quercetin stimulates osteoblast differentiation via the activation of Smad (Yamaguchi et al., 2011) and P38 MAPK (Zhou et al., 2015) and the inhibition of the WNT/-Catenin signaling pathways (Casado-díaz et al., 2016). Quercetin inhibits RANKL-induced osteoclastogenesis via inhibiting NF-B activation, as observed by Yamaguchi et al. (Yamaguchi et al., 2011).

Aside from that, it has been shown that supplementing with quercetin may reduce bone loss in a mouse model of ovarian cancer. (Tsuji et al., 2009). All of these studies tend to indicate that quercetin has a significant potential for use as a bone-health supplement in the near future.

6.8.3.4 GENISTEIN

Genistein, a form of phytoestrogen, accounts for about two-thirds of the isoflavones found in soy-derived products. In animal models, genistein has been shown to have hypolipidemic, anticancer, and antioxidant properties (Lv et al., 2019). A large number of *in vitro* and *in vivo* studies have shown that genistein has substantial promise for bone loss prevention in both humans and animals. (An et al., 2016; King et al., 2015; Zhang et al., 2016). Numerous processes have been postulated along the same propositions. For instance, it was shown that isoflavones increase osteoblast development by upregulating PPARγ (Zhang et al., 2016), BMP2, Wnt/-catenin (Kushwaha et al., 2014), Smad5, Runx2 (Dai et al., 2013), and oestrogen and oestrogen receptors (Liao et al., 2014). Isoflavones also inhibit osteoclastogenesis by inhibiting a range of transcription factors that are implicated in bone loss, such as NF-B and AP-1 (Liao et al., 2014), as well as cytokines linked with bone loss, such as IGF-1 and TGF (Joo et al., 2004). Following the findings of this study, it is blatantly obvious that utilizing soy products as a dietary supplement may be beneficial in terms of avoiding bone deterioration.

6.8.4 MARINE SOURCES

With its incredible variety, the marine environment also serves as the source of an abundance of bioactive substances that have enormous promise in the food and pharmaceutical industries, as well as other fields. Water covers more than 70% of the earth's surface, and it is also home to approximately 70% of all organisms. Several recent investigations have established that the vast majority of these species contain active chemicals, but that they are currently neglected as supplies for a multitude of distinct uses. A variety of bioactive peptides derived from diverse fish sources have been discovered in recent investigations. When ingested *in vivo*, they include short amino acid sequences ranging in length from 2 to 20 amino acids in length, which provide physiological advantages. They may be created by proteolytic hydrolysis utilizing commercially available enzymes such as trypsin, chymotrypsin, and alcalase, as well as through the use of proteolytic microbes and fermentation processes (Vercruysse et al., 2005). A study found that fish peptides have the ability to increase the rate of calcium absorption. According to the findings of the study, fish protein hydrolysates include peptides and the growth factors that help in the acceleration of calcium absorption. (Fouchereau-Peron et al., 1999; Jung et al., 2005). These peptides have the capacity to connect to osteoclast cell surface receptors and are largely involved in calcium metabolism by decreasing the amount of osteoclasts in the bone matrix. By discovering these peptides, the prospect of treating the PDB has been enhanced significantly, according to the researchers. Furthermore, gelatin peptides have been found to increase calcium bioavailability in animal models by boosting calcium absorption. It has also been reported that fish bone peptides (FBP) II may suppress the production of insoluble calcium ions at neutral pH. The administration of FBP II to ovariectomized rats throughout the study period resulted in improved calcium retention and reduced bone mineral loss. The FBP II diet also significantly increased the levels of femoral total calcium, bone mineral density, and strength, bringing them at levels that were equivalent to those seen in the casein phosphopeptide diet group (Giri et al., 2012; Jung et al., 2006; Senevirathne et al., 2012).

6.9 REGULATORY ASPECTS

When it comes to foods, nutraceuticals is a wide umbrella word that refers to any product generated from food sources that provides additional health advantages on top of the fundamental nutritional content present in food. The name "nutraceutical" is derived from the combination of the terms "nutrient," meaning a nutritive dietary component, and "pharmaceutical," meaning a therapeutic medicine. Stephen De Felice, the founder and head of the Foundation for Innovation in Medicine, which is based in Cranford, New Jersey, created this term in 1989.

Nutraceuticals are often classified as either functional foods or dietary supplements in India, with the former being more common. Nutraceuticals, in contrast to medicines, are normally not subject to the regulatory oversight of the "Drugs and Cosmetics Act of 1940." As a result, a number of legislative proposals have been enacted to regulate nutraceuticals.

The Food Safety and Standards Act (FSSA) aims to standardize the country's food safety laws in order to facilitate systematic and scientific growth in the food processing industry and to accomplish a fundamental change away from a regulatory regime towards self-compliance. FSSA had two principal objects, first was to establish a unified food statute and the second was to advance the scientific development of the food processing industry.

The Food Safety and Standards Authority of India (FSSAI) is the regulatory authority for food and nutrition in India and is responsible for establishing laws governing the nutraceutical sector. The FSSAI defines nutraceuticals as functional foods or foods for special dietary purposes."

FSSAI 2006 requires registration and licensing for manufacturing. To import, a NOC from FSSAI and Import-Export Certificate (IEC) is required. Except for those indicated in Schedule I, FSS rules 2010 and 2011 requires registration of the production site and the acquisition of a manufacturing license in India from the state licensing authority. A central licensing authority-issued import license is required for the import.

In 1994, the United States enacted legislation to regulate the manufacturing and marketing of nutraceuticals, which was termed as "Dietary Supplement Health and Education Act" (DSHEA). The DSHEA regulates in a way distinct from the regulations governing traditional foods and drugs. Under the 1994 DSHEA, the manufacture of a dietary product or dietary ingredient is responsible for ensuring the safety of the product prior to marketing, and the Food and Drug Administration (FDA) is responsible for taking actions against any unsafe dietary supplement product, once it is one the market. When the Federal Good Manufacturing Practices (GMP) standard (21 CFR, part 111) was implemented in 2007, the FDA and the Department of Health and Human Services were responsible for GMP enforcement. As per the regulations, all domestic and foreign companies that manufacture, label, package, or store the nutraceuticals in the US, as well as those involved in quality control, testing, and distribution of nutraceuticals are responsible for ensuring the safety of the food products before as well as after they are introduced into the market.

The European Food and Safety Authority (EFSA) controls the food regulations in the European Union (EU), with a particular emphasis on "food supplements." In Europe, food supplements are defined as concentrated supplies of nutrients and other compounds with positive nutritional effects. Directive 2002/46/EC is the European Union rule that governs dietary supplements and other nutritional products. There is a great deal of variation in product claims and recommended daily allowances (RDAs) for various nutraceuticals across the European Union. Following this development, the European food supplement and nutraceutical industries joined the forces to form the Food Supplements Europe (FSE) coalition, which works with the regulators to ensure the quality of products and create a better regulatory environment for the future.

As a viable and more effective therapy for a number of bone-related illnesses, nutraceuticals are garnering more and more awareness. While the regulations that are presently available have addressed the vast majority of problems, there is always space for improvement and updating in light of the constantly changing dynamics of the environment (H. et al., 2003; Ray et al., 2016).

6.10 CONCLUSION

PDB is a kind of bone disease that develops as a consequence of the aging process in the body's bones. Despite the fact that the general frequency of the condition is decreasing, research into PDB is still crucial. Given the enormous therapeutic potential of nutraceuticals in a wide variety of diseases, it is vital to have a thorough knowledge of the role that nutraceuticals may play in bone-related disorders, and PDB is no exception to that.

KEYWORDS

- **compound annual growth rate**
- **interleukin-6**
- **osteoprotegerin**
- **Paget's disease of the bone**
- **receptor activator of nuclear factor κB ligand**
- **zoledronic acid**

REFERENCES

Aggarwal, S., Ichikawa, H., Takada, Y., Sandur, S. K., Shishodia, S., & Aggarwal, B. B., (2006). Curcumin (diferuloylmethane) down-regulates expression of cell proliferation and antiapoptotic and metastatic gene products through suppression of IκBα kinase and Akt activation. *Molecular Pharmacology, 69*(1), 195–206.

An, J., Yang, H., Zhang, Q., Liu, C., Zhao, J., Zhang, L., & Chen, B., (2016). Natural products for treatment of osteoporosis: The effects and mechanisms on promoting osteoblast-mediated bone formation. *Life Sciences, 147*(6), 46–58.

Barker, D. J. P., & Gardner, M. J., (1974). Distribution of Paget's disease in England, Wales and Scotland and a possible relationship with vitamin D deficiency in childhood. *British Journal of Preventive and Social Medicine, 28*(4), 226–232.

Basle, M. F., Fournier, J. G., Rozenblatt, S., Rebel, A., & Bouteille, M., (1986). Measles virus RNA detected in Paget's disease bone tissue by in situ hybridization. *Journal of General Virology, 67*(5), 907–913.

Baur, J. A., & Sinclair, D. A., (2006). Therapeutic potential of resveratrol: The *in vivo* evidence. *Nature Reviews Drug Discovery, 5*(6), 493–506.

Baxter-Jones, A. D. G., Faulkner, R. A., Forwood, M. R., Mirwald, R. L., & Bailey, D. A., (2011). Bone mineral accrual from 8 to 30 years of age: An estimation of peak bone mass. *Journal of Bone and Mineral Research, 26*(8), 1729–1739.

Bharti, A. C., Takada, Y., & Aggarwal, B. B., (2004). Curcumin (diferuloylmethane) inhibits receptor activator of NF-κB ligand-induced NF-κB activation in osteoclast precursors and suppresses osteoclastogenesis. *The Journal of Immunology, 172*(10), 5940–5947.

Birch, M. A., Taylor, W., Fraser, W. D., Ralston, S. H., Hart, C. A., & Gallagher, J. A., (1994). Absence of paramyxovirus RNA in cultures of pagetic bone cells and in pagetic bon. *Journal of Bone and Mineral Research, 9*(1), 11–16.

Casado-díaz, A., Anter, J., Dorado, G., & Quesada-gómez, J. M., (2016). Effects of quercetin, a natural phenolic compound, in the differentiation of human mesenchymal stem cells (MSC) into adipocytes and osteoblasts. *The Journal of Nutritional Biochemistry, 32*, 151–162.

Chapuy, M. C., Arlot, M. E., Duboeuf, F., Brun, J., Crouzet, B., Arnaud, S., Delmas, P. D., & Meunier, P. J., (1992). Vitamin D3 and calcium to prevent hip fractures in elderly women. *N. Engl. J. Med.., 327*(23), 1637–1662.

Cooper, C., Harvey, N. C., Dennison, E. M., & Van, S. T. P., (2006). Update on the epidemiology of Paget's disease of bone. *Journal of Bone and Mineral Research, 22*(suppl. 2), 10–15.

Corral-Gudino, L., Borao-Cengotita-Bengoa, M., Del Pino-Montes, J., & Ralston, S., (2013). Epidemiology of Paget's disease of bone: A systematic review and meta-analysis of secular changes. *Bone, 55*(2), 347–352.

Cundy, T., (2018). Paget's disease of bone. *Metabolism: Clinical and Experimental, 80*, 5–14.

Cushing, F. R., & Bone, H. G., (2002). Radiographic diagnosis and laboratory evaluation of Paget's disease of bone. *Clinical Reviews in Bone and Mineral Metabolism, 1*(2), 115–134.

Dai, J., Li, Y., Zhou, H., Chen, J., Chen, M., & Xiao, Z., (2013). Genistein promotion of osteogenic differentiation through BMP2/SMAD5/RUNX2 signaling. *International Journal of Biological Sciences, 9*(10), 1089–1098.

Dawson-Hughes, B., Harris, S. S., Krall, E. A., & Dallal, G. E., (1997). Effect of calcium and vitamin D supplementation on bone density in men and women 65 years of age or older. *New England Journal of Medicine, 337*(10), 670–676.

DeLuca, H. F., (2008). Evolution of our understanding of vitamin D. *Nutrition Reviews, 66*(2), S73–S87.

DeLuca, H. F., (2014). History of the discovery of vitamin D and its active metabolites. *BoneKEy Reports, 3*(1), 1–8.

Dosier, C. R., Erdman, C. P., Park, J. H., Schwartz, Z., Boyan, B. D., & Guldberg, R. E., (2012). Resveratrol effect on osteogenic differentiation of rat and human adipose derived stem cells in a 3-D culture environment. *Journal of the Mechanical Behavior of Biomedical Materials, 11*, 112–122.

Douglas, D. L., Duckworth, T., Kanıs, J. A., Jefferson, A. A., Martın, T. J., & Russel, R. G. G., (1981). Spınal cord dysfunction in Paget's disease of bone. *British Society of Bone and Joint Surgery, 63 B*(4), 495–503.

Fouchereau-Peron, M., Duvail, L., Michel, C., Gildberg, A., Batista, I., & Le Gal, Y., (1999). Isolation of an acid fraction from a fish protein hydrolysate with a calcitonin-gene-related-peptide-like biological activity. *Biotechnology and Applied Biochemistry, 29*(1), 87–92.

Giri, A., & Ohshima, T., (2012). Bioactive marine peptides. *Advances in Food and Nutrition Research, 65*, 73–105.

Glaser, D. L., & Kaplan, F. S., (2002). Orthopedic surgery considerations in Paget's disease of bone. *Clinical Reviews in Bone and Mineral Metabolism, 1*(2), 159–165.

Global Nutraceuticals Market Report, 2021: Market to Reach, (2021). Retrieved from: https://www.globenewswire.com/en/news-release/2021/10/06/2309742/28124/en/Global-Nutraceuticals-Market-Report-2021-Market-to-Reach-441-7-Billion-by-2026-Functional-Food-to-Account-for-217-2-Billion.html (accessed on 25 July 2023).

Good, D. A., Busfield, F., Fletcher, B. H., Lovelock, P. K., Duffy, D. L., Kesting, J. B., Andersen, J., & Shaw, J. T. E., (2004). Identification of SQSTM1 mutations in familial Paget's disease in Australian pedigrees. *Bone, 35*(1), 277–282.

Gordon, M. T., Anderson, D. C., & Sharpe, P. T., (1991). Canine distemper virus localized in bone cells of patients with Paget's disease. *Bone, 12*(3), 195–201.

Gordon, M. T., Mee, A. P., Anderson, D. C., & Sharpe, P. T., (1992). Canine distemper virus transcripts sequenced from pagetic bone. *Bone and Mineral, 19*(2), 159–174.

Grahame, R., (2000). Heritable disorders of connective tissue. *Bailliere's Best Practice and Research in Clinical Rheumatology, 14*(2), 345–361.

Haddaway, M. J., Davie, M. W. J., McCall, W., & Howdle, S., (2007). Effect of age and gender on the number and distribution of sites in Paget's disease of bone. *British Journal of Radiology, 80*(955), 532–536.

Hansen, M. F., Seton, M., & Merchant, A., (2007). Osteosarcoma in Paget's disease of bone. *Journal of Bone and Mineral Research, 22*(SUPPL. 2), 58–63.

Hocking, L. J., Lucas, G. J. A., Daroszewska, A., Cundy, T., Nicholson, G. C., Donath, J., Walsh, J. P., et al., (2004). Novel UBA domain mutations of SQSTM1 in Paget's disease of bone: Genotype phenotype correlation, functional analysis, and structural consequences. *Journal of Bone and Mineral Research, 19*(7), 1122–1127.

Johnson-Pais, T. L., Wisdom, J. H., Weldon, K. S., Cody, J. D., Hansen, M. F., Singer, F. R., & Leach, R. J., (2003). Three novel mutations in SQSTM1 identified in familial Paget's disease of bone. *Journal of Bone and Mineral Research, 18*(10), 1748–1753.

Joo, S., Won, T., Kang, H., & Lee, D., (2004). Isoflavones extracted from *Sophorae* fructus upregulate IGF-I and TGF- and inhibit osteoclastogenesis in rat born marrow. *Archives of Pharmacal Research, 27*(1), 99–105.

Joshi, S. R., Ambhore, S., Butala, N., Patwardhan, M., Kulkarni, M., Pai, B., & Karne, R., (2006). Paget's disease from Western India. *Journal of Association of Physicians of India, 54*(JULY), 535–538.

Jung, W. K., Lee, B. J., & Kim, S. K., (2006). Fish-bone peptide increases calcium solubility and bioavailability in ovariectomized rats. *British Journal of Nutrition, 95*(1), 124–128.

Jung, W. K., Park, P. J., Byun, H. G., Moon, S. H., & Kim, S. K., (2005). Preparation of Hoki (*Johnius belengerii*) bone oligophosphopeptide with a high affinity to calcium by carnivorous intestine crude proteinase. *Food Chemistry, 91*(2), 333–340.

Kakkar, V., Modgill, N., & Kumar, M., (2016). *From Nutraceuticals to Nanoceuticals,* In: *Nanoscience in Food and Agriculture 3.* Ranjan, S., Dasgupta, N., & Lichtfouse, E., (eds.). Springer International Publishing Switzerland. 183–198.

Kannan, S., Mahadevan, S., Satya, A., & Sriram, U., (2008). A tale of three diseases of the bone. *Singapore Medical Journal, 49*(10), 263–265.

Kaushik, D., Dureja, H., & Kumar. V., (2003). Developments in nutraceuticals. *Indian Journal of Pharmacology, 35*(6), 363–372.

Keservani, R. K., & Sharma, A. K., (2014). Flavonoids: Emerging trends and potential health benefits. *Journal of Chinese Pharmaceutical Sciences, 23*(12), 815.

Keservani, R. K., Kesharwani, R. K., Sharma, A. K., Vyas, N., & Chadoker, A., (2010b). Nutritional supplements: An overview. *International Journal of Current Pharmaceutical Review and Research, 1*(1), 59–75.

Keservani, R. K., Kesharwani, R. K., Vyas, N., Jain, S., Raghuvanshi, R., & Sharma, A. K., (2010a). Nutraceutical and functional food as future food: A review. *Der Pharmacia Lettre, 2*(1), 106–116.

Keservani, R. K., Sharma, A. K., & Kesharwani, R. K., (2016a). Nutraceutical and functional foods for cardiovascular health. In: *Food Process Engineering* (pp. 257–278, 291–312). Apple Academic Press, CRC Press. ISBN: 9781771884020.

Keservani, R. K., Sharma, A. K., & Kesharwani, R. K., (2016b). Medicinal effect of nutraceutical fruits for the cognition and brain health. *Scientifica*, 10. Article ID 3109254. doi: 1 0.1155/2016/3109254.

Keservani, R. K., Sharma, A. K., & Kesharwani, R. K., (2017). An overview and therapeutic applications of nutraceutical and functional foods. *Recent Advances in Drug Delivery Technology*, 160–201. ISBN:9781522507543.

Keservani, R. K., Sharma, A. K., & Kesharwani, R. K., (2020). *Nutraceuticals and Dietary Supplements: Applications in Health Improvement and Disease Management*. CRC Press. ISBN: 9781771888738.

Kesterke, M. J., & Judd, M. A., (2019). A microscopic evaluation of Paget's disease of bone from a byzantine monastic crypt in Jordan. *International Journal of Paleopathology, 24*, 293–298.

Kim, H. N., Han, L., Iyer, S., De Cabo, R., Zhao, H., O'Brien, C. A., Manolagas, S. C., & Almeida, M., (2015). Sirtuin1 suppresses osteoclastogenesis by deacetylating foxos. *Molecular Endocrinology, 29*(10), 1498–1509.

Kim, J. H., Gupta, S. C., Park, B., Yadav, V. R., & Aggarwal, B. B., (2011). Turmeric (*Curcuma longa*) inhibits inflammatory nuclear factor (NF)-κB and NF-κB-regulated gene products and induces death receptors leading to suppressed proliferation, induced chemosensitization, and suppressed osteoclastogenesis. *Molecular Nutrition and Food Research, 56*(3), 454–465.

King, T. J., Shandala, T., Lee, A. M., Foster, B. K., Chen, K., Howe, P. R., & Xian, C. J., (2015). *Potential Effects of Phytoestrogen Genistein in Modulating Acute Methotrexate Chemotherapy-Induced Osteoclastogenesis and Bone Damage in Rats*., 18293–18311.

Kristensen, E. B., (1971). Ocular manifestations in Paget's Disease (Osteitis Deformans). *Acta Ophthalmologica, 49*(5), 741–746.

Kushwaha, P., Khedgikar, V., Gautam, J., Dixit, P., Chillara, R., Verma, A., Thakur, R., Mishra, D. P., Singh, D., & Maurya, R., (2014). *A Novel Therapeutic Approach with Caviunin-based Isoflavonoid that en routes Bone Marrow Cells to bone Formation via BMP2 / Wnt- b -Catenin Signaling*. *5*(9), e1422–1416.

Langston, A. L., & Ralston, S. H., (2004). Management of Paget's disease of bone. *Rheumatology, 43*(8), 955–959.

Langston, A. L., Campbell, M. K., Fraser, W. D., MacLennan, G. S., Selby, P. L., & Ralston, S. H., (2010). Randomized trial of intensive bisphosphonate treatment versus symptomatic management in Paget's disease of bone. *Journal of Bone and Mineral Research, 25*(1), 20–31.

Langston, A. L., Campbell, M. K., Fraser, W. D., MacLennan, G., Selby, P., & Ralston, S. H., (2007). Clinical determinants of quality of life in Paget's disease of bone. *Calcified Tissue International, 80*(1), 1–9.

Lanham-New, S. A., Buttriss, J. L., Miles, L. M., Ashwell, M., Berry, J. L., Boucher, B. J., Cashman, K. D., et al., (2011). Proceedings of the rank forum on vitamin D. *British Journal of Nutrition, 105*(1), 144–156.

Lawrence, G., Raisz, M. D., Barbara, E., & Kream, P. D., (1975). Regulation of bone formation. *Pharmasciences, 309*(2), 2020.

Liao, M., Tai, Y., Cherng, Y., Liu, S., Chang, Y., Lin, P., & Chen, R., (2014). *Genistein Induces Oestrogen Receptor- a Gene Expression in Osteoblasts Through the Activation of Mitogen-Activated Protein Kinases / NF- k B / Activator Protein-1 and Promotes Cell Mineralization*, 55–63.

Lv, Z. P., Yan, S. J., Li, G., Liu, D., & Guo, Y. M., (2019). Genistein improves the reproductive performance and bone status of breeder hens during the late egg-laying period. *Poultry Science, 98*(12), 7022–7029.

Manna, S. K., Mukhopadhyay, A., & Aggarwal, B. B., (2000). Resveratrol suppresses TNF-induced activation of nuclear transcription factors NF-κB, activator protein-1, and apoptosis: Potential role of reactive oxygen intermediates and lipid peroxidation. *The Journal of Immunology, 164*(12), 6509–6519.

Matkovic, V., Gael, P. K., Badenhop-Stevens, N. E., Landoll, J. D., Li, B., Ilich, J. Z., Skugor, M., et al., (2005). Calcium supplementation and bone mineral density in females from childhood to young adulthood: A randomized controlled trial. *American Journal of Clinical Nutrition, 81*(1), 175–188.

Mays, S., (2010). Archaeological skeletons support a northwest European origin for Paget's disease of bone. *Journal of Bone and Mineral Research, 25*(8), 1839–1841.

McCloskey, E. V., & Kanis, J. A., (2002). Neurological complications of Paget's disease. *Clinical Reviews in Bone and Mineral Metabolism, 1*(2), 135–143.

Melton, L. J., Tiegs, R. D., Atkinson, E. J., & O'Fallon, W. M., (2000). Fracture risk among patients with Paget's disease: A population-based cohort study. *Journal of Bone and Mineral Research, 15*(11), 2123–2128.

Monsell, E. M., (2002). Emerging concepts of hearing loss in Paget's disease of bone. *Clinical Reviews in Bone and Mineral Metabolism, 1*(2), 145–147.

Morales-Piga, A. A., Moya, J. L., Bachiller, F. J., Muñoz-Malo, M. T., Benavides, J., & Abraira, V., (2000). Assessment of cardiac function by echocardiography in Paget's disease of bone. *Clinical and Experimental Rheumatology, 18*(1), 31–37.

Morales-Piga, A. A., Rey-Rey, J. S., Corres-González, J., Garc ÍA-Sagredo, J. M., & LÓPez-Abente, G., (1995). Frequency and characteristics of familial aggregation of Paget's disease of bone. *Journal of Bone and Mineral Research, 10*(4), 663–670.

Nakamura, M., Udagawa, N., Matsuura, S., Mogi, M., Nakamura, H., Horiuchi, H., Saito, N., et al., (2003). Osteoprotegerin regulates bone formation through a coupling mechanism with bone resorption. *Endocrinology, 144*(12), 5441–5449.

Nieves, J. W., Melsop, K., Curtis, M., Kelsey, J. L., Bachrach, L. K., Greendale, G., Sowers, M. F., & Sainani, K. L., (2010). Nutritional factors that influence change in bone density and stress fracture risk among young female cross-country runners. *PM and R., 2*(8), 740–750.

Palacios, C., (2006). The role of nutrients in bone health, from A to Z. *Critical Reviews in Food Science and Nutrition, 46*(8), 621–628.

Pandey, M., Verma, R. K., & Saraf, S. A., (2010). Nutraceuticals: New era of medicine and health. *Asian Journal of Pharmaceutical and Clinical Research, 3*(1), 11–15.

Parvizi, J., Klein, G. R., & Sim, F. H., (2006). Surgical management of Paget's disease of bone. *Journal of Bone and Mineral Research, 21*(S2), P75–P82.

Prince, R. L., Devine, A., Dhaliwal, S. S., & Dick, I. M., (2006). Effects of calcium supplementation on clinical fracture and bone structure: Results of a 5-year, double-blind, placebo-controlled trial in elderly women. *Archives of Internal Medicine, 166*(8), 869–875.

Pronicka, E., Rowińska, E., Kulczycka, H., Łukaszkiewicz, J., Lorenc, R., & Janas, R., (1997). Persistent hypercalciuria and elevated 25-hydroxyvitamin D3 in children with infantile hypercalcemia. *Pediatric Nephrology, 11*(1), 2–6.

Rabjohns, E. M., Hurst, K., Ghosh, A., Cuellar, M. C., Rampersad, R. R., & Tarrant, T. K., (2021). Paget's disease of bone: Osteoimmunology and osteoclast pathology. *Current Allergy and Asthma Reports, 21*(4).

Ralston, S. H., (2008). Pathogenesis of Paget's disease of bone. *Bone, 43*(5), 819–825.

Ralston, S. H., (2013). Paget's disease of bone. *New England Journal of Medicine, 368*(7), 644–650.

Ralston, S. H., Corral-Gudino, L., Cooper, C., Francis, R. M., Fraser, W. D., Gennari, L., Guañabens, N., et al., (2019). Diagnosis and management of Paget's disease of bone in adults: A clinical guideline. *Journal of Bone and Mineral Research, 34*(4), 579–604.

Ralston, S. H., Digiovine, F. S., Gallacher, S. J., Boyle, I. T., & Duff, G. W., (1991). Failure to detect paramyxovirus sequences in Paget's disease of bone using the polymerase chain reaction. *Journal of Bone and Mineral Research, 6*(11), 1243–1248.

Ray, A., Joshi, J., & Gulati, K., (2016). Regulatory aspects of nutraceuticals. In: *Nutraceuticals* (PP. 941–946).

Reddy, S. V., Singer, F. R., & Roodman, G. D., (1995). Bone marrow mononuclear cells from patients with Paget's disease contain measles virus nucleocapsid messenger ribonucleic acid that has mutations in a specific region of the sequence. *Journal of Clinical Endocrinology and Metabolism, 80*(7), 2108–2111.

Reddy, S. V., Singer, F. R., Mallette, L., & Roodman, G. D., (1996). Detection of measles virus nucleocapsid transcripts in circulating blood cells from patients with Paget disease. *Journal of Bone and Mineral Research, 11*(11), 1602–1607.

Reid, I. R., Lyles, K., Su, G., Brown, J. P., Walsh, J. P., Del Pino-Montes, J., Miller, P. D., et al., (2011). A single infusion of zoledronic acid produces sustained remissions in Paget disease: Data to 6.5 years. *Journal of Bone and Mineral Research, 26*(9), 2261–2270.

Rogers, J., Jeffrey, D. R., & Watt, I., (2002). Paget's disease in an archeological population. *Journal of Bone and Mineral Research, 17*(6), 1127–1134.

Roodman, G. D., (2010). Insights into the pathogenesis of Paget's disease. *Annals of the New York Academy of Sciences, 1192*, 176–180.

Roodman, G. D., Kurihara, N., Ohsaki, Y., Kukita, A., Hosking, D., Demulder, A., Smith, J. F., & Singer, F. R., (1992). Interleukin 6. A potential autocrine/paracrine factor in Paget's disease of bone. *Journal of Clinical Investigation, 89*(1), 46–52.

Sabharwal, R., Gupta, S., Sepolia, S., Panigrahi, R., Mohanty, S., Subudhi, S. K., & Kumar, M., (2014). An insight in to Paget's disease of bone. *Nigerian Journal of Surgery: Official Publication of the Nigerian Surgical Research Society, 20*(1), 9–15.

Sakai, E., Shimada-Sugawara, M., Yamaguchi, Y., Sakamoto, H., Fumimoto, R., Fukuma, Y., Nishishita, K., et al., (2013). Fisetin inhibits osteoclastogenesis through prevention of RANKL-induced ROS production by Nrf2-mediated up-regulation of phase II antioxidant enzymes. *Journal of Pharmacological Sciences, 121*(4), 288–298.

Seitz, S., Priemel, M., Von, D. C., Beil, F. T., Barvencik, F., Amling, M., & Rueger, J. M., (2008). The second most common bone disease: A review on Paget's sisease of bone. *European Journal of Trauma and Emergency Surgery, 34*(6), 549–553.

Senevirathne, M., & Kim, S. K., (2012). Development of bioactive peptides from fish proteins and their health promoting ability. *Advances in Food and Nutrition Research, 65*(1), 235–248.

Seton, M., Choi, H. K., Hansen, M. F., Sebaldt, R. J., & Cooper, C., (2003). *Disease of Bone—The New England Registry for Paget ' s Disease of Bone, 18*(8), 18–20.

Shaker, J. L., (2009). Paget's disease of bone: A review of epidemiology, pathophysiology and management. *Therapeutic Advances in Musculoskeletal Disease, 1*(2), 107–125.

Shaw, B., Burrell, C. L., Green, D., Navarro-Martinez, A., Scott, D., Daroszewska, A., Van 'T Hof, R., et al., (2019). Molecular insights into an ancient form of Paget's disease of bone. *Proceedings of the National Academy of Sciences of the United States of America, 116*(21), 10463–10472.

Singer, F. R., Bone, H. G., Hosking, D. J., Lyles, K. W., Murad, M. H., Reid, I. R., & Siris, E. S., (2014). Paget's disease of bone: An endocrine society clinical practice guideline. *Journal of Clinical Endocrinology and Metabolism, 99*(12), 4408–4422.

Siris, E. S., (1998). Paget's disease of bone. *Journal of Bone and Mineral Research, 13*(7), 1061–1065.

Siris, E. S., Ottman, R., Flaster, E., & Kelsey, J. L., (1991). Familial aggregation of Paget's disease of bone. *Journal of Bone and Mineral Research, 6*(5), 495–500.

Takata, S., Hashimoto, J., Nakatsuka, K., Yoshimura, N., Yoh, K., Ohno, I., Yabe, H., et al., (2006). Guidelines for diagnosis and management of Paget's disease of bone in Japan. *Journal of Bone and Mineral Metabolism, 24*(5), 359–367.

Tou, J. C., (2014). Resveratrol supplementation affects bone acquisition and osteoporosis: Pre-clinical evidence toward translational diet therapy. *Biochimica et Biophysica Acta – Molecular Basis of Disease, 1852*(6), 1186–1194.

Tsuji, M., Yamamoto, Æ. H., Sato, Æ. T., & Mizuha, Æ. Y., (2009). *Dietary Quercetin Inhibits Bone Loss Without Effect on the Uterus in Ovariectomized Mice*, 673–681.

Van, S. T. P., Selby, P., Leufkens, H. G. M., Lyles, K., Sprafka, J. M., & Cooper, C., (2002). Incidence and natural history of Paget's disease of bone in England and Wales. *Journal of Bone and Mineral Research, 17*(3), 465–471.

Vercruysse, L., Van, C. J., & Smagghe, G., (2005). ACE inhibitory peptides derived from enzymatic hydrolysates of animal muscle protein: A review. *Journal of Agricultural and Food Chemistry, 53*(21), 8106–8115.

Williams, E. D., Barr, W. T., & Rajan, K. T., (1981). Relative vitamin D deficiency in Paget's disease. *Lancet, 1979*–1980.

Yamaguchi, M., & Weitzmann, M. N., (2011). *Quercetin, a Potent Suppressor of NF- κ B and Smad Activation in Osteoblasts*, 521–525.

Zainabadi, K., Liu, C. J., & Guarente, L., (2017). *SIRT1 is a Positive Regulator of the Master Osteoblast Transcription Factor, RUNX2.*, 1–14.

Zhang, L. Y., Xue, H. G., Chen, J. Y., Chai, W., & Ni, M., (2016). *Genistein Induces Adipogenic Differentiation in Human Bone Marrow Mesenchymal Stem Cells and Suppresses Their Osteogenic Potential by Upregulating PPAR γ*, 1853–1858.

Zhao, L., Wang, Y., Wang, Z., Xu, Z., Zhang, Q., & Yin, M., (2015). Effects of dietary resveratrol on excess-iron-induced bone loss via antioxidative character. *Journal of Nutritional Biochemistry, 26*(11), 1174–1182.

Zhou, Y., Wu, Y., Jiang, X., Zhang, X., Xia, L., & Lin, K., (2015). *The Effect of Quercetin on the Osteogenesic Differentiation and Angiogenic Factor Expression of Bone Marrow-Derived Mesenchymal Stem Cells*, 1–21.
Zhu, K., & Prince, R. L., (2012). Calcium and bone. *Clinical Biochemistry, 45*(12), 936–942.

CHAPTER 7

HEALTH BENEFITS OF NUTRACEUTICALS IN GOUT PATIENTS

SURESH KUMAR,[1] BHAWNA CHOPRA,[2] RAMESHWAR DASS,[2] and ASHWANI K. DHINGRA[2]

[1]*Department of Pharmacy, School of Medical and Allied Sciences, GD Goenka University, Gurugram, Haryana, India*

[2]*Guru Gobind Singh College of Pharmacy, Yamuna Nagar, Haryana, India*

ABSTRACT

The monosodium urate (MSU) crystals get deposited into the joints associated with increased uric acid in the blood usually leads to GOUT. Gout is also characterized as inflammatory arthritis, triggered by the interactions between MSU crystals and tissue during purine catabolism by xanthine oxidase. The oxidative hydroxylation induced by the xanthine oxidase enzyme is results in the formation of xanthine from hypoxanthine followed by uric acid synthesis, thereby causing painful inflammation. Natural dietary supplements (NPS) are very essential for the treatment of gout. Despite that, the complete knowledge about the use of that agents is generally less or not well organized. Nutraceuticals or natural products like nuts, vegetables, fruits, legumes, and whole grains, are involved in the onset of hyperuricemia and gout. Xanthine oxidase inhibitors are currently being investigated as potential therapeutics for disorders such as gout, hyperuricemia, ulcers, cancer, and oxidative damage. Non-purine xanthine oxidase inhibitors have rarer side effects as compared to purine analogs, making them appropriate for advanced study, which is noticeable by articles and patents being published. In addition, colchicine or corticosteroids, nonsteroidal anti-inflammatory

Nutraceuticals and Bone Health. Deepak Sharma, Madan Mohan Gupta, Anil K. Sharma, Raj K. Keservani, Rajesh K. Kesharwani (Eds.)

medications (NSAIDs), and IL-1 inhibitors should be used to treat acute gout attacks.

7.1 INTRODUCTION

The formation of uric acid crystals is caused by a significant increase in serum uric acid (SUA) contents above that range. Hyperuricemia is the primary pathogenic defect in gout; nevertheless, several persons who might have hyperuricemia don't get gout or even generate UA crystals. Therefore, the usual thought is that other factors like genetic predisposition involve gout pathophysiology (Dalbeth, et al., 2016; Emmerson, 1996). The deposition of MSU crystals in tissues, particularly, around the joints, results in painful attacks and warm and swollen joints. Gout is diagnosed primarily by the significant accumulations of crystals beneath the skin known as tophi or identifying pathognomonic MSU crystals through joint fluid aspiration. Presently, to manage gout, serum uric acid lowering drugs or dietary modification is used (Keservani et al., 2017, 2020), reducing the SUA levels by dissolving the MSU crystals and preventing further attacks (Pascual & Sivera et al., 2007; Singh et al., 2014).

Gout is also categorized as inflammatory arthritis, caused by the interactions between MSU crystals and tissue (Dhingra & Chopra, 2020; Dalbeth, et al., 2014) while xanthine oxidase catabolizes purines (Nile, et al., 2013). Xanthine oxidase enzyme is responsible for the oxidative hydroxylation of hypoxanthine to xanthine to uric acid, causing painful inflammation (Chopra, et al., 2015; Nile & Khobragade, et al., 2011). Uricase is another enzyme that translates uric acid into highly soluble allantoin, further excreted in the urine (Figure 7.1). Moreover, uricase is not a functional human enzyme that results in hyperuricemia in humans (Gliozzi, et al., 2016). Gout has also been associated with joint abnormalities, tophi, and kidney stones, among other factors (Teh et al., 2014).

The kidney is the primary regulator of SUA levels, where renal urate excretion is characterized by the balance between urate secretion and reabsorption. Two urate transporters mainly mediate renal urate reabsorption, comprising urate transporter 1 (URAT1) and glucose transporter 9 (GLUT9) (Enomoto et al., 2002; Matsuo, et al., 2008). The mechanisms involved in plasma uric acid levels reduction consists of urate reabsorption inhibition *via* renal mRNA, protein levels of URAT1, GLUT9, organic carnitine and cation transporters (OCTN1/2, OCT1/2) along with organic anion transporter 1 (OAT1) (Sungthong, et al., 2016). The increased SUA levels (>0.42 mmol/L) result in hyperuricemia) (Stamp, et al., 2007). As a result, the primary strategy for managing gout is to minimize uric acid levels, with target SUA levels less than 0.36 mmol/L (Pillinger, et al., 2007; Falasca, 2006).

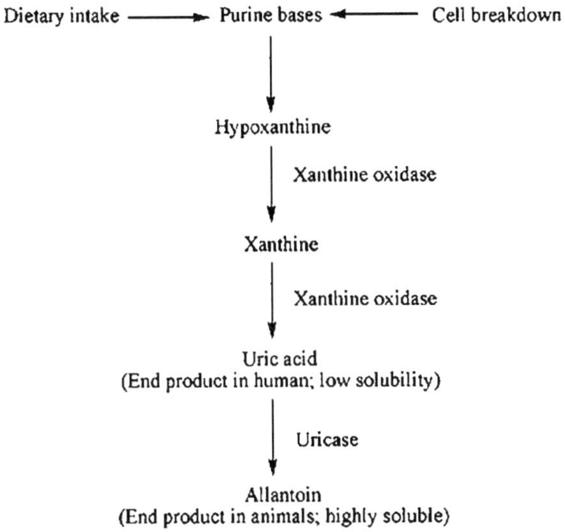

FIGURE 7.1 Representation of the purine catabolism by xanthine oxidase.

Several risk factors accompanying gout development include age, alcohol consumption, cholesterol level, chronic renal disease, dietary factors, usage of diuretics, genetic factors, hyperuricemia, hypertension, obesity, and metabolic syndrome (Roddy & Doherty, 2010). It is believed that men have a four- to the nine-fold enlarged risk of producing gout as compared to women. However, they are so much more prone to develop gout after menopause, as estrogen's uricosuric action is no longer present (Tausche, et al., 2009). Furthermore, genetics and race factors may also take part in the development of gout (Mohd, et al., 2011).

7.2 PATHOGENESIS OF HYPERURICEMIA

Uric acid exists in its ionized form urate in the body, and it is a weak acid with a pH of 5.8. The deposition of urate crystals triggered an increased SUA level than the standard pathological threshold (6.8 mg/dL) (McCarty & Hollander, 1961). Various factors responsible for uric acid's solubility in the joint include electrolytes level, water concentration, synovial fluid pH, and other synovial components like collagen and proteoglycans SUA level in the body is characterized by the balance between its formation (either from endogenous or purine intake) and excretion through kidneys and GIT. Its renal under-excretion causes 90% of cases of gout, while the remaining 10% is caused by increased production of UA (Mandal & Mount, 2015).

Factors that affect the SUA levels include gender and age. Levels are low among children and young adults, and men possess higher SUA levels than women. Postmenopausal women's SUA levels, however, rise to match those of males. This explains why postmenopausal women, middle-aged, and older men are more likely to develop gout. Rarely, it may occur in kids and young adults due to rare inborn purine metabolism abnormalities. The enzymatic abnormalities cause an increase in the formation of SUA and UA crystals in the kidneys and joints (Kamei, et al., 2014).

7.3 OVERPRODUCTION OF URIC ACID

Lack of purine metabolism-related enzymes is the root cause of the excessive supply of UA. For example, Lesch-Nyhan syndrome is a metabolic disorder characterized by deficiency of hypoxanthine-guanine phosphoribosyl-transferase, involved in UA metabolism. It's an X-linked recessive genetic disorder with various levels of severity depending on the type of mutation. The clinical symptoms of the disease includes renal stones, dystonia, chorea, cognitive impairment, obsessive injurious behavior, self-mutilation, and articular symptoms. If left untreated, it can lead to tophi formation and renal failure (Torres & Puig, 2007).

In addition, super activity of one more enzyme *i.e.*, phosphoribosyl pyro-phosphate synthetase results in the development of gout in the young. The clinical picture includes neurological abnormalities like hypotonia, ataxia and sensorineural hearing loss, in severe form. There are two clinical forms of this X-linked dominant hereditary disorder: a severe early-onset type (in children) and a mild late juvenile type. Therefore, Urate cause kidney stones and arthritis, both are mild symptoms. However, these enzymatic deficiencies account for less than 10% of all cases of urate overproduction (Reginato & Olsen, 2007).

7.4 ENDOGENOUS URATE PRODUCTION

Increased cellular regeneration leads to excessive endogenous uric acid production, such as in malignancies hematological and diseases leading to inflammation. Chemotherapy and tissue damage may lead to increased purine production. Moreover, obesity and increased body weight promote the uric acid formation, which is raising the risk of hyperuricemia. The hormone leptin has been found to increase the blood urate levels. Therefore, exercise and weight loss helps to lower the SUA levels and reduce the risk of gout (Mahmoud et al., 1998; Emmerson, 1998; Bedir et al., 2003; Dessein et al., 2000).

7.5 DECREASED EXCRETION OF URIC ACID

The maximum amount of urate excretion occurs through the kidneys, while the remaining is excreted through GIT. Reduced transporter secretory function ABCG2 causes a decrease in uric acid excretion through the GI tract, leading to a rise in uric acid blood levels and increased renal excretion (Mandal & Mount, 2015; Ichida et al., 2012). Uric acid crystals are not soluble, requiring specific membrane transporters to cross cell membranes which includes the urate transporter/channel (URAT), namely URAT1, OAT1 and OAT3 transporters (Mandal & Mount, 2015; Enomoto & Endou, 2005; Bobulescu & Moe, 2012).

7.6 MANAGEMENT OF ACUTE GOUTY ARTHRITIS (CURRENT APPROACHES)

Xanthine oxidase inhibitors are considered promising in treatment/mitigation of disorders such as gout, hyperuricemia, ulcers, cancer, and oxidative damage. Non-purine xanthine oxidase inhibitors are observed to have minimum side effects. In addition, colchicine or corticosteroids, nonsteroidal anti-inflammatory medications (NSAIDs), and IL-1 inhibitors should be used to treat acute gout attacks (Dhingra et al., 2015).

7.7 XANTHINE OXIDASE INHIBITORS

Gout is largely caused by the oxidation of xanthine and hypoxanthine to uric acid, which is catalyzed by the enzyme xanthine oxidase (XOD) (Groff, et al., 1990). During the reoxidation of XO, molecular oxygen performances as an electron acceptor, which produces superoxide radicals and hydrogen peroxide (Clive, 2000). These reactions can be represented as follows (Schlesinger, et al., 2002).

$$\text{Xanthine} + 2O_2 + H_2O \rightarrow \text{Unic acid} + 2O_2 + 2H$$
$$+ \text{Xanthine} + O_2 + H_2O \rightarrow \text{Unic acid} + H_2O_2.$$

XOD is a highly versatile flavoprotein enzyme. It leads to the production of superoxide radicals, which along with other reactive oxygen species (ROS), is the leading cause of oxidative stress and are involved in many pathological conditions like atherosclerosis, aging, cancer, inflammation, etc., (Halliwell et al., 1992). In addition, the hydroxylation of purines is catalyzed by XO, particularly the production of uric acid from xanthine is responsible for gout, renal diseases, and stone formation in the urinary system. The enzyme catalyzes the oxidative hydroxylation of purine substrates followed by subsequent reduction of O_2 at the flavin center with generation ROS, either superoxide anion radical or hydrogen peroxide. Allopurinol is a potent inhibitor of the XO enzyme, clinically used in treating gout is a xanthine oxidase inhibitor (Borges, et al., 2002). Xanthine oxidase inhibitors are much helpful for treating hyperuricemia and related medical conditions, including gout, as they possess lesser side effects than uricosuric and anti-inflammatory agents (Muthuswamy, et al., 2012). Xanthine oxidase inhibitors as purine analogs include allopurinol, oxypurinol, and tisopurine (Iwanaga et al., 2005), whereas other non-purine analogs include febuxostat and inositols (Becker, et al., 2005).

The prototypical xanthine oxidase (XO) inhibitor allopurinol has been used in the clinical management of gout and conditions associated with hyperuricemia for several decades (Pacher, et al., 2006). Allopurinol is usually well-tolerated; however, the adverse effect includes allopurinol hypersensitivity syndrome (AHS) (1%), gastrointestinal events (2%), rash (3%), fever (1%), and musculoskeletal events (1%) (Schlesinger, et al., 2002; Schlesinger, 2006). Another drug, febuxostat, is associated with atrial fibrillation, cardiovascular toxicity, and cardiovascular thromboembolic events. However, if one XOI is ineffective in achieving the target level of SUA or intolerance develops, then the second XOI may be substituted. Apart from this, pegloticase therapy (polyethylene glycol uricase preparation) adverse reactions includes infusion-related reactions, gout flares, nausea (7%) and headache (11%) (Sundy, et al., 2011).

7.8 NUTRITIONAL STATUS AND DIETARY MANIPULATION AMONG GOUT PATIENTS

7.8.1 VITAMIN C

Ascorbic acid (vitamin C) represents a uricosuric activity that inhibits uric acid synthesis and reduces uric acid levels. However, it did not show significant action on the SUA level compared to Allopurinol (Stamp, et al., 2013).

7.8.2 DAIRY PRODUCTS

A randomized clinical study performed in New Zealand exhibited that the intake of skimmed milk powder supplemented with G600 milk fat extract and glycomacropeptide (GMP) for 3-months decreased the probability of gout attacks. In addition, the anti-inflammatory potential of milk reduced the inflammation characterized by the formation of monosodium urate crystals in the joint (Dalbeth & Palmano, 2011). Additionally, the Third National Health and Nutrition Examination Survey showed that yogurt significantly reduced the SUA levels compared to cheese (Choi, et al., 2005; Choi, et al., 2004).

7.8.3 CHERRIES

Literature reveals that the consumption of cherry juice for more than four months significantly reduced gout attacks. Cherries have ample nutrients as it contains quercetin, anthocyanins, and vitamins (E, C, and A) responsible for the antioxidant and anti-inflammatory activity (Schlesinger, et al., 2013; Collins, et al., 2019). Clinical studies performed on 633 patients showed that the cherry extract combined with allopurinol remarkably decreased the probability of acute gout attack (Zhang, et al., 2012). Another cross-over, placebo-controlled trial on obese patients meant a significant reduction in the SUA levels (19.2%) by the intake of 100% Tart Cherry Juice (TCJ) (Martin & Coles, 2019). However, none of the studies confined the exact dose of cherry tart required for the anti-inflammatory activity against acute gout attacks (Stamp, et al., 2020).

7.8.4 GREEN TEA

Green tea is a rich fount of catechins, primarily Epigallocatechin (EGCG), responsible for its antioxidant properties. It showed significant free-radical scavenging, xanthine oxidase inhibitory, and peroxidation of anti-lipids. Studies concluded that the consumption of green tea significantly reduced the SUA and uric acid clearance (Zhang, et al., 2017; Khan & Mukhtar, 2018). The polyphenols (Keservani et al., 2016a, 2016b) in green tea remarkably reduced the SUA levels in hyperuricemic mice (Chen, et al., 2015). In contrast, black tea's antioxidant and urate-lowering potential are due to the presence of flavonoids, particularly quercetin (Shi & Williamson, 2016).

7.8.5 COFFEE

Coffee (Kesharwani et al., 2016) is a widely consumed beverage, which abundantly contains alkaloids, antioxidants, carbohydrates, caffeine, kahweol, and lipids (Zhang, et al., 2016; Zhang & Zhang, 2018). Studies suggested that long-term coffee consumption decreases the SUA level, reducing the risk of gout among men (Choi, et al., 2007). Furthermore, postmenopausal women represented high levels of SUA as compared to premenopausal women, which suggests the protective effect of estrogen against gout (Park, et al., 2016). Studies conducted on women exhibited that the consumption of coffee reduced the risk of gout (Choi & Curhan, 2010; Bae, et al., 2015). However, the type of coffee, brewing methods, size, and the number of cups intake may contribute to the confounding factors (Park, et al., 2016).

7.8.6 OMEGA-3 FATTY ACIDS

Omega-3 fatty acids, mainly DHA (docosahexaenoic acid) and EPA (eicosapentaenoic acid) possess significant anti-inflammatory potential. Mackerel, salmon, and tuna contain both DHA and EPA abundantly. In addition, omega-6 fatty acids such as arachidonic and linoleic acid showed significantly higher concentrations of pro-inflammatory activity because of thromboxanes, leukotrienes, and prostaglandins. Fish consumption has omega-3 fatty acids associated with reduced gout flares (Zhang, et al., 2019; Saito, et al., 2020). The mechanism involved is elucidated as the reduction of IL-1β secretion and caspase-1 activation along with the NLRP3 inflammasome activation inhibition and chemotaxis (Yan, et al., 2013).

7.9 PLANT POSSESSING ANTI-GOUT ACTIVITY:

Medicinal plants have served as a prominent alternative to complementary medicines against gout pathology because of bioactive compounds possessing antioxidants and anti-inflammatory activity (Chopra & Dhingra, 2021). Several bioactive compounds like colchicine and curcumin have been reported to possess significant anti-gout activity. They are used as a first-line treatment for the management of acute gout attacks along with oral NSAIDs.

Table 7.1 Classifies medicinal plants that have been used traditionally to manage gout disease.

TABLE 7.1 The Medicinal Plants with Anti-gout Activity

S/n	Common name	Part used	Scientific name	Family	Phytoconstituents	Mechanism of action/use	Ref
1.	Crown Wood-Fern	Rhizome	*Dryopteris crassirhizoma*	Dryopteridaceae	Flavonoids (flavaspidic acid AB, flavaspidic acid AP, flavaspidic acid BB and flavaspidic acid PB)	Inhibit xanthine oxidase (XO) enzyme	Yuk, et al., 2021
2.	Chandramallika	Flower	*Chrysanthemum indicum* Linn	Asteraceae	Terpenoids, flavonoids, and DKB114	Anti-hyperuricemia, Anti-inflammatory activity in acute and chronic cutaneous inflammation	Park, et al., 2020
3	Cassia	Bark	*Cinnamomum cassia*	Lauraceae	Essential oil, and DKB114	Anti-hyperuricemia	Park, et al., 2020
4	Alfalfa, lucerne		*Medicago sativa*	Fabaceae	--	Inhibit xanthine oxidase (XO) enzyme	Hsu, et al., 2021
5.	Chilli	Fruit	Capsicum species (*C. frutescens*, *C. annuum*, *C. pubescens*, *C. baccatum*, and *C. chinense*)	Solanaceae	Carotenoids (lutein, carotene, cryptoxanthin, capsanthin, violaxanthin, zeaxanthin, and capsorubin), phenolic compounds (quercetin, luteolin, and phenolic acids), vitamins (C and E) and capsaicinoids	Anti-inflammatory, antioxidant	Villa-Rivera, et al., 2020
6.	White mulberry	Fruit, leaves	*Morus alba*	Moraceae	Proteins	anti-gouty arthritis	Oh, et al., 2021
7.	Ananas	Stem	*Ananas comosus*	Bromeliaceae	Bromelain	Anti-inflammatory, anti-gouty arthritis, sinusitis	Ali, et al., 2020
8	Guava	Leaves	*Psidium guajava*	Myrtaceae	Rich vitamin-C, flavonoids (kaempferol, quercetin, rutin, catechin, quercitrin, epicatechin, luteolin) and phenolic acids	Inhibitors of ACE and XO.	Irondi, et al., 2016
9	Milk thistle	Capsule	*Silybum marianum*	Asteraceae	Silybin, Silymarin	anti-inflammatory	Ali, et al., 2017; Simanek, et al., 2001

TABLE 7.1 *(Continued)*

S/n	Common name	Part used	Scientific name	Family	Phytoconstituents	Mechanism of action/use	Ref
10	Dandelion	Root	*Taraxacum officinale*	Asteraceae	Lactones, triterpenes, sterols, flavonoids, and phenolic acids	Kidney problem, Antiuricemia	Amim, et al., 2013
11	Turmeric	Rhizome	*Curcuma longa*	Zingiberaceae	Curcumin	Anti-inflammatory, gout	Chen, et al., 2019
12	Cherry	Fruit	*Prunus avium (Sweet), Prunus cerasus (Sour)*	Rosaceae	Ascorbic acid and phenolic compounds	Cherries may reduce acute and chronic inflammation by inhibiting cytokines and COX-I, and COX-II pathways.	Collins, et al., 2019; Kelley, et al., 2018; Zhang, et al., 2012
13	Corn	Stigma	*Zea mays*	Graminae	Phenolic acids, flavonoids, stilbenes, coumarins, and tannins	Inhibition of xanthine oxidase	Wang, et al., 2012; Sheng, et al., 2018
14	Cabra	Fruit & leaves	*Capparis spinosa*	Capparidaceae	Phenolic compounds (catechin, caffeic acid, coumarin, chlorogenic acid, kaempferol, ferulic acid, quercetin, luteolin, rutin, resveratrol, vanillic acid, and syringic acid)	Inhibition of xanthine oxidase	Ameni, et al., 2012; Mollica, et al., 2017
15	Colchicum	Seeds	*Colchicum autumnale*	Liliaceae	Alkaloid (colchicine)	Antigout	Fernandez-Cuadros, et al., 2019; Finkelstein, et al., 2010
16	Clove	Flowers	*Caryophyllus aromaticus*	Myrtaceae	Essential oil	Inhibition of xanthine oxidase	Daoudi, et al., 2020
17	*Ginkgo*	Leaves	*Ginkgo biloba*	Ginkgoaceae	Flavonoids, glycosides, and ginkgolides	Inhibition of xanthine oxidase	Chen, et al., 2017
18	*Hyssop*	Flowers	*Hyssopus officinalis*	Lamiaceae	Essential oils and flavonoids	Inhibition of xanthine oxidase	Daoudi, et al., 2020
19	*Lavender*	Flowers	*Lavandula angustifolia*	Lamiaceae	Essential oils	Inhibition of xanthine oxidase	Daoudi, et al., 2020
20	Mentha	Leaves	*Mentha spicata*	Lamiaceae	Essential oils	Inhibition of xanthine oxidase	Daoudi, et al., 2020

KEYWORDS

- **colchicine**
- **gout**
- **hyperuricemia**
- **monosodium urate**
- **xanthine oxidase**

REFERENCES

Ali, K., Yeter, D., & Serkan, Y., (2017). Protective effect of *Silybum marianum* and *Taraxacum officinale* extracts against oxidative kidney injuries induced by carbon tetrachloride in rats. *Ren Fail, 39*(1), 1–6.

Ali, M. M., Hashim, N., Aziz, S. A., & Lasekan, O., (2020). Pineapple (*Ananas comosus*): A comprehensive review of nutritional values, volatile compounds, health benefits, and potential food products. *Food Res Int., 137*, 109675.

Ameni, B. D., & Boumerfeg, S., (2012). Studies of antioxidants and xanthine oxidase inhibitory potentials of root and aerial parts of medicinal plant *Capparis spinosa* L. *Am. J. M. Sc., 2*(1), 25–32.

Amim, M. M., Sawhney, S. S., & Jassl, M. S., (2013). Comparative antioxidant power determination of *Taraxacum officinale* by FRAP and DTPH method. *Pharm. Anal. Acta, 4*, 221–223.

Bae, J., Park, P. S., Chun, B. Y., et al., (2015). The effect of coffee, tea, and caffeine consumption on serum uric acid and the risk of hyperuricemia in Korean multi-rural communities cohort. *Rheumatol. Int., 35*(2), 327–336.

Becker, M. A., Schumacher, H. R., Wortmann, R. L., et al., (2005). Febuxostat, a novel non-purine selective inhibitor of xanthine oxidase: A twenty-eight-day, multicenter, phase II, randomized, double-blind, placebo-controlled, dose-response clinical trial examining safety and efficacy in patients with gout. *Arthr. Rheumat., 52*, 916–923.

Bedir, A., Topbas, M., Tanyeri, F., Alvur, M., & Arik, N., (2003). Leptin might regulate serum uric acid concentrations in humans. *Jpn Heart J., 44*(4), 527–536.

Bobulescu, I. A., & Moe, O. W., (2012). Renal transport of uric acid: Evolving concepts and uncertainties. *Adv. Chronic Kidney Dis., 19*(6), 358–371.

Borges, F., Fernandes, E., & Roleira, F., (2002). Progress towards the discovery of xanthine oxidase inhibitors. *Curr. Med. Chem., 9*(2), 195–217.

Chen, B., Li, H., Ou, G., Ren, L., Yang, X., & Zeng, M., (2019). Curcumin attenuates MSU crystal-induced inflammation by inhibiting the degradation of IκBα and blocking mitochondrial damage. *Arthritis Res. Ther., 21*(1), 193.

Chen, G., Tan, M. L., Li, K. K., Leung, P. C., & Ko, C. H., (2015). Green tea polyphenols decreases uric acid level through xanthine oxidase and renal urate transporters in hyperuricemic mice. *J. Ethnopharmacol., 175*, 14–20.

Chen, W., Liu, Y., Wei, M., Shi, L., Wu, Y., Liu, Z. Y., Liu, S., et al., (2017). Studies on the effect of *Ginkgo biloba* L. leaves in acute gout with hyperuricemia model rats by using UPLC-ESI-Q-TOF/MS metabolomics approach. *RSC Adv., 7*(68), 42964–42972.

Choi, H. K., & Curhan, G., (2010). Coffee consumption and risk of incident gout in women: The nurses' health study. *Am. J. Clin. Nutr., 92*(4), 922–927.

Choi, H. K., Atkinson, K., Karlson, E. W., Willett, W., & Curhan, G., (2004). Purine-rich foods, dairy and protein intake, and the risk of gout in men. *N Engl. J. Med., 350*(11), 1093–1103.

Choi, H. K., Liu, S., & Curhan, G., (2005). Intake of purine-rich foods, protein, and dairy products and relationship to serum levels of uric acid: The third national health and nutrition examination survey. *Arthritis Rheum., 52*(1), 283–289.

Choi, H. K., Willett, W., & Curhan, G., (2007). Coffee consumption and risk of incident gout in men: A prospective study. *Arthritis Rheum., 56*(6), 2049–2055.

Chopra, B., & Dhingra, A. K., (2021). Natural products: A lead for drug discovery and development. *Phytother. Res., 35*(9), 4660–4702.

Chopra, B., Dhingra, A. K., Kapoor, R. P., & Parsad, D. N., (2015). Microwave assisted synthesis of some 5-substituted imidazolone analogs as a new class of non-purine xanthine oxidase inhibitors. *Der Pharma Chemica., 7*(9), 145–152.

Clive, D. M., (2000). Renal transplant-associated hyperuricemia and gout. *J. Am. Soc. Nephrol., 11*(5), 974–979.

Collins, M. W., Saag, K. G., & Singh, J. A., (2019). Is there a role for cherries in the management of gout. *Ther. Adv. Musculoskelet Dis., 11*, 1759720X19847018.

Dalbeth, N., & Palmano, K., (2011). Effects of dairy intake on hyperuricemia and gout. *Curr. Rheumatol. Rep., 13*(2), 132–137.

Dalbeth, N., Lauterio, T. J., & Wolfe, H. R., (2014). Mechanism of action of colchicine in the treatment of gout. *Clin. Ther., 36*(10), 1465–1479.

Dalbeth, N., Merriman, T. R., & Stamp, L. K., (2016). Gout. *The Lancet., 388*(10055), 2039–2052.

Daoudi, N. E., Bouhrim, M., Ouassou, H., & Bnouham, M., (2020). Medicinal plants as a drug alternative source for the antigout therapy in Morocco. *Scientifica (Cairo), 2020*, 8637583.

Dessein, P. H., Shipton, E. A., Stanwix, A. E., Joffe, B. I., & Ramokgadi, J., (2000). Beneficial effects of weight loss associated with moderate calorie/carbohydrate restriction, and increased proportional intake of protein and unsaturated fat on serum urate and lipoprotein levels in gout: A pilot study. *Ann. Rheum. Dis., 59*(7), 539–543.

Dhingra, A. K., & Chopra, B., (2020). Inflammation as a therapeutic target for various deadly disorders: A review. *Curr. Drug Targets, 21*(6), 582–588.

Dhingra, A. K., Chopra, B., Dass, R., & Mittal, S. K., (2015). An update on anti-inflammatory compounds: A review. *Anti-Inflamm. Anti-Allergy Agents Med. Chem., 14*(2), 81–97.

Emmerson, B. T., (1996). The management of gout. *N Engl. J. Med., 334*(7), 445–451.

Emmerson, B., (1998). Hyperlipidemia in hyperuricemia and gout. *Ann Rheum Dis., 57*(9), 509, 510.

Enomoto, A., & Endou, H., (2005). Roles of organic anion transporters (OATs) and a urate transporter (URAT1) in the pathophysiology of human disease. *Clin. Exp. Nephrol., 9*(3), 195–205.

Enomoto, A., Kimura, H., Chairoungdua, A., Shigeta, Y., Jutabha, P., Cha, S. H., et al., (2002). Molecular identification of a renal urate–anion exchanger that regulates blood urate levels. *Nature, 417*, 447–452.

Falasca, G. F., (2006). Metabolic diseases: Gout. *Clin. Dermatol., 24*(6), 498–508.

Fernández-Cuadros, M. E., Goizueta-San-Martin, G., Varas-De-Dios, B., et al., (2019). Colchicine-induced rhabdomyolysis: Clinical, biochemical, and neurophysiological features and review of the literature. *Clin. Med. Insights Arthritis Musculoskelet Disord., 12*, 1179544119849883.

Finkelstein, Y., Aks, S. E., & Hutson, J. R., (2010). Colchicine poisoning: The dark side of an ancient drug. *Clin. Toxicol., 48*(5), 407–414.

Gliozzi, M., Malara, N., Muscoli, S., & Mollace, V., (2016). The treatment of hyperuricemia. *Int. J. Cardiol., 213*, 23–27.

Groff, G. D., Franck, W. A., & Raddatz, D. A., (1990). Systemic steroid therapy for acute gout: A clinical trial and review of the literature. *Semin Arthritis Rheum., 19*(6), 329–336.

Halliwell, B., Gutteridge, J. M., & Cross, C. E., (1992). Free radicals, antioxidants, and human disease: Where are we now CE cross. *J. Lab Clin. Med., 119*(6), 598–620.

Hsu, S. J., Verpoorte, R., Lin, S. M., & Lee, C. K., (2021). Fast dereplication of xanthine oxidase-inhibiting compounds in alfalfa using comparative metabolomics. *Food Res. Int., 141*, 110170.

Ichida, K., Matsuo, H., Takada, T., Nakayama, A., Murakami, K., & Shimizu, T., (2012). Decreased extra-renal urate excretion is a common cause of hyperuricemia. *Nature Commun., 3*, 764.

Irondi, E. A., Agboola, S. O., Oboh, G., Boligon, A. A., Athayde, M. L., & Shode, F. O., (2016). Guava leaves polyphenolics-rich extract inhibits vital enzymes implicated in gout and hypertension *in vitro. J. Intercult. Ethnopharmacol., 5*(2), 122–130.

Iwanaga, T., Kobayashi, D., Hirayama, M., & Maeda, T., (2005). Involvement of uric acid transporter in increased renal clearance of the xanthine oxidase inhibitor oxypurinol induced by a uricosuric agent, benzbromarone. *Drug Metab. Dispos., 33*(12), 1791–1795.

Kamei, K., Konta, T., Hirayama, A., Suzuki, K., Ichikawa, K., & Fujimoto, S., (2014). A slight increase within the normal range of serum uric acid and the decline in renal function: Associations in a community-based population. *Nephrol, Dial. Transplant., 29*(12), 2286–2292.

Kelley, D. S., Adkins, Y., & Laugero, K. D., (2018). A review of the health benefits of cherries. *Nutrients, 10*(3), 368.

Keservani, R. K., Sharma, A. K., & Kesharwani, R. K., (2016a). Nutraceutical and functional foods for cardiovascular health. In: *Food Process Engineering* (pp. 257–278, 291–312). Apple Academic Press, CRC Press. ISBN: 9781771884020.

Keservani, R. K., Sharma, A. K., & Kesharwani, R. K., (2016b). Medicinal effect of nutraceutical fruits for the cognition and brain health. *Scientifica*, 10. Article ID 3109254. doi: 1 0.1155/2016/3109254.

Keservani, R. K., Sharma, A. K., & Kesharwani, R. K., (2017). An overview and therapeutic applications of nutraceutical and functional foods. *Recent Advances in Drug Delivery Technology*, 160–201. ISBN:9781522507543.

Keservani, R. K., Sharma, A. K., & Kesharwani, R. K., (2020). *Nutraceuticals and Dietary Supplements: Applications in Health Improvement and Disease Management*. CRC Press. ISBN: 9781771888738.

Kesharwani, R. K., Singh, P., & Keservani, R. K., (2016). Green coffee bean extract and chlorogenic acids: Chemistry and novel antioxidant benefits; Chapter 1. In: Bagchi, D., Moriyama, H., & Swaroop, A., (eds.), *Green Coffee Bean Extract in Human Health* (pp. 1–17). Boca Raton, Taylor and Francis, USA. ISBN: 9781498716376.

Khan, N., & Mukhtar, H., (2018). Tea polyphenols in promotion of human health. *Nutrients, 11*(1), 39.

Mahmoud, H. H., Leverger, G., Patte, C., Harvey, E., & Lascombes, F., (1998). Advances in managing malignancy-associated hyperuricemia. *Br J. Cancer, 77*(4), 18–20.

Mandal, A. K., & Mount, D. B., (2015). The molecular physiology of uric acid homeostasis. *Annu Rev. Physiol., 77*, 323–345.

Martin, K. R., & Coles, K. M., (2019). Consumption of 100% tart cherry juice reduces serum urate in overweight and obese adults. *Curr. Dev. Nutr., 3*(5), nzz011.

Matsuo, H., Chiba, T., Nagamori, S., Nakayama, A., Domoto, H., Phetdee, K., et al., (2008). Mutations in glucose transporter nine gene SLC2A9 cause renal hypouricemia. *Am. J. Hum. Genet., 83*(6), 744–751.

McCarty, D. J., & Hollander, J. L., (1961). Identification of urate crystals in gouty synovial fluid. *Ann. Intern. Med., 54*, 452–460.

Mohd, A., Das, G. E., Loh, Y. L., Gandhi, C. D., Souza, B., & Gun, S. C., (2011). Clinical characteristics of gout: A hospital case series. *Malaysian Fam. Phys., 6*(2, 3), 72–73.

Mollica, G., Zengin, M., Locatelli, et al., (2017). Anti-diabetic and anti-hyperlipidemic properties of *Capparis spinosa* L.: *In vivo* and *in vitro* evaluation of its nutraceutical potential. *J. Funct. Foods, 35*, 32–42.

Muthuswamy, U., Prabhu, R., Asokkumar, K., Sivashanmugam, T., Subhadradevi, V., Jagannath, P., & Madeswaran, A., (2012). In silico docking studies and *in vitro* xanthine oxidase inhibitory activity of commercially available terpenoids. *Int. J. Phytopharm Res., 2*, 135–142.

Nile, S. H., & Khobragade, C. N., (2011). *In vitro* anti-inflammatory and xanthine oxidase inhibitory activity of *Tephrosia purpurea* shoot extract. *Nat. Prod. Commun., 6*(10), 1437–1440.

Nile, S. H., Kumar, B., & Park, S. W., (2013). *In vitro* evaluation of selected benzimidazole derivatives as an antioxidant and xanthine oxidase inhibitors. *Chem. Biol. Drug Des., 82*(3), 290–295.

Oh, K. K., Adnan, M., & Cho, D. H., (2021). Network pharmacology study o*n Morus alba* l. leaves: Pivotal functions of bioactives on RAS signaling pathway and its associated target proteins against gout. *Int. J. Mol. Sci., 22*(17), 9372.

Pacher, P., Nivorozhkin, A., & Szabo, C., (2006). Therapeutic effects of xanthine oxidase inhibitors: Renaissance half a century after the discovery of allopurinol. *Pharmacol. Rev., 58*(1), 87–114.

Park, K. Y., Kim, H. J., Ahn, H. S., et al., (2016). Effects of coffee consumption on serum uric acid: Systematic review and meta-analysis. *Semin Arthritis Rheum., 45*(5), 580–586.

Park, Y. H., Kim, D. H., Lee, J. S., Jeong, H. I., Lee, K. W., & Kang, T. H., (2020). A 12-week, multicenter, randomized, double-blind, placebo-controlled clinical trial for evaluation of the efficacy and safety of dkb114 on reduction of uric acid in serum. *Nutrients, 10*(12), 3794.

Pascual, E., & Sivera, F., (2007). Time required for disappearance of urate crystals from synovial fluid after successful hypouricemic treatment relates to the duration of gout. *Ann. Rheum. Dis., 66*(8), 1056–1058.

Pillinger, M. H., Rosenthal, P., & Abeles, A. M., (2007). Hyperuricemia and gout: New insights into pathogenesis and treatment. *Bull. NYU Hosp. Jt. Dis., 65*(3), 215–221.

Reginato, A. M., & Olsen, B. R., (2007). Genetics and experimental models of crystal-induced arthritis. Lessons learned from mice and men: Is it crystal clear? *Curr. Opin. Rheumatol., 19*(2), 134–145.

Roddy, E., & Doherty, M., (2010). Gout. Epidemiology of gout. *Arthritis Res. Ther., 12*, 223.

Saito, H., Toyoda, Y., Takada, T., et al., (2020). Omega-3 polyunsaturated fatty acids inhibit the function of human urat1, a renal urate re-absorber. *Nutrients, 12*(6), 1601.

Schlesinger, N., & Schlesinger, M., (2013). Previously reported prior studies of cherry juice concentrate for gout flare prophylaxis: Comment on the article by Zhang et al. *Arthritis Rheum., 65*(4), 1135–1136.

Schlesinger, N., (2006). Response to application of ice may help differentiate between gouty arthritis and other inflammatory arthritides. *J. Clin. Rheumatol., 12*(6), 275–276.

Schlesinger, N., Detry, M. A., Holland, B. K., Baker, D. G., Beutler, A. M., Rull, M., et al., (2002). Local ice therapy during bouts of acute gouty arthritis. *J. Rheumatol., 29*, 331–334.

Sheng, S., Tong, L., & Liu, R. H., (2018). Corn phytochemicals and their health benefits. *Food Sci. Hum. Wellness, 7*(3), 185–119.

Shi, Y., & Williamson, G., (2016). Quercetin lowers plasma uric acid in pre-hyperuricaemic males: A randomized, double-blinded, placebo-controlled, cross-over trial. *Br. J. Nutr., 115*(5), 800–806.

Simanek, V., Walterova, D., Vicar, J., et al., (2001). Silymarin, extract from the milk thistle (*Silybum marianum*) – medicine or a nutraceutical. *Ceska a Slovenska Farmacie., 50*, 66–69.

Singh, J. A., (2014). Challenges faced by patients in gout treatment: A qualitative study. *J. Clin. Rheumatol., 20*(3), 172–174.

Stamp, L. K. O., Donnell, J. L., & Chapman, P. T., (2007). Emerging therapies in the long-term management of hyperuricemia and gout. *Intern. Med. J., 37*(4), 258–266.

Stamp, L. K., Chapman, P., & Frampton, C., (2020). Lack of effect of tart cherry concentrate dose on serum urate in people with gout. *Rheumatology (Oxford), 59*(9), 2374–2380.

Stamp, L. K., O'Donnell, J. L., Frampton, C., Drake, J. M., Zhang, M., & Chapman, P. T., (2013). Clinically insignificant effect of supplemental vitamin C on serum urate in patients with gout: A pilot randomized controlled trial. *Arthritis Rheum., 65*(6), 1636–1642.

Sundy, J. S., Baraf, H. S., & Yood, R. A., (2011). Efficacy and tolerability of pegloticase for the treatment of chronic gout in patients refractory to conventional treatment: Two randomized controlled trials. *JAMA, 306*(7), 711–720.

Sungthong, B., Manok, S., Sato, H., & Sato, V. H., (2016). Effects of *Aquilaria Crassna* on xanthine oxidase activity *in vitro* and hyperuricemic mice. *Indian J. Pharm. Sci., 78*(4), 547–552.

Tausche, A. K., Jansen, T. L., Schröder, H. E., Bornstein, S. R., Aringer, M., & Ladner, U. M., (2009). Gout—Current diagnosis and treatment. *Dtsch. Arztebl. Int., 106*(34, 35), 549–555.

Teh, C. L., Chew, K. F., & Ling, G. R., (2014). Acute gout in hospitalized patients in Sarawak general hospital. *Med. J. Malaysia., 69*(3), 126–128.

Torres, R. J., & Puig, J. G., (2007). Hypoxanthine-guanine phosphoribosyltransferase (HPRT) deficiency: Lesch-Nyhan syndrome. *Orphanet J. Rare Dis., 2*, 48.

Villa-Rivera, M. G., & Ochoa-Alejo, N., (2020). Chili pepper carotenoids: Nutraceutical properties and mechanisms of action. *Molecules, 25*, 5573.

Wang, G. O., Xu, T., Bu, X. M., & Liu, B. Y., (2012). Anti-inflammation effects of corn silk in a rat model of carrageenin-induced pleurisy. *Inflammation, 35*(3), 822–827.

Yan, Y., Jiang, W., Spinetti, T., et al., (2013). Omega-3 fatty acids prevent inflammation and metabolic disorder through inhibition of NLRP3 inflammasome activation. *Immunity, 38*(6), 1154–1163.

Yuk, H. J., Kim, J. Y., Sung, Y. Y., & Kim, D. S., (2021). Phloroglucinol derivatives from *Dryopteris crassirhizoma* as potent xanthine oxidase inhibitors. *Molecules, 26*(1), 122.

Zhang, M., Zhang, Y., Terkeltaub, R., Chen, C., & Neogi, T., (2019). Effect of dietary and supplemental omega-3 polyunsaturated fatty acids on risk of recurrent gout flares. *Arthritis Rheumatol., 71*(9), 1580–1586.

Zhang, Y., & Zhang, D. Z., (2018). Is coffee consumption associated with a lower level of serum C-reactive protein? A meta-analysis of observational studies. *Int. J. Food Sci. Nutr., 69*(8), 985–994.

Zhang, Y., Cui, Y., & Li, X. A., (2017). Is tea consumption associated with the serum uric acid level, hyperuricemia or the risk of gout? A systematic review and meta-analysis. *BMC Musculoskelet Disord., 18*(1), 95.

Zhang, Y., Neogi, T., Chen, C., et al., (2012). Cherry consumption and decreased risk of recurrent gout attacks. *Arthritis Rheum., 64*, 4004–4011.

Zhang, Y., Yang, T., Zeng, C., Wei, J., Li, H., Xiong, Y. L., Yang, Y., & Ding, X., (2016). Is coffee consumption associated with a lower risk of hyperuricemia or gout? A systematic review and meta-analysis. *BMJ, 6*(7), e009809.

CHAPTER 8

ROLE OF NUTRACEUTICALS IN PREVENTION AND MANAGEMENT OF BURSITIS

PRAGYA PANDEY,[1] SAVVY SHARMA,[1] UTSAV RATHORE[1] and MADAN MOHAN GUPTA,[2]

[1]*Department of Food Technology and Nutrition, School of Agriculture Lovely Professional University, Phagwara, Punjab, India*

[2]*School of pharmacy, Faculty of Medical Sciences, The University of the West Indies St Augustine Trinidad and Tobago. West Indies*

ABSTRACT

Bursitis is a typical reason for outer muscle torment and frequently prompts muscular interview. Bursitis should be recognized from joint inflammation, crack, tendinitis, and nerve pathology. Normal kinds of bursitis incorporate prepatellar, olecranon, trochanteric, and retrocalcaneal. Inflammation is a part of the ordinary host guard instrument that gives assurance from contamination and different affronts. Irritation starts the killing of microorganisms and is associated with the cycles of tissue fix as supporting rebuilding of homeostasis at tainted or harmed destinations. The fiery reaction includes cooperation among numerous cell types and the creation of, and reactions to, various compound. These in-betweens are harming to microbes however may likewise make harm to tissues. It is the inundation of cells into the site of fiery movement and the presence of the provocative go between delivered accordingly that are liable for the cardinal indications of irritation: redness, expanding, hotness, torment and loss of capacity. One of the self-improvement solutions to joint fix is monitoring muscle to fat ratio, getting movement

Nutraceuticals and Bone Health. Deepak Sharma, Madan Mohan Gupta, Anil K. Sharma, Raj K. Keservani, Rajesh K. Kesharwani (Eds.)

and work out, and keeping up with great nutritive dietary plans. Although ongoing drug progresses have further developed treatment of joints, patients with related issues frequently ask about dietary mediations/sources for further development, as they see quick changes in their indications after utilization of specific food varieties. Many nutraceuticals have been proven to have anti-inflammatory effect which can in turn help in the management of bursitis. In this chapter we have discussed some of the significant nutraceutical and their role in the prevention and treatment of bursitis.

8.1 INTRODUCTION

Bursitis is irritation of bursae (when bursae become inflamed). The bursa serves as a cushion between muscles, ligaments, and bones, allowing structures to glide and slide past one another with little friction. These sacs are present all-around human body loaded up with liquid that facilitates friction and rubbing between tissues like bones, muscles, ligaments, and skin (Pathak, 2021). The major bursae are situated close to the ligaments and to the enormous joints, for example, in the shoulders, hips, knees, elbows. Bursitis is generally a transitory condition. It might restrict or reduce the movement of joints. Bursitis for most of the part disappears in half a month with treatment but it can occur again. It develops gradually or be unexpected and serious, particularly on the off chance when an individual has calcium stores. At the point when bones and ligament never again get the important supplements, they become delicate and fragile. Besides, the acids from food tend to gather in the joints, compounding irritation. Stress adds to ferment the body while abundance weight can bother what is going on. Agony of the knee, elbow, hip, and heel is among the most well-known outer muscle or skeletal problem. Septic as well as aseptic bursitis is normal reasons for torment, and they should be separated from joint pain, tendinitis, break, ligament or tendon injury, contamination, and neoplasm (Aaron et al., 2011).

Bursitis emerges from irresistible and non-infectious etiologies and recognizing between the two can challenge. Bursitis emerges from numerous fiery peculiarities, yet disease is the essential concern. Non-infectious etiologies of bursitis incorporate injury as gout, sarcoid, idiopathic calcification, and calcinosis, Raynaud peculiarity, esophageal dysmotility, sclerodactyly, and telangiectasia (CREST) condition (Aaron et al., 2011). A careful history and actual assessment are expected for determination. Subordinate tests are useful in deciding the analysis. Most patients with bursitis can be effectively treated with non-surgical treatment using diet therapy which includes anti-inflammatory and anti-oxidant food sources as at the point when imbalance happens among oxidants and immune system, occurrence of oxidative pressure takes place. This oxidative pressure in cells results in extreme metabolic dysfunctions, including loss of cell unity, compound capacity, genomic security, which eventually lead to pathogenesis of numerous human sicknesses (e.g., irritation, inflammations) (Menon & Sudheer, 2007). The present chapter discusses about various types of bursitis and nutraceuticals can be an effective approach in its management.

8.2 TYPES OF BURSITIS

Bursitis could be classified into six categories on the basis of their location of the body (Figure 8.1).

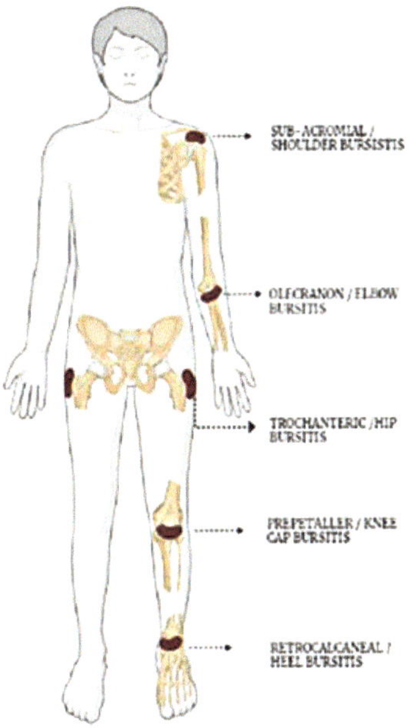

FIGURE 8.1 Types of bursitis.

8.2.1 PREPATELLAR BURSITIS/KNEE CAP

This category of bursitis is average in individuals who are on their knees for work as a great deal. Around 80% of occurrences of infected prepatellar bursitis are conveyed about by *Staphylococcus aureus*. The dermal prepatellar bursa and the low infrapatellar bursa are the two standard bursae about the knee joint. They are typically indicated to all things considered as the prepatellar bursa which lies amid the skin and the patella, and the low infrapatellar bursa lies amid the skin and the tibial tubercle. Non-infectious etiologies of bursitis integrate injury as gout, Raynaud peculiarity, idiopathic calcification, and calcinosis, sarcoid, esophageal dysmotility, sclerodactyly, and telangiectasia (CREST) condition. The method of infection is recognized to be instantaneous immunization, conceivable due to the low blood supply to the bursa, for example, cover layers and handymen (Reid et al., 2010).

8.2.2 OLECRANON BURSITIS/ELBOW BURSITIS

It is conveyed about by the annoyance of the bursa positioned amid the skin and bones of the elbow. Elbow bursitis can be conveyed almost by injury or unswerving strain on the elbow (for instance, while inclining toward a hard surface (Rinkel et al., 2013). Olecranon bursitis is the most widely recognized bursitis. Liquid assortment inside and irritation around the bursa are brought due to damaging, provocative, and enticing procedures. Olecranon bursitis is commonly noncontagious in beginning; infected bursitis represents around 20% of all cases. The olecranon bursa structures after the age of 7 years. Tension from the hard olecranon and shearing powers pragmatic to the covering skin during action might add to bursa arrangement. Serious olecranon bursitis might bring about a thoughtful radiation of the hidden elbow joint. Emanation regularly settles with the executives of the bursitis (Aaron et al., 2011).

8.2.3 TROCHANTERIC BURSITIS/HIP BURSITIS

It is recurrently the consequence of injury, abuse, spinal anomalies, joint pain, or medical procedure. This sort of bursitis is more normal in ladies and moderately matured and more established individuals. Patients with trochanteric bursitis commonly contemporaneous with sidelong hip agony, which might emanate to the buttock, crotch, or low back. Side effects might be exacerbated by ambulation, strolling uphill, step climbing, and ascending from a situated position. Actual assessment might uncover ordinary scope of movement at the hip joint (Lustenberger et al., 2011).

8.2.4 SUB-ACROMIAL BURSITIS

Sub-acromial bursitis is a typical etiology of shoulder torment. It results from aggravation of the bursa, a sac of tissue present under the acromion cycle of the shoulder. It is generally achieved by redundant upward exercises or injury. Sub-acromial bursitis is a disorder brought about by aggravation of the bursa that isolates the prevalent superficial of the supra-spinatus ligament (one of the four ligaments of the rotator sleeve) from the super imposing coraco-acromial tendon, acromion, and coracoid (the acromial curve) and from the profound superficial of the deltoid muscle (Kashid et al., 2020).

8.2.5 KNEE BURSITIS

Knee bursitis also known as Goosefoot bursitis or Pes Anserine bursitis, is a condition of swelling and inflammation of the superficial pre- or infra-patellar bursa sac. The prevalent factors for knee bursitis are microtrauma caused by repetitive or prolonged kneeling on or resting against the knee. Infection or general illnesses could also be contributing factor towards knee bursitis (Draghi et al., 2015).

8.2.6 RETROCALCANEAL BURSITIS

Inflammation of the retrocalcaneal bursa can restrict capacity and cause torment. The Achilles ligament and its hard inclusion might be engaged with extreme cases. This range of sickness has been given many names, including Haglund condition, Albert infection; calcaneus altus, siphon knock, winter heel, and achillodynia (Aaron et al., 2011). There are two types of retrocalcaneal bursitis.

8.2.6.1 RETROMALLEOLAR LIGAMENT BURSITIS

Retromalleolar ligament bursitis is also known as Albert illness. This type of bursitis can result due to injury, disease, or shoes with stiff back support. The bursa where the tendon connects to the heel can get inflamed when excessive pressure in applied to the lower part of the Achilles tendon (Canoso et al., 1998).

8.2.6.2 POSTERIOR ACHILLES TENDON BURSITIS

The bursa between the skin of the heel and the Achilles ligament is affected by this kind of bursitis, also known as Haglund disfigurement. Back Achilles ligament connects the muscles of the lower leg to the heel. It is aggravated by excess and continuous pressure of soft heel tissue towards hard back support of a shoe during (Wiegerinck & colleagues, 2012).

8.3 SIGNS AND SYMPTOMS

Tormentis the most well-known manifestation of bursitis it could develop gradually or be unexpected and serious; It develops gradually or be unexpected and serious, particularly the major chances is when an individual have calcium stores around the bursae forming regions. Most likely when an individual stretch or broaden the joint, the torment starts to begin. Symptoms like expanding of joints, swelling in joints, joints become tender, shows redness, restricted movement, hurts when individual move joints, muscle or press on it, throbbing or firmness occurs in the joint that deteriorates when individual move the joint. Other symptoms such as individual gets handicap, unexpected powerlessness to move joint, inordinate enlarging or rash appears on the impacted region, sudden sharp torment in the effected region. The aggravation might start at the same time or develop after some time (Pathak, 2021; Roberts, 2002).

8.4 ETIOLOGY

Anybody could develop bursitis; however certain variables can increase individual's susceptibility towards it. Some these factors are discussed below.

8.4.1 AGE

The occurrence of bursitis increases with aging, particularly after the age of 40. As person ages, the capacity of ligaments to withstand the pressure decreases. Ligaments become less flexible and more straightforward to tear (Pathak, 2021).

8.4.2 OCCUPATIONS OR SIDE INTERESTS

The occupation of an individual could be a huge factor of bursitis. During work, there could be more strain for a continuous period of time on specific bursae, individual's possibility of creating bursitis escalate in works such as tile setting, planting, painting and playing an instrument.

8.4.3 OTHER AILMENTS

Certain foundational infections and conditions – like rheumatoid joint pain, gout and diabetes – increment individual's possibility of creating bursitis. Being overweight can expand individual's possibility of creating hip and knee bursitis. It's typically brought about by persistent tension on apart or by utilizing a joint disproportionately. High hazard exercises incorporate cultivating, raking, carpentry, scooping, scouring, tennis, golf, skiing, and tossing. Individual can likewise get bursitis by sitting or standing the incorrect way for quite a while at work or home, or by not extending adequate before work out. Unexpected injury can some of the time cause bursitis. Assuming there's an issue with the construction of a bone or joint, (for example, legs that are various lengths or joint inflammation in a joint), that can place more weight on a bursa, causing bursitis (Roberts, 2002).

8.5 DIAGNOSIS

Bursitis could be diagnosed by various tests. Total clinical history and actual test, analytic tests for bursitis might include, X-beam, Magnetic resonance imaging (MRI), ultrasound and aspiration technique (involving the utilization of a slim needle to aspirate liquid from the enlarged bursa to check for infection) and Blood tests (C-reactive protein (CRP) and erythrocyte sedimentation rate (ESR)) (Aaron et al., 2011).

8.6 ROLE OF NUTRACEUTICALS

Nutraceutical is a broad term that refers to foods and food products providing additional health benefits. Nutraceuticals are aimed at preventing rather than curing diseases. This product category has gotten a lot of attention because of its safe medicinal effects in addition to its nutritional benefits. They have the potential to prevent a variety of life-threatening diseases, including diabetes, cancer, and neurodegenerative disorders. Nutraceuticals (Keservani et al., 2010a, 2010b, 2017, 2020) include dietary supplements (proteins, vitamins, and minerals), bioactive compounds from natural sources. Inflammation is the biological response to a detrimental stimulus. While inflammation is beneficial in quenching the foreign substance it also damages body's own cells. As bursitis is an inflammatory condition, nutraceutical having anti-inflammatory effect could play a significant role in their management. Some of the important nutraceuticals has been discussed below for inflammation prevention and therapy in bursitis.

8.6.1 OMEGA-3 UNSATURATED FATS

Fatty fish are an incredible source of protein and the long-chain omega-3 unsaturated fats such as eicosapentaenoic corrosive (EPA) and docosahexaenoic corrosive (DHA). These fats can be found in slick fish and fish oil supplements. Omega-3 unsaturated fats can be found in a variety of seafood, including anchovies, mackerel, sardines, salmon, and herring. Fish oil or flaxseed oil are two of the common sources. These unsaturated fats have the ability to suppress several aspects of discomfort. Resolvins and protectins, which are produced by the EPA and DHA, help to reduce irritants. The properties of n-3 unsaturated fats include changing cell layer phospholipid, unsaturated fat structure, lipid pontoon disturbance, inhibition of inflammation, and aiding in the maintenance of atomic variable kappa B for reducing the articulation of fiery components of joint inflammation. Meta-studies back up preliminary clinical findings of fish oil in patients with joint problems (Calder, 2013).

8.6.2 GLUCOSAMINE SULFATE

Ligament, the tissue that covers the ends of the bones in a joint, contains glucosamine. It may help treat osteoarthritis aggravation, according to some evidence. It may also aid in the relief of bursitis pain. Glucosamine sulfate, like non-steroidal palliative medications, has been shown to have interesting moderating effects. Oral dose of glucosamine sulfate is used to treat a painful condition caused by ligament irritation, degradation, and probable (Rovati et al., 2016).

8.6.3 VITAMIN B12

Cyanocobalamin may have anti-inflammatory or moderating properties. They are useful in the pursuit of a "bursa well-disposed" diet. It can be found in food sources and functions as an anti-inflammatory substance for the benefit of bone, joints, and ligaments. Broccoli, spinach, and bananas are examples of vitamin B-rich foods that can be used to alter one's diet.

8.6.4 VITAMIN C

Vitamin C, also known as L-ascorbic acid, is fundamental for the development, growth and collagen synthesis of all body tissues. It is engaged with many body physiological functions, including arrangement of collagen, retention of iron, wound healing, and plays an important role of supporting ligament, bones and teeth. L-ascorbic acid with flavonoids assist in protecting connective tissue (like ligament). Foods high in L-ascorbic acid, as well as Omega 3-rich foods, have been reported to help strengthen the bursa sac. It also acts as a co-factor in the collagen matrix, which is the main protein found in joint tissue and bone. L-ascorbic acid aids in the fight against infection and may help to manage disease-related aggravation, which can cause joint inflammation eruptions. In addition, it appears to direct the immune system response in rheumatoid joint arthritis and aid in the prevention of continuous condition deterioration (Maleki et al., 2019).

8.6.5 VITAMIN E

There are four tocopherols and four tocotrienols in the vitamin E family. Tocopherol (T) is the most common form of vitamin E found in tissues and organs. Recent research shows that other forms of vitamin E, such as γ-tocopherol (γT), -tocopherol, and γ-tocotrienol, have exceptional cell reinforcing and reposeful effect that outperform αT in the prevention and treatment of persistent infections. Other vitamin E structures are used in conjunction with these vitamin E compounds to suppresspro-inflammatory signaling. Long-chain carboxychromanols, especially 130-carboxychromanols, have been shown to have stronger moderating effects than non-metabolized nutrients. Recent research suggests that γT and tocotrienols may be beneficial in the treatment of irritation-related illnesses and inflammations (Jiang, 2014).

8.6.6 BROMELAIN

A catalyst derived from pineapples reduces irritability. Bromelain appears to help with osteoarthritis by reducing pain and improving joint capacity when combined with trypsin and rutin. Bromelain has also been shown to help with knee pain. Bromelain may increase the level of anti-microbial efficacy in the body when combined with anti-infection drugs, which could be dangerous. Hence, turmeric is also combined with bromelain in a low dose, which strengthens the effects of bromelain (Varilla, 2021).

8.6.7 VITAMIN D

The most well-known aspects of the vitamin D metabolic flagging system are its conventional effects on outer muscle health. Vitamin D is linked to calcium absorption, immunity, and the protection of bone, muscle, and heart health. It is naturally available in food sources, and the body can also produce it when the skin is exposed to sunlight. A working vitamin D metabolite is involved in reducing irritation by controlling incendiary and restraining the expansion of supportive fiery cells, both of which are important and might be found in patients. Vitamin D deficiency appears to affect bacterial and viral illness sensitivity and severity (Bishop et al., 2021).

8.6.8 SPICES

Spices are a well-known method of strengthening the body and healing illness. It has been established that people who consume spices have a significant impact on the improvement of enduring infections. The functional elements in these flavors can inhibit provocative pathways that lead to continued aggravation, adding to the flavors' inherent qualities. Spices, on the other hand, can cause side effects and interact with other spices, enhancers, or prescriptions.

8.6.9 BOSWELLIA (BOSWELLIA SERRATE)

It is frequently used in Ayurveda, the traditional Indian medicine. In the Western world, Boswellia is known as frankincense. This compound contains anti-arthritic, pain-relieving, and moderating effects. Osteoarthritis, rheumatoid joint inflammation, bursitis, and tendonitis have all been treated with it. Boswellia tar contains compounds that may reduce irritation and increase the safe response. (Siddiqui, 2011).

8.6.10 TURMERIC (CURCUMA LONGA)

Turmeric is frequently used with bromelain because it enhances the effects of bromelain. Turmeric contains curcumin, a powerful relaxing bioactive compound. Turmeric has been demonstrated in studies to help with joint inflammation, diabetes, and a variety of illnesses (Menon&Sudheer, 2007).

8.6.11 WHITE WILLOW (SALIX ALBA)

White willow (Salix alba) serves as an anti-inflammatory agent. Willow bark has traditionally been used as a remedy for pain, migraines, and inflammatory conditions such as bursitis and tendinitis. Salicin, a compound found in white willow bark, is used to promote anti-inflammatory therapy. It's usually turned into a tea by boiling half a teaspoon (2 grams) of bark in eight ounces of water. Drink up to 5 cups of water every day. Turmeric and white willow can also be used to prevent enlargement (Shara & Stohs, 2015).

8.6.12 GREEN TEA

Green tea and its extracts have a broad range of beneficial activities in the treatment of osteoarthritis and joint pain by altering the synthetic compounds in a damaged joint to point toward mending and closing down harmful variables. It is this line of inquiry that researchers are considering. Green tea concentrate appears to be used as an additional treatment for both pain relief and improving knee joint capacity in adults (Hashempur et al., 2018).

8.7 CONCLUSION

Bursitis is an inflammation of the liquid-filled sacs (bursae) that are commonly found near joints. It is excruciatingly painful and restricts the movement. An excess of strain on one of these sacs can cause discomfort (a bursa). There are approximately 100 bursae in the human body, with many of them located near joints. Bursitis can also be caused by an infection, which is usually bacterial. Anti-inflammatory foods are helpful in reducing irritation. The most well-known symptom of bursitis is pain. It is critical to adjust a healthy lifestyle to help kill acids (particularly extending, strolling and swimming and keeping up with healthy eating routine). When someone is in pain and continues to move, their condition will worsen. Enhancements and treatments, such as fat fish, such as herring, mackerel, salmon, fish, and sardines, contain undeniable amounts of Omega 3. These lipids have a relaxing effect on the joints, and other mitigating and preventative agents, such as vitamin (A, B, C, D, E) and flavors, such as turmeric and green tea, play a key role in reducing aggravation. Thus, properties like anti-inflammatory; anti- oxidative plays a major role in the treatment for bursitis. (Bustamante et al., 2020). The ability of nutraceuticals to improve the bursitis condition may plausibly transfer to the ability of these compounds to be used as dietary strategies for prevention and and/or treatment of infections in bursitis.

KEYWORDS

- **C-reactive protein**
- **docosahexaenoic**
- **eicosapentaenoic**
- **erythrocyte sedimentation rate**
- **magnetic resonance imaging**

REFERENCES

Aaron, D. L., Patel, A., Kayiaros, S., & Calfee, R., (2011). Four common types of bursitis: Diagnosis and management. *American Academy of Orthopaedic Surgeon, 19*, 359–367.

Bishop, L. E., Ismailova, A., Dimeloe, S., Hewison, M., & White, J. H., (2020). Vitamin D and immune regulation: Antibacterial, antiviral, anti-inflammatory. *JBMR Plus, 5.*

Bustamante, M. F., Agustín-Perez, M., Cedola, F., Coras, R., Narasimhan, R., Golshan, S., & Guma, M., (2020). Design of an anti-inflammatory diet (ITIS diet) for patients with rheumatoid arthritis. *Contemporary Clinical Trials Communications, 17*, 100524.

Calder, P. C., (2013). Omega-3 polyunsaturated fatty acids and inflammatory processes: Nutrition or pharmacology? *British Journal of Clinical Pharmacology, 75*, 645–662.

Canoso, J. J., Liu, N., Traill, M. R., & Runge, V. M., (1988). Physiology of the retrocalcaneal bursa. *Annals of the Rheumatic Diseases, 47*, 910–912.

Draghi, F., Corti, R., Urciuoli, L., Alessandrino, F., & Rotondo, A., (2015). Knee bursitis: A sonographic evaluation. *Journal of Ultrasound, 18*, 251–257.

Hashempur, M. H., Sadrneshin, S., Mosavat, S. H., & Ashraf, A., (2018). Green tea (*Camellia sinensis*) for patients with knee osteoarthritis: A randomized open-label active-controlled clinical trial. *Clinical Nutrition, 37*, 85–90.

Jiang, Q., (2014). Natural forms of vitamin E: Metabolism, antioxidant, and anti-inflammatory activities and their role in disease prevention and therapy. *Free Radical Biology and Medicine, 72*, 76–90.

Jimenez-Palop, M., (2002). Olecranon bursitis due to *Candida parapsilosis* in an immunocompetent adult. *Annals of the Rheumatic Diseases, 61*, 279–281.

Kashid, M., Rai, S. K., & Chakraborty, B., (2020). Rare presentation of subacromial bursitis mimics neoplasm in a case of rheumatoid arthritis. *J. Orthopcase Rep.*, *9*(6), 23–26.

Keservani, R. K., Kesharwani, R. K., Sharma, A. K., Vyas, N., & Chadoker, A., (2010b). Nutritional supplements: An overview. *International Journal of Current Pharmaceutical Review and Research, 1*(1), 59–75.

Keservani, R. K., Kesharwani, R. K., Vyas, N., Jain, S., Raghuvanshi, R., & Sharma, A. K., (2010a). Nutraceutical and functional food as future food: A review. *Der Pharmacia Lettre, 2*(1), 106–116.

Keservani, R. K., Sharma, A. K., & Kesharwani, R. K., (2017). An overview and therapeutic applications of nutraceutical and functional foods. *Recent Advances in Drug Delivery Technology*, 160–201. ISBN:9781522507543.

Keservani, R. K., Sharma, A. K., & Kesharwani, R. K., (2020). "*Nutraceuticals and Dietary Supplements: Applications in Health Improvement and Disease Management*. CRC Press. ISBN: 9781771888738.

Lustenberger, D. P., Ng, V. Y., Best, T. M., & Ellis, T. J., (2011). Efficacy of treatment of trochanteric bursitis: A systematic review. *Clinical Journal of Sport Medicine, 21*, 447–453.

Maleki, S. J., Crespo, J. F., & Cabanillas, B., (2019). Anti-inflammatory effects of flavonoids. *Food Chemistry, 299*, 125124.

McFarland, E. G., Mamanee, P., Queale, W. S., & Cosgarea, A. J., (2000). Olecranon and prepatellar bursitis: Treating acute, chronic, and inflamed treatment varies with bursitis, patient activities, and presence of infection. *The Physician and Sportsmedicine*.

Menon, V. P., & Sudheer, A. R., (2007). Antioxidant and anti-inflammatory properties of curcumin. *Advances in Experimental Medicine and Biology*, 105–125.

Pathak, N. (2023). *Bursitis: Types, Causes, Symptoms, Treatment, and Prevention* https://www.webmd.com/pain-management/arthritis-bursitis (accessed on 25 July 2023).

Reid, C. R., Bush, P. M. C., Cummings, N. H., McMullin, D. L., & Durrani, S. K., (2010). A review of occupational knee disorders. *Journal of Occupational Rehabilitation, 20*, 489–501.

Rinkel, W. D., Schreuders, T. A. R., Koes, B. W., & Huisstede, B. M., (2013). Current evidence for effectiveness of interventions for cubital tunnel syndrome, radial tunnel syndrome, instability, or bursitis of the elbow. *The Clinical Journal of Pain, 29*, 1087–1096.

Rovati, L. C., Girolami, F., D'Amato, M., & Giacovelli, G., (2016). Effects of glucosamine sulfate on the use of rescue non-steroidal anti-inflammatory drugs in knee osteoarthritis: Results from the pharmaco-epidemiology of gonarthrosis (Pegasus) study. *Seminars in Arthritis and Rheumatism, 45*.

Shara, M., & Stohs, S. J., (2015). Efficacy and safety of white willow bark (*Salix alba*) extracts. *Phytotherapy Research, 29*, 1112–1116.

Siddiqui, M. Z., & Boswellia, S., (2011). A potential anti-inflammatory agent: An overview. *Indian J. Pharm Sci.*, *73*(3), 255–226.
Vallianou, N. G., (2014). Honey and its anti-inflammatory, anti-bacterial and anti-oxidant properties. *General Medicine: Open Access, 2.*
Varilla, C., Marcone, M., Paiva, L., & Baptista, J., (2021). Bromelain, a group of pineapple proteolytic complex enzymes (*Ananas comosus*) and their possible therapeutic and clinical effects: A summary. *Foods, 10*, 2249.
Wiegerinck, J. I., Kerkhoffs, G. M., Van, S. M. N., Sierevelt, I. N., & Van, D. C. N., (2012). Treatment for insertional Achilles tendinopathy: A systematic review. *Knee Surgery, Sports Traumatology, Arthroscopy, 21*, 1345–1355.

CHAPTER 9

CARPAL TUNNEL SYNDROME MANAGEMENT BY NUTRACEUTICALS

RUHUL AMIN, FARUK ALAM, DURGAPRASAD KEMISETTI, DHRUBAJOYTI SARKAR and BIPLAB KUMAR DEY

Faculty of Pharmaceutical Science, Assam Down Town University, Panikhaiti, Gandhinagar, Guwahati, Assam, India

ABSTRACT

The term *nutraceutical* has emerged from the words Nutritional and Pharmaceutical terms. Earlier nutritional diet was preferred for patients and drugs were prescribed. As of now, the trend has been changed from earlier pattern and started including the active constituents in food, which serve the purpose of avoiding or treating certain diseased conditions or abolish certain conditions like Obesity. Carpel Tunnel Syndrome (CTS), a disorder of hand numbness, inflammation. Usage of certain analgesics or anti-inflammatory drugs provided certain relief temporarily. Anatomy, Mechanism, Pathophysiology, Epidemiology of CTS was discussed in detail in the main topic under their respective headings. Few of the risk factors which are linked with CTS were enlightened like obesity, diabetes, gender, hypothyroidism, rheumatoid arthritis, workplace factors, connective tissue diseases, etc. For treatment of CTS, some of the following have to be incorporated in preparation of foods like Red bell peppers, spinach, salmon, walnuts, pineapple, turmeric. The active constituents present in the above, when included in food at certain concentration may relieve from CTS. The chemical constituents present in them include omega-3-fatty acids, vitamin B6, curcumin, α-lipoic acid, eicosapentaenoic acid, docosahexaenoic acid, capsaicin, α-linonlenic acid, vitamin B12 and bromelain. Avoid junk food and include an appropriate or

Nutraceuticals and Bone Health. Deepak Sharma, Madan Mohan Gupta, Anil K. Sharma, Raj K. Keservani, Rajesh K. Kesharwani (Eds.)

required quantities of the chemical constituents in food and get relieved from CTS. Hence, it can be concluded that nutraceuticals play an important role and serve as prophylactic treatment of certain diseases.

9.1 INTRODUCTION

Nutraceuticals have been used by humans for a long time, but very lately has been scientifically proven nutritional and medical data permitted them to emerge as potentially useful (Dillard & German, 2000). Non-specific biological treatments, such as nutraceuticals, are used to boost overall health, regulate symptoms, and prevent malignant processes. Nutritional and Pharmaceutical terminology had been combined to form the phrase "Nutraceutical," which refers to as a dietary supplement that contains both nutrients and pharmacologically active ingredients (Keservani et al., 2010a, 2010b, 2017, 2020). In the year 1989, Stephen DeFelice, the founder and chairman of the Foundation for Innovation in Medicine in Cranford, New Jersey, had named Nutraceutical.

A new diet health paradigm has emerged during these years, emphasizing the beneficial elements of nutrition. With more modern way of living trend, a shift in people's fundamental eating pattern also had changed. One of the most common causes of illness is poor nutrition, which has led to a rise in the consumption of junk foods. Obesity had been increasingly acknowledged as a worldwide problem. In most developing nations, heart disease remains the lead cause of mortality, followed by cancer, osteoporosis, arthritis, carpal tunnel syndrome (CTS), and a host of other conditions. Modern medicine's price, high-tech disease-treatment strategy is driving consumers to seek out complementary and alternative foods, and carefully managed type of care makes nutraceuticals especially enticing. The 2,500-year-old dictum of Hippocrates, "Let food be the medicine and medicine not be the food," is unquestionably still relevant today. Emerging natural compounds known as nutraceuticals help blur the border between food and medicine use (Adelaja & Schilling, 1999).

In the future, the food industry would be able to develop new food items using nutraceuticals derived from plants and animals. A broad range of stakeholders, including consumers, healthcare practitioners, regulators, food manufacturers, and distributors, have an interest in the function of nutraceuticals in the human diet. Anti-arthritic, carpal tunnel syndrome (CTS), cough, cold, insomnia, digestion, prevention of few malignancies, osteoporosis, hypertension, managing cholesterol levels, analgesic, depression, diabetes are just a few of the many therapeutic areas covered by nutraceuticals (Pandey, Verma & Saraf, 2010; Rao & Naidu, 2016).

CTS, a disorder characterized by discomfort, numbness, and inflammation in hands and wrists, as well as the arms. Splints and analgesics may provide relief from the symptoms of CTS (Alfonso, Jann, Massa & Torreggiani, 2010). Nutraceuticals, on the other hand, may be able to alleviate the symptoms of conditions like CTS if added to the diet, and in selected foods and consumed.

By compressing the median nerve, it travels through the carpal tunnel and cause carpal tunnel syndrome (CTS). Common symptoms associated with median nerve include Pain, abnormal sensations like tingling and pricking, sometimes weakness also (Genova, Dix, Saefan, Thakur & Hassan, 2020).

9.2 ANATOMY

Transverse carpel ligament (flexor retinaculum) supports the carpal bones which creates the carpal tunnel. Median nerve and nine flexor tendons of the forearm muscle must be traversed through this anatomical tube. When the median nerve is squeezed, ischemia and mechanical damage results (Jillapalli & Shefner, 2005; Keir & Rempel, 2005).

The median nerve arises from the brachial plexus and travels via the C6-C7-C8-T1 nerve roots to reach its final destination. There are median sensory fibers that detect the fleshy portion of your thumb and the first three and a half fingers of your hand from these roots. The median nerve's motor fibers are supplied via the C8 and T1 roots in the forearm and hand (Maleki et al., 2014). Those nerve fibers that were supposed to be part of the median nerve are diverted to the brachial plexus due to the aforementioned factors. The brachial plexus fibers C5-C6 join to produce the median nerve, which does not have any branches in the upper arm. It comes within a few millimeters of the brachial artery on the medial side as it passes through the antecubital fossa area. The pronator teres, flexor carpi radialis, palmaris longus, and flexor digitorum superficialis are all supplied by the median nerve in the upper forearm. The anterior interosseous nerve must be developed from the median nerve in order to feed the wrist's pronator quadratus muscle. Flexor pollicis longus and the medial heads of the flexor digitorum profundus I and II muscles are likewise supplied by the anterior interosseous nerve. In the forearm, the anterior interosseous nerve and the median nerve run side by side. The palmar cutaneous sensory branch of the median nerve branches off at the wrist to give feeling on the palm's lateral side. Because the median nerve's palmer branch travels above the carpal tunnel rather than through it, pain or numbness in the thenar eminence might go unnoticed in people with carpal tunnel syndrome (Ghasemi-Rad et al., 2014).

The carpal tunnel is the passageway through which the median nerve and the tendons of the nine flexor muscles of the wrist travel. The median nerve is most often inflamed and compressed in the carpal tunnel. During rare circumstances, medial entrapment may also develop closer to elbow or away from the elbow. The median nerve leaves the carpal tunnel and feeds the hand with both motor and sensory nerves. These hand muscles: abductor pollicis brevis, superficial head flexor pollicis, opponens pollicis, and 1st and 2nd lumbricals are all supplied with sensory input via the median nerve. Despite the fact that these muscles may be affected, the lumbrical may be spared in persons with severe or longstanding CTS. The anatomy described above is most common, however anomalous innervations in the hand may complicate diagnosis in certain instances (Oh et al., 2008).

9.3 PATHOPHYSIOLOGY

The etiology of CTS is unclear. Clincial CTS is associated with elevated intracarpal canal pressure (Maleki et al., 2014). Experimental research suggests that anatomic compression or inflammation may be important factors in CTS, while the actual etiology is uncertain. This condition may be triggered by increased pressure in the carpal tunnel itself, as well as by axonal transport problems or venous compression in the perineurium. The carpal tunnel houses the median nerve as well as nine flexor tendons, all of which may be inflamed or swollen. Subsynovial connective tissue around flexor tendons may become inflamed and induce anatomic compression (Duncan, Bhate & Mustaly, 2017). Inborn with a decreased the anatomical space, mass lesions like cyst, tumor, or persistent median artery, and inflammation induced by systemic illness (such as rheumatoid arthritis) are all possible sources of compression. Carpal tunnel syndrome may be exacerbated or averted if the wrist is held in an unnatural posture. Lowest pressure in carpal tunnel is observed, when wrist is in a neutral or slightly flexed posture, and the pressure increases as the wrist moves away from its position (Keir, Wells, Ranney & Lavery, 1997; Seradge, Jia & Owens, 1995; Szabo & Chidgey, 1989; Weiss, Gordon, Bloom, So & Rempel, 1995). Endoneurial and perineurial edema and thickening of vessel walls, non-inflammatory synovial fibrosis and proliferation of vessels, fibrillation, myelin thinned out and degeneration and regeneration of nerve fibers have all been seen in human compression (Donato et al., 2009; Mackinnon, Dellon, Hudson & Hunter, 1986). There have been surgical reports of a "hourglass" shaped abnormality of the median nerve in the carpal tunnel region. Nerve weakness and swelling in the distal and proximal portions of the transverse carpal tunnel ligament may be the cause of this deformity. The increased expression of prostaglandin E2 and vascular endothelial growth factor (VEGF) has been associated with vascular proliferation and fibrosis. However, these factors are uncertain about their role in CTS (Abate et al., 2009; Duncan et al., 2017).

9.4 EPIDEMIOLOGY

Carpal tunnel syndrome affects large percentage of adults. Among 1,00,000 population, 324 to 542 women, and 125 to 303 men are expected to suffer from CTS each year (Atroshi, Englund, Turkiewicz, Tägil & Petersson, 2011; Gelfman et al., 2009). CTS is rare among children (Davis & Vedanarayanan, 2014; Eun, Kim, Hwang & Choi, 2019), however the frequency remains unknown. As per the information provided, CTS affects in between 1% to 5% of general population (Atroshi et al., 1999; De Krom, Kester, Knipschild & Spaans, 1990). CTS, is more common in women (0.7 to 9.2%) than in men (0.4 to 2.1%) (Atroshi et al., 1999; De Krom et al., 1992; Stevens, Sun, Beard, O'fallon & Kurland, 1988; Stevens, Witt, Smith & Weaver, 2001). Women who are overweighed or obese are more likely to suffer from CTS than men who are slim or average in weight (Werner, Albers, Franzblau & Armstrong, 1994). In two different studies among larger groups (Atroshi et al., 1999), female-to-male ratio for CTS prevalence was around 3:1.

- Risk Factors:

CTS has been linked with few of the following parameters:

- Obesity
- Diabetes
- Female gender
- Rheumatoid arthritis
- Pregnancy
- Aromatase inhibitor use
- Hypothyroidism
- Pre-existing median mononeuropathy
- Connective tissue diseases

- Genetic predisposition
- Workplace factors

There is an increased risk of median neuropathy at the wrist and symptoms of Chronic Traumatic Encephalopathy, if a person is overweighed or obese (CTS). On the other hand, several studies have shown that no link exists between the two (Kurt et al., 2008). According to many studies, CTS is more frequent in women than in males. (For further information on epidemiology, refer above section). Gender inequity may be explained as physical differences between men and women. Compared to men, women's proximal carpal tunnels are smaller in size (Keir & Rempel, 2005). This suggests that women with CTS had smaller cross-sectional areas than those without the condition. A larger proportion of people with diabetics (Albers, Brown, Sima & Greene, 1996; Leach & Odom Jr, 1968), rheumatoid arthritis patients and those with connective tissue disorders (e.g., lupus), pregnant women (Padua et al., 2001; Luca Padua et al., 2010), and thyroid disease patients suffer from CTS. CTS was shown to be linked with diabetes (OR 2.2, 95% CI 1.5–3.1), hypothyroidism (OR 1.4, 95% CI 1–2), and rheumatoid arthritis in a systematic study in the year 2003 (Van Dijk, Reitsma, Fischer & Sanders, 2003). In 2014, meta-analysis of five studies, hypothyroidism was revealed to be associated with CTS (effect size 1.44, 95% CI 1.27–1.63) (Shiri, 2014). Pregnancy-induced CTS, is very certainly brought on by fluid retention (L Padua et al., 2001). According to one study, up to one third of those, with acromegaly experienced bilateral median nerve compressions. These outcomes are a consequence of soft tissue expansion (synovial edoema and tendon hyperplasia).

Patients with a pre-existing median mononeuropathy have a higher likelihood of acquiring CTS, as shown by widely accepted electrodiagnostic tests (i.e., a prolonged median sensory delay). Pressure palsy is a risk factor for CTS in patients with a hereditary neuropathy. According to few studies, there may be a genetic component to CTS. According to a study of twins, up to half of women's risk for CTS is due to their genes. People who have bilateral CTS are more prone to pass the ailment down to their family as a hereditary (Alford, Weiss & Akelman, 2004). There may be a genetic anatomical difference in the size of the tunnel or a family history of medical issues, when the carpal tunnel is different in size (such as type 2 diabetes). Hereditary neuropathy with pressure palsy (HNPP), an episodic demyelinating neuropathy, occurs very often. As the most prevalent symptom, nerve palsies are the most common. Nerves such as the brachial plexus and the axillary and medial branches, as well as the plexus itself, are often damaged.

Single-nerve palsy symptoms might last anywhere from a few days to few month's time. "Charcot-Marie-Tooth disease: Genetics, clinical features, and diagnosis" discusses 'Hereditary neuropathy with vulnerability to pressure palsy.' Arthralgias and CTS may be worsened, if you use aromatase inhibitors (Nishihori et al., 2008; Spagnolo, Sestak, Howell, Forbes & Cuzick, 2016). Breast cancer patients on anastrozole, an aromatase inhibitor, have shown to have significantly more occurrences of CTS in comparison to those patients using tamoxifen (Spagnolo et al., 2016). The majority of CTS sufferers did not need any kind of treatment. Another case-control study found that using aromatase inhibitors was associated with significantly increased tendon thickness on sonograms when compared to those who did not take the drug. The mechanism through which aromatase inhibitors induce CTS seems to be by thickening of the tendon.

9.5 BEST FOODS FOR CTS

Carpal tunnel syndrome (CTS) symptoms may be improved by eating right foods. As a suggestion, we should begin to include in our diets from now.

9.5.1 RED BELL PEPPERS

In order to cure carpal tunnel syndrome, it is necessary to decrease inflammation, and foods, high in antioxidants accomplish this objective. Red peppers and other colorful food, such as tomatoes, carrots, yellow peppers, and leafy greens, should be a regular part of your diet. Try to include a rainbow of veggies into your diet every day.

9.5.2 SPINACH

Supplemental vitamin B6 treatment may provide alleviation for those with carpal tunnel syndrome (Ryan-Harshman & Aldoori, 2007). However, B6 may also be present in a variety of foods, including spinach. Don't take more than 200 mg a day of vitamin B6 since it also acts as an analgesic. Carpal tunnel syndrome sufferers say it helps them feel better. Fresh spinach salad, spinach pie, or an omelet stuffed with spinach are all good sources of vitamin K. Citrus fruits and vegetables including oranges, bananas, cauliflower and cantaloupe all contain the vitamin.

9.5.3 *SALMON*

Carpal tunnel syndrome and other inflammatory conditions, such as rheumatoid arthritis, may make symptoms worse, but salmon can assist by reducing the inflammation. Salmon and other fatty fish, such as tuna and sardines, are richest sources of two omega-3 fats termed as **Eicosapentaenoic acid** (EPA) and Docosahexaenoic acid (DHA), which may help to reduce inflammation. Use of fish oil supplements may cause a benefit for Carpal tunnel syndrome.

9.5.4 *WALNUTS*

For their higher content of omega-3 fatty acids (ALA), walnuts may also reduce inflammation of the body and alleviate symptoms of carpal tunnel syndrome, although probably not to the same degree as foods rich in **Eicosapentaenoic acid** (EPA) and Docosahexaenoic acid (DHA). 'Flaxseed and chia seeds, as well as other foods rich in ALA, may help in reducing inflammation.' Make a salad with walnuts, stack them with fruit and fat-free or low-fat Greek yogurt for a parfait, or crush and use to 'crust' chicken or fish filets before baking.

9.5.5 *PINEAPPLE*

Fresh pineapple, chopped, diced, or sliced, may help prevent the symptoms of carpal tunnel syndrome. Bromelain, an enzyme found in pineapple, has anti-inflammatory properties. If you're looking for an extra health benefit, pineapple is also an excellent provider of vitamins and digestive help. Combine pineapple, papaya, coconut nectar, and blueberries in a fruit salad. Fresh ginger, which has anti-inflammatory effects, adds a burst of flavor to the dish.

9.5.6 *TURMERIC*

Curcumin, an anti-inflammatory molecule found in this unusual and tasty spice, may help alleviate pain. It may be used to curries and mixed into a pilaf or used as a seasoning for grilled meats, fish or vegetarian grain and bean meals. With the addition of black pepper and ginger, the anti-inflammatory qualities are enhanced.

9.6 BENEFICIAL FOODS FOR CTS

Carpal tunnel syndrome symptoms may be alleviated by lowering inflammation, according to a study published by Kiecolt-Glaser (2010).

9.6.1 *OMEGA-3 FATTY ACIDS*

Omega-3 fatty acids reduce inflammation and protect the brain from damage. When it comes to carpal tunnel syndrome, researchers discovered that consuming omega-3 fatty acids decreased the numbness and discomfort. Fatty fish such as salmon, tuna, mackerel and mackerel oil, sardines, herring, and herring oil all are rich in omega-3 fatty acids.

9.6.2 *ANTIOXIDANTS*

Carpal tunnel syndrome may be alleviated by consuming foods which have high content of antioxidants. Fruits and vegetables rich in phytonutrients include: berries including blueberries, strawberries, and goji berries.

Vitamin C-rich foods include citrus fruits like grapefruits and tangerines; red bell peppers; herbs (Rane et al., 2020a; Rane et al., 2020b; Rane et al., 2020c; Sharma et al., 2020) such as parsley and turmeric; walnuts; pistachios; salmon; and other fatty fish.

9.6.3 *VITAMIN B6*

Vitamin B6 supplementation for carpal tunnel syndrome is a continuous therapy option. Apart from its potential to alleviate carpal tunnel syndrome symptoms, it is sometimes advised by doctors for overall health improvement.

B6 may serve as an analgesic by boosting the pain threshold in persons with carpal tunnel syndrome, but the exact mechanism by which it does so is unknown (Talebi et al., 2013). Salmon, pork, poultry (chicken and turkey), bananas, soybeans, wheat germ, spinach, chickpeas, and potatoes are some of the foods that have a high amount of B6.

9.6.4 *VITAMIN B12*

When it comes to alleviating the symptoms of nerve injury, vitamin B12 supplementation may be of use. Clams, liver, salmon, tuna, eggs, and trout are all good sources of this vitamin.

9.6.5 TURMERIC

Curcumin, an antioxidant, anti-inflammatory, and neuroprotective compound in turmeric is found to be of use as spice (Pajardi, Bortot, Ponti & Novelli, 2014). Carpal tunnel syndrome and other inflammatory disorders, such as arthritis, have been shown to get benefit from usage of turmeric, according to its users.

Turmeric and black pepper may be combined for improved results. Increased bioavailability of curcumin may be achieved by adding a little quantity of pepper to turmeric. Carpal tunnel syndrome may be alleviated by consuming turmeric tea, which has anti-inflammatory properties.

9.7 SUPPLEMENTARY FOOD FOR CTS

Consuming supplements such as vitamin B12 and B6, fish oil capsules rich in omega-3 fatty acids and antioxidants, and turmeric, or curcumin, may give some assistance, if your regular diet is short in vitamins and minerals among the foods mentioned above, Supplements are an option. If your diet isn't providing you the nutrients you need, Supplements should be provided. Structures of, α-lipoic acid have been shown to alleviate symptoms of carpal tunnel syndrome (Di Geronimo, Caccese, Caruso, Soldati & Passaretti, 2009). Lower quantities of it may be found in various meals, such as spinach and organ meat. There are many compounds in hot red peppers that have been shown to have pain-relieving qualities, including capsaicin (Fattori, Hohmann, Rossaneis, Pinho-Ribeiro & Verri, 2016).

9.8 FOODS TO BE AVOIDED DURING CTS

Carpal tunnel syndrome is a debilitating ailment that occurs when the median nerve is swollen and inflamed in the wrist and the tendons of the fingers. Pain, tingling, and numbness are all symptoms of a compressed median nerve caused by bulging tendons. Sometimes, carpal tunnel syndrome is referred to as a condition resulting from repeated motion. Inflammation-reducing foods and supplements may be helpful in treating this illness. In certain cases, a healthy diet may be enough to reduce symptoms and prevent the need for further medical care. Medical therapy may benefit from it in certain cases. Inflammation in the body may be exacerbated by certain meals, which can lead to further symptoms.

It is possible that a heavy diet in processed foods might lead to an increase in cytokines, which are pro-inflammatory proteins. An significant cytokine in CTS is tumor necrosis factor alpha (TNFα), which is linked to pain (BURNS, 2005). TNFα, an inflammatory protein, has been linked to diets high in processed foods.

CTS may be exacerbated by foods that cause inflammation. Refined starches and sugars are among them. In addition to being referred to as 'bad carbs,' refined starches are also known as "simple carbohydrates." Grains that have been stripped of fiber, minerals and bran are included in this category of products. Desserts and soft drinks are also part of the menu.

The term 'refined carbohydrates and sugars' encompasses foods such as sweet morning cereals; white bread; ordinary pasta; and white rice. It is not uncommon for refined carbohydrates to include significant levels of saturated or trans-fat. Among them, is a variety of sweets, including pie and cake. Beef, pork, and processed lunch on meats such as bologna, margarine, microwave popcorn, and cake frosting are examples of other foods rich in saturated and trans-fat. Trans fat and salt may be found in fried meals. Fast food items such as French fries, fried chicken and tacos are all included. Foods high in salt may exacerbate swelling and strain on the median nerve, since salt can increase water retention. Avoid foods that have high salt content, such as pre-packaged and commercially prepared ones. High quantity of Table salt should also be avoided in meals. Toxic effects of alcohol include increased levels of inflammation in the body (Guan et al., 2018). It has been shown that frequent drinking may raise the likelihood of getting carpal tunnel syndrome over time.

KEYWORDS

- **carpal tunnel syndrome**
- **nutraceuticals**
- **salmon**
- **spinach**
- **walnuts**

REFERENCES

Abate, M., Silbernagel, K. G., Siljeholm, C., Di Iorio, A., De Amicis, D., Salini, V., & Paganelli, R., (2009). Pathogenesis of tendinopathies: Inflammation or degeneration? *Arthritis Research & Therapy, 11*(3), 1–15.

Adelaja, A. O., & Schilling, B. J., (1999). Nutraceuticals: Blurring the line between food and drugs in the twenty-first century. *Choices, 14*(4), 35–39.

Albers, J. W., Brown, M. B., Sima, A. A., & Greene, D. A., (1996). Frequency of median mononeuropathy in patients with mild diabetic neuropathy in the early diabetes intervention trial (EDIT). *Muscle & Nerve: Official Journal of the American Association of Electrodiagnostic Medicine, 19*(2), 140–146.

Alfonso, C., Jann, S., Massa, R., & Torreggiani, A., (2010). Diagnosis, treatment and follow-up of the carpal tunnel syndrome: A review. *Neurological Sciences, 31*(3), 243–252.

Alford, J. W., Weiss, A., & Akelman, E., (2004). The familial incidence of carpal tunnel syndrome in patients with unilateral and bilateral disease. *Am. J. Orthop., 33*(8), 397–400.

Atroshi, I., Englund, M., Turkiewicz, A., Tägil, M., & Petersson, I. F., (2011). Incidence of physician-diagnosed carpal tunnel syndrome in the general population. *Archives of Internal Medicine, 171*(10), 941–954.

Atroshi, I., Gummesson, C., Johnsson, R., Ornstein, E., Ranstam, J., & Rosén, I., (1999). Prevalence of carpal tunnel syndrome in a general population. *JAMA, 282*(2), 153–158.

BURNS, T. M., (2005). Mechanisms of acute and chronic compression neuropathy. *Peripheral Neuropathy,* 1391–1402.

Davis, L., & Vedanarayanan, V., (2014). Carpal tunnel syndrome in children. *Pediatric Neurology, 50*(1), 57–59.

De Krom, M., Kester, A., Knipschild, P., & Spaans, F., (1990). Risk factors for carpal tunnel syndrome. *American Journal of Epidemiology, 132*(6), 1102–1110.

De Krom, M., Knipschild, P., Kester, A., Thijs, C., Boekkooi, P., & Spaans, F., (1992). Carpal tunnel syndrome: Prevalence in the general population. *Journal of Clinical Epidemiology, 45*(4), 373–376.

Di Geronimo, G., Caccese, A. F., Caruso, L., Soldati, A., & Passaretti, U., (2009). Treatment of carpal tunnel syndrome with alpha-lipoic acid. *Eur. Rev. Med. Pharmacol. Sci., 13*(2), 133–139.

Dillard, C. J., & German, J. B., (2000). Phytochemicals: Nutraceuticals and human health. *Journal of the Science of Food and Agriculture, 80*(12), 1744–1756.

Donato, G., Galasso, O., Valentino, P., Conforti, F., Zuccalà, V., Russo, E., & Amorosi, A., (2009). Pathological findings in subsynovial connective tissue in idiopathic carpal tunnel syndrome. *Clinical Neuropathology, 28*(2), 129–135.

Duncan, S. F., Bhate, O., & Mustaly, H., (2017). Pathophysiology of carpal tunnel syndrome. In: *Carpal Tunnel Syndrome and Related Median Neuropathies* (pp. 13–29). Springer.

Eun, Y., Kim, H., Hwang, H., & Choi, S. A., (2019). Idiopathic median mononeuropathy in a previously healthy child: The usefulness of magnetic resonance imaging in making the diagnosis. *Annals of Child Neurology., 27*(1), 29–31.

Fattori, V., Hohmann, M. S., Rossaneis, A. C., Pinho-Ribeiro, F. A., & Verri, W. A., (2016). Capsaicin: Current understanding of its mechanisms and therapy of pain and other pre-clinical and clinical uses. *Molecules, 21*(7), 844.

Gelfman, R., Melton, L. R., Yawn, B., Wollan, P., Amadio, P., & Stevens, J., (2009). Long-term trends in carpal tunnel syndrome. *Neurology, 72*(1), 33–41.

Genova, A., Dix, O., Saefan, A., Thakur, M., & Hassan, A., (2020). Carpal tunnel syndrome: A review of literature. *Cureus, 12*(3).

Ghasemi-Rad, M., Nosair, E., Vegh, A., Mohammadi, A., Akkad, A., Lesha, E., & Maleki-Miyandoab, T., (2014). A handy review of carpal tunnel syndrome: From anatomy to diagnosis and treatment. *World Journal of Radiology, 6*(6), 284.

Guan, W., Lao, J., Gu, Y., Zhao, X., Rui, J., & Gao, K., (2018). Case-control study on individual risk factors of carpal tunnel syndrome. *Experimental and Therapeutic Medicine, 15*(3), 2761–2766.

Jillapalli, D., & Shefner, J. M., (2005). *Electrodiagnosis in Common Mononeuropathies and Plexopathies.* Paper presented at the Seminars in neurology.

Keir, P. J., & Rempel, D. M., (2005). Pathomechanics of peripheral nerve loading: Evidence in carpal tunnel syndrome. *Journal of Hand Therapy, 18*(2), 259–269.

Keir, P. J., Wells, R. P., Ranney, D. A., & Lavery, W., (1997). The effects of tendon load and posture on carpal tunnel pressure. *The Journal of Hand Surgery, 22*(4), 628–634.

Keservani, R. K., Kesharwani, R. K., Sharma, A. K., Vyas, N., & Chadoker, A., (2010b). Nutritional supplements: An overview. *International Journal of Current Pharmaceutical Review and Research, 1*(1), 59–75.

Keservani, R. K., Kesharwani, R. K., Vyas, N., Jain, S., Raghuvanshi, R., & Sharma, A. K., (2010a). Nutraceutical and functional food as future food: A review. *Der Pharmacia Lettre, 2*(1), 106–116.

Keservani, R. K., Sharma, A. K., & Kesharwani, R. K., (2017). An overview and therapeutic applications of nutraceutical and functional foods. *Recent Advances in Drug Delivery Technology*, 160–201. ISBN:9781522507543.

Keservani, R. K., Sharma, A. K., & Kesharwani, R. K., (2020). *Nutraceuticals and Dietary Supplements: Applications in Health Improvement and Disease Management.* CRC Press. ISBN: 9781771888738.

Kiecolt-Glaser, J. K., (2010). Stress, food, and inflammation: Psychoneuroimmunology and nutrition at the cutting edge. *Psychosomatic Medicine, 72*(4), 365.

Kurt, S., Kisacik, B., Kaplan, Y., Yildirim, B., Etikan, I., & Karaer, H., (2008). Obesity and carpal tunnel syndrome: Is there a causal relationship? *European Neurology, 59*(5), 253–257.

Leach, R. E., & Odom, Jr. J. A., (1968). Systemic causes of carpal tunnel syndrome. *Postgraduate Medicine, 44*(2), 127–131.

Mackinnon, S., Dellon, A., Hudson, A., & Hunter, D., (1986). Chronic human nerve compression–a histological assessment. *Neuropathology and Applied Neurobiology, 12*(6), 547–565.

Maleki, N., Azami, A., Anari, H., Iranparvar, A. M., Tavosi, Z., & Hajaty, S., (2014). Value of ultrasonography in the diagnosis of carpal tunnel syndrome confirmed by nerve conduction study. *Scientific Journal of Kurdistan University of Medical Sciences, 19*(4), 58–66.

Nishihori, T., Choi, J., DiGiovanna, M. P., Thomson, J. G., Kohler, P. C., McGurn, J., & Chung, G. G., (2008). Carpal tunnel syndrome associated with the use of aromatase inhibitors in breast cancer. *Clinical Breast Cancer, 8*(4), 362–365.

Oh, S., Ettema, A. M., Zhao, C., Zobitz, M. E., An, K. N., Amadio, P. C., & Wold, L. E., (2008). Dextrose-induced subsynovial connective tissue fibrosis in the rabbit carpal tunnel: A potential model to study carpal tunnel syndrome? *Hand, 3*(1), 34–40.

Padua, L., Aprile, I., Caliandro, P., Carboni, T., Meloni, A., Massi, S., & Murasecco, D., (2001). Symptoms and neurophysiological picture of carpal tunnel syndrome in pregnancy. *Clinical Neurophysiology, 112*(10), 1946–1951.

Padua, L., Pasquale, A. D., Pazzaglia, C., Liotta, G. A., Librante, A., & Mondelli, M., (2010). Systematic review of pregnancy-related carpal tunnel syndrome. *Muscle & Nerve, 42*(5), 697–702.

Pajardi, G., Bortot, P., Ponti, V., & Novelli, C., (2014). Clinical usefulness of oral supplementation with alpha-lipoic acid, curcumin phytosome, and B-group vitamins in patients with carpal tunnel syndrome undergoing surgical treatment. *Evidence-Based Complementary and Alternative Medicine, 2014*.

Pandey, M., Verma, R. K., & Saraf, S. A., (2010). Nutraceuticals: New era of medicine and health. *Asian J. Pharm. Clin. Res, 3*(1), 11–15.

Rane, B. R., Bharath, M. S., Patil, R. R., Keservani, R. K., & Jain, A. S., (2020b). Novel approaches in nutraceuticals; Chapter 11. In: Rajesh, K. K., Raj, K. K., & Anil, K. S., (eds.), *Enhancing the Therapeutic Efficacy of Herbal Formulations Through Novel Drug Delivery Systems* (pp. 241–266). IGI Global International Publisher, Pennsylvania, USA. ISBN13: 9781799844532.

Rane, B. R., Patil, A. S., Keservani, R. K., & Jain, A. S., (2020a). Novel approaches in herbal formulation; Chapter 2. In: Rajesh, K. K., Raj, K. K., & Anil, K. S., (eds.), *Enhancing the Therapeutic Efficacy of Herbal Formulations Through Novel Drug Delivery Systems* (pp. 43–68). IGI Global International Publisher, Pennsylvania, USA. ISBN13: 9781799844532.

Rane, B., R., Tadavi, S., A., & Keservani, R. K., (2020c). Naturopathy; Chapter 12. In: Anil, K. S., Raj, K. K., & Surya, P. G., (eds.), *Herbal Product Development* (pp. 321–347). Apple Academic Press, CRC Press, Taylor & Francis Group. ISBN: 9781771888776.

Rao, P. J., & Naidu, M. M., (2016). Nanoencapsulation of bioactive compounds for nutraceutical food. In: *Nanoscience in Food and Agriculture 2* (pp. 129–156). Springer.

Ryan-Harshman, M., & Aldoori, W., (2007). Carpal tunnel syndrome and vitamin B6. *Canadian Family Physician, 53*(7), 1161–1162.

Seradge, H., Jia, Y. C., & Owens, W., (1995). *In vivo* measurement of carpal tunnel pressure in the functioning hand. *The Journal of Hand Surgery, 20*(5), 855–859.

Sharma, A. K., Keservani, R. K., & Gautam, S. P., (2020). *Herbal Product Development*. Apple Academic Press, CRC Press, Taylor & Francis Group. ISBN: 9781771888776.

Shiri, R., (2014). Hypothyroidism and carpal tunnel syndrome: A meta-analysis. *Muscle & Nerve, 50*(6), 879–883.

Spagnolo, F., Sestak, I., Howell, A., Forbes, J. F., & Cuzick, J., (2016). Anastrozole-induced carpal tunnel syndrome: Results from the international breast cancer intervention study II prevention trial. *Journal of Clinical Oncology*.

Stevens, J. C., Sun, S., Beard, C., O'fallon, W., & Kurland, L., (1988). Carpal tunnel syndrome in Rochester, Minnesota, 1961 to 1980. *Neurology, 38*(1), 134–134.

Stevens, J. C., Witt, J. C., Smith, B. E., & Weaver, A. L., (2001). The frequency of carpal tunnel syndrome in computer users at a medical facility. *Neurology, 56*(11), 1568–1570.

Szabo, R. M., & Chidgey, L. K., (1989). Stress carpal tunnel pressures in patients with carpal tunnel syndrome and normal patients. *The Journal of Hand Surgery, 14*(4), 624–627.

Talebi, M., Andalib, S., Bakhti, S., Ayromlou, H., Aghili, A., & Talebi, A., (2013). Effect of vitamin B6 on clinical symptoms and electrodiagnostic results of patients with carpal tunnel syndrome. *Advanced Pharmaceutical Bulletin, 3*(2), 283.

Van, D. M. A., Reitsma, J. B., Fischer, J. C., & Sanders, G. T., (2003). Indications for requesting laboratory tests for concurrent diseases in patients with carpal tunnel syndrome: A systematic review. *Clinical Chemistry, 49*(9), 1437–1444.

Weiss, N. D., Gordon, L., Bloom, T., So, Y., & Rempel, D. M., (1995). Position of the wrist associated with the lowest carpal-tunnel pressure: Implications for splint design. *The Journal of Bone and Joint Surgery American, 77*(11), 1695–1699.

Werner, R. A., Albers, J. W., Franzblau, A., & Armstrong, T. J., (1994). The relationship between body mass index and the diagnosis of carpal tunnel syndrome. *Muscle & Nerve, 17*(6), 632–636.

PART III

Inflammatory Deformities and Rickets

CHAPTER 10

ROLE OF NUTRACEUTICALS IN MANAGING OSTEOMYELITIS

JOYETA GHOSH,[1] SUDRITA ROY CHOUDHURY,[2] and SAMARPITA KONER[2]

[1]*Department of Dietetics and Applied Nutrition, Amity University Kolkata, India*

[2]*Department of Dietetics and Nutrition, NSHM Knowledge Campus-Kolkata, West Bengal. India*

ABSTRACT

Osteomyelitis, the inflammatory bone deformities, mainly caused by infection, leading to the destruction or necrosis of bones. These infection is prevalent among all age groups, involving any bone, can become chronic disease, causing persistent morbidity. In most of the cases *S. aureus* is the causative agent, although other bacteria or fungi are also involved. Current literature supports that the osteomyelitis treatment is most challenging, especially while the causative agents become multi drug resistance (MDR), and bacterial biofilm already established abscess and bone deformities. Furthermore the microbial drug resistance or antibiotic resistance is emerging globally as a serious human threat and the MDR strains are exhibiting major hurdles to any fatality related to infections. Thus the newer development of antibiotic-resistance mechanisms contributes to inefficiency to the existing drugs with prolonged illness and increased expenditure. Besides, conventional antibiotic therapy has several adverse effects on the patient's health, resulting in poor immune health development. There are many more nutraceuticals and its successful application against various MDR pathogens (especially

Nutraceuticals and Bone Health. Deepak Sharma, Madan Mohan Gupta, Anil K. Sharma, Raj K. Keservani, Rajesh K. Kesharwani (Eds.)

osteomyelitis causing pathogens) observed lately. Among them curcumin, resveratrol, quercetin, fatty acids, eugenol, allicin, kaempferol, etc., are found to be promising. Not only that nutraceuticals proofs to be beneficial in better bone health and improved immune functions. Thus application of nutraceuticals in osteomyelitis treatment may not only help the host to kill the pathogenic causative bacteria efficiently, but also it will improve the bone health and immune health of the host, most required criteria to recover from any bone infection and its associated health complications. Although very few phytocompounds are still now implemented in therapeutic treatment of infections, the commercial application of such phytochemicals is still low and urgently needs one drastic improvement regarding the correct application of these natural compounds against treating infection. Thus, there is need for fast-tracking more research, clinical trials and applications of these nutraceuticals to battle against osteomyelitis and its complications.

10.1 INTRODUCTION

Osteomyelitis is one typical bone infection. Archeological findings of animal fossils proves the existence of this ancient old disease in human history (Schmitt et al., 2017). Over the years this disease got various names until Nelaton introduced the term 'Osteomyelitis' in 1844 (Schmitt et al., 2017). Most importantly it is one chronic or acute process of inflammation, involving the bone and its structures. The pyogenic organisms like bacteria, mycobacteria and fungi are the causative factors behind the incidence of osteomyelitis. The traumatic or hematogenous sources are the crucial source of osteomyelitis development (Lew et al, 1997; Calhoun et al., 2009; Tong et al., 2015; Walter et al., 2012). However most orthopedic surgeries are the comparative threat of developing osteomyelitis through the traumatic route, as it possess huge risk of bacterial translocation from the skin to the surgical site intraoperatively (Calhoum et al., 2009). According to its histopathological findings osteomyelitis is categorized as a chronic or acute form of bone infection. Where acute osteomyelitis is accompanied with bone inflammation caused by pathogenic bacteria and the typical symptoms persist within two weeks after infection. The necrosis of the bone is common in chronic osteomyelitis, the symptoms may not continue until six weeks after onset of infection (Mylona et al., 2009). Another classification of osteomyelitis is according to the presumed mechanism of infection. For example: the chronic overlying open wound or hematogenous soft tissue infection, etc. (Waldvogel et al., 1970). Another complex classification is according to Cierny-Mader classification system, developed with the aim to guide the surgical management mostly, thus it is commonly less implemented during primary care of the diseases (Cierny et al., 2003). Earlier the common management of osteomyelitis was surgery including extensive debridement, saucerization, wound packing following which the affected area used to left for the healing by secondary intention (Schmitt et al., 2017). These ultimately leads to high mortality due to septicemia. Recently the availability of antibiotics, use of modern surgical techniques improved recovery rates from osteomyelitis. Although the arrival of multi drug resistance (MDR) microorganisms increases the treatment cost and morbidity/mortality day by day (Schmitt et al., 2017). Thus, again septicemia remains a common consequence of osteomyelitis, and in spite of advanced antibiotic development, it all too often remains refractory to chemotherapeutic treatment alone (Schmitt et al., 2017).

The antibiotic resistance occurred while bacteria/fungi evolved the ability to escape drugs, targeted to kill them, and continued its growth and survivability spontaneously in spite of drug treatment. A complex interaction was initiated while antimicrobial drugs were applied, evolving the drug resistance determinants. The imprudent application of antibiotics for humans led to the fast emergence, spread and persistence of such MDR, where microbes are developing resistance against a wide range of conventional antibiotics (Yu et al., 2020). In the US itself the highest health care cost was seen in hospital onset invasive infections estimated 4.6 billion dollars in 2017 (Calhoun et al., 2009). Thus, MDR associated infections are becoming the leading burden to any health care system. Such a situation put the medical community with a much-shrunken pool of therapeutic options against treatment of those infections. Thus, it is significantly associated with substantial mortality and morbidity along with a huge economic burden on global public health (Centers for Disease Control and Prevention, 2019). Thus, there is urgent need for finding alternative approaches in treating osteomyelitis effectively. Present chapter illustrate the gainful attempts of the applications of nutraceuticals in the fight against osteomyelitis.

10.2 OSTEOMYELITIS-CURRENT CONCEPT

Healthy bones are protective against infections. Though the bone becomes vulnerable to diseases while it is exposed to an inoculum of bacteria due to presence of foreign body, ischemia or trauma (Lew et al., 1997). The most common pathogen of osteomyelitis is *Staphylococcus aureus,* responsible for both acute and chronic hematogenous osteomyelitis in children and adults. The next most in children were Group A *Streptococcus, Kingella kingae* and

Streptococcus pneumonia. Primarily the group B streptococcal infection occurs in newborns (Kaplan et al., 2005). *S. aureus* is the most common causative agent in bone and prosthetic joint infections among adults. In the current scenario while MDR is common among infections, the methicillin-resistant *S. aureus* (MRSA) is increasingly isolated from various osteomyelitis patients (Aragón-Sánchez et al., 2009). *Staphylococcus aureus* most commonly cling to the bone through the expression of specific receptors, called adhesions. This is present in some components of bone matrix, including collagen, laminin, bone sialoglycoprotein and fibronectin. In recent discovery it is also observed that collagen-binding adhesion was expressed by *S. aureus,* predispose strong attachment to bone cartilage while fibronectin-binding adhesion initiate the attachment of bacteria to surgically implanted devices (Gornitzky et al., 2020). In cultured osteoblasts *S. aureus* capable to survive intracellularly after being internalized. Other bacteria are capable of creating protective biofilm in its surroundings and underlying surfaces. Such bacteria tend to adhere the surgically implanted devices or to bone and expressed resistance to antibiotic therapy (Schmitt et al., 2017). *Pseudomonas aeruginosa*, *Staphylococcus epidermidis, Escherichia coli* and *Serratia marcescens,* etc., are the example of some chronic osteomyelitis. Overall fungal or mycobacterial infections are also observed in osteomyelitis, these are common among immunocompromised patients (Kohli et al, 2005).

Patients with different osteomyelitis execute symptoms like erythema, fistula and or sinus, swelling, pain purulent drainage, wound breakdown, and mild rise in local temperature (McNally et al, 2010; Govaert et al., 2020). Additionally there are high erythrocyte sedimentation rate, leukocyte counts, and C-reactive protein levels, existence of organisms in more than one deep specimen verified with microbiological cultures and signs of bone destruction and the development of sequestra and purulent collection in soft tissue assessed through imaging (Govaert et al., 2020). In general histopathological examinations of deep tissue samples help us to confirm the disease status (mostly used specific staining techniques) (Metsemakers et al., 2018). While culture-negative patients were diagnosed using histological signs of active bone resorption and remodeling & presence of acute and chronic inflammatory cells (McNally et al., 2010). The most complicated infection of bones and joints are termed chronic osteomyelitis, prevalent among adults and predisposes symptoms after several weeks. Hematogenous spread is common in acute cases, while chronic osteomyelitis developed due to open wounds, fractures, orthopedic implants and contiguous soft-tissue infections. Especially in peripheral vascular disease or diabetes mellitus (DM), as because of poor immune response patient become highly susceptible to develop chronic osteomyelitis through infections of contiguous tissues. Bone necrosis is common in these cases (Hatzenbuehler et al., 2011).

10.2.1 PATHOPHYSIOLOGY

The hematogenous route of infection through microbial (bacteria, fungi) seeding of bone is the pertinent route of bone infection. Further the source can be contiguous from surrounding tissue and joints, or can be from a distant source of infection, or can be the direct inoculation of bone during surgery or trauma attack (Schmitt et al., 2017; Lew et al., 1997). The incidence rate of hematogenous osteomyelitis is high among children than adults, while long bones are usually affected more frequently than others (*Diagnosis and Management of Osteomyelitis – American Family Physician*, n.d.). The reason behind this high incidence rate among children may be due to common exposure of metaphyseal (growing) regions of the bones especially long bones, highly vascular and susceptible to minimal trauma. Study also shows that the highly prevalent age group is below 5 years of age (Gutierrez, 2005). Fever, irritability, local swelling, erythema, tenderness over the bone are some of the common symptoms present in patients within two weeks of disease (Saavedra-Lozano et al., 2008) (Figure 10.1).

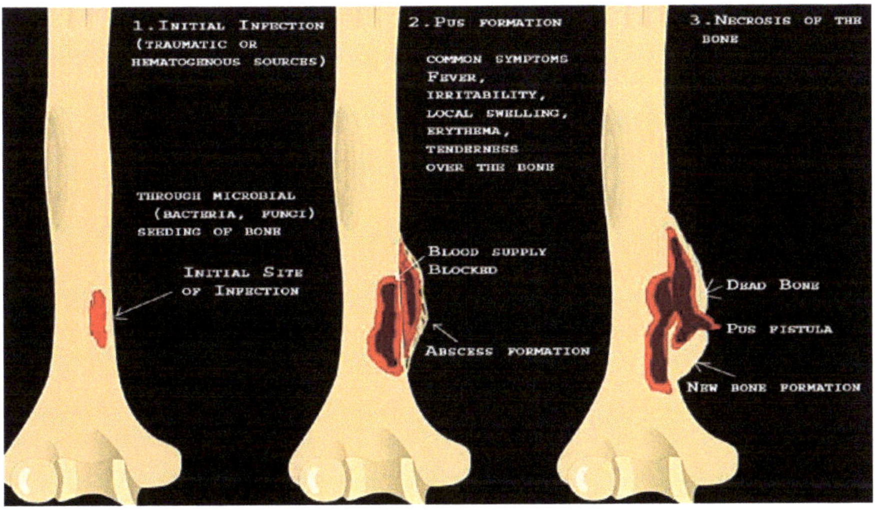

FIGURE 10.1 Development of osteomyelitis.

Source: Adapted from Schmitt et al., 2017; Saavedra-Lozano et al., 2008; Lew et al., 1997.

Chronic state of infection is less prevalent among children and mostly causes due to contiguous soft tissue infection, bacteremia or open fractures (Auh et al., 2004). In case of adult patients, the vertebrae is most commonly infected and most common occurrence during related surgery or trauma (Schmitt et al., 2017). Whereas in the case of elderly, the osteomyelitis is typically correlated with joint arthroplasties, decubitus ulcers. Diabetes mellitus is one important risk factor of osteomyelitis associated with vascular insufficiency. Vertebral osteomyelitis patients often have existing complications like chronic renal disease, cancer, diabetes mellitus or history of intravenous drug use (Zimmerli, 2010). Of late, the high incidence rate of Chronic osteomyelitis from contiguous soft tissue infection is common due to the increasing prevalence of peripheral vascular disease and diabetic foot infections. According to the statistical data around 50% of diabetic patients develop peripheral neuropathy at a later stage and increases the risk of unrecognized infections (Barrett et al., 2007). However, among diabetic patients the peripheral vascular disease is also common, responsible for delay in wound healing and thus contributes to the generation of soft tissue infection (Abdulrazak et al., 2005).

Contiguous osteomyelitis is highly prevalent among bedbound or wheelchair bound patients, commonly developing pressure-related skin ulcerations. These typically colonized by polymicrobial flora and soon spread over the soft tissue and underlying bone (Schmitt et al., 2017). Trauma leads by infected exposed skin and soft tissue are another causative reason of contiguous osteomyelitis (Schmitt et al., 2017; Lew et al., 1997).

10.2.2 COMMON PATHOGEN AND ITS PROTECTIVE MECHANISM: WHY DO ANTIBIOTICS FAILS SO OFTEN

As previously discussed the most common pathogen of osteomyelitis is *Staphylococcus aureus*, which tends to arrive at the bone surface via the hematogenous or iatrogenic routes that can easily adhere to soft tissue, metal implants or bone (Lew et al., 2004). This microbes adapt such functionality via binding to extracellular matrix proteins through the adhesive matrix molecules present at the microbial surface, such as collagen binding proteins and the sialoprotein of bone (Ellington et al., 2006). Here this specific microbe adapts multiple strategies to survive immune attack and the continued therapy (Figure 10.2). The infected *S. aureus* then multiply, aggregate and form typical colonies called microcolonies (Szafranska et al., 2014a), it is also

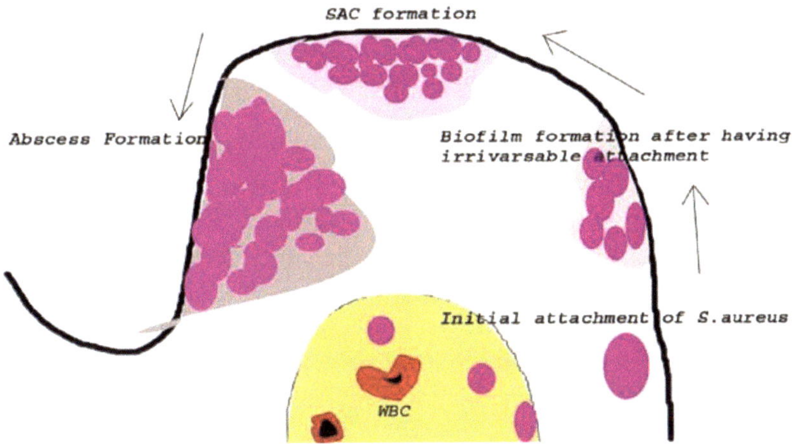

FIGURE 10.2. *Staphylococcus aureus* and its mechanism of action in osteomyelitis.

Source: Adapted from Brandt et al., 2018; Horst et al., 2012; Cheng et al., 2011; Giavaresi et al., 2008).

termed as staphylococcal abscess communities (SACs) (Qin et al., 2019). Other than bones presence of SACs also observed in kidney, skin, brain or renal tissues (Monaco et al., 2016). Specifically this microbe develops fibrin through initiating polymerization of fibrinogen. It also initiate the secretion of enzymes, such as coagulase and von Willebrand factor binding protein, responsible for the activation of endogenous prothrombin & leading to the development of fibrin (González-Martín et al., 2022). Such network of fibrin (SACs) ultimately protects the bacteria from immune attack responsible for the clearance of the bacteria from the infection site, as a result the accumulation of immune cells to the specific bacterial nidus took place (Masters et al., 2021). These developmental structure become more complex when *Staphylococcus aureus* adhering to implanted devices or sequestra (Inzana et al., 2015). Antibiotics are less penetrable against bacteria with biofilm. It is due to several factors including less oxygen tension and altered metabolic activity. Interesting the residing *S. aureus* beneath the biofilm, capable to secrete extracellular polymeric matrix consisting of proteins, polysaccharides and possibly extracellular DNA from the dead cells of microbes, such matrix create strong barrier to immune cell infiltration (W. Zhou et al., 2020). Besides SACs activity *S. aureus* also interacts with bones and involved in osteocyte-lacunae canaliculi networks (De Mesy Bentley et al., 2018). Invasion of *S. aureus* in canalicular network may lead to the persistence and chronic infection, and escape immune clearance permanently (De Mesy Bentley et al., 2018). Other than these, chronic infection may also predisposed by *Pseudomonas aeruginosa, Staphylococcus epidermidis, Escherichia coli, Serratia marcescens, Streptococcus spp.* Again the MRSA is the most prevalent pathogen in chronic osteomyelitis (Hatzenbuehler et al., 2011).

Another observed defense mechanism of *S. aureus* is the capability to release a wide range of toxins targeted to host cells. These includes pore-forming toxins, superantigens and exfoliative toxins (Oliveira et al., 2018). The shock syndrome toxin 1(superantigen toxic) is associated with further bone infection, responsible to promote osteoclastogenesis and the *in vitro* bone resorption activity of osteoclasts (Kavanagh et al., 2018). *S. aureus* released pore forming toxins can be categorized into hemolysin-a, hemolysin-b, phenol-soluble modulins and leukotoxins; mainly responsibly to reduce host cell membrane integrity (Oliveira et al., 2018). Precisely the phenol-soluble modulins exhibits cytotoxic effect on *in vitro* osteoblasts, where as hemolysin-a exhibit cellular (*in vitro*) death of both osteoblast and osteoclast (Kavanagh et al., 2018). The other pore forming toxin Hemolysin-b exhibits phagosomal escape, shown *in vitro* in polymorphonuclear cells (PMN) and *in vivo* stimulates biofilm formation (Oliveira et al., 2018). Leukotoxin Panton-Valentine leucocidin exhibits more aggressive efforts and thus is difficult to treat (Dohin et al., 2007, Kavanagh et al., 2018). *S. aureus* has the capability to invade and survive intracellularly during the exclusive phagocytosis performed by host immune cells (Kubica et al., 2008; Szafrańska et al., 2014b). Additionally it capable to trigger phagocytic internalization through expressing fibronectin-binding proteins (A and B), and adheres to the fibronectin, remains connected to a_5b_1 integrins on neutrophils or macrophages (Smith et al., 2011; Yang et al., 2018). *S. aureus* evades cell death after successful internalization by remains persisting within vacuoles or by inhibiting phagolysosomal fusion (Bernard et al., 2010). Furthermore *S. aureus* can survive within non-professional phagocytes like primary human osteoblasts (*in vitro*) (Kalinka et al., 2014), human osteocytes (*in vivo* and *ex vivo*) (Kalinka et al., 2014) mouse osteoclasts (*in vitro* and *in vivo*) (de Barros et al., 2010; Krauss et al., 2019). Thus *S. aureus* continue to survive within the osteoblasts, matured to become osteocytes and still remain infected (Strobel et al., 2016). Another interesting observation by Tuchscherr et al. (2010) was that to survive internally the *S. aureus* adopt frequently such dormant small colony variant that predispose slow growth and decreased metabolic activity. Thus numerous mechanistic approach of *S. aureus* make them survive within the bone niche for prolonged period and exhibits chronic and recurrent osteomyelitis.

10.2.3 BACTERIAL INTERACTIONS WITH SKELETAL CELLS

Bones can be defined as the mineralized organic matrix consisting of bone forming osteoblasts, osteocytes and bone-resorbing osteoclasts. *S. aureus* has a strong impact against all three cells. Basically, in osteoblasts staphylococcal protein A (SpA) binds to TNF receptor-1(Tumor necrosis factor) and initiates apoptosis. It also decreases the cell differentiation and calcium deposition of the osteoblasts (Claro et al., 2013; Widaa et al., 2012; Josse et al., 2015a). However osteoblasts viability and functionality got affected due to *S. aureus* mediated internalization through integrin bridging by fibronectin-binding protein A/Bea5b1(Claro et al., 2011). Thus both of these interaction reduces bone formation and inhibited matrix mineralization (Claro et al., 2013; Widaa et al., 2012). Thus as demonstrated in rabbits, *S. aureus* increases periosteal bone formation in osteoblasts (Claro et al., 2013). In case of osteoclasts the SpA secreted by *S. aureus* up-regulate bone resorption capacity through initiating TNF and epidermal growth factor (Thompson et al., 2019). Such activity leads to lacunae formation and formation of necrotic bone pieces, commonly observed in biopsies osteomyelitis patients (Gaida et al., 2012). The osteolysis activity was increased indirectly by osteoclasts through osteocytes, osteoblasts and PMNs, responsible for secreting receptor activator of NF-kB ligand (RANK-L), initiates osteoclastogenesis and stimulate osteoclasts for bone resorption. The osteocytes repeat the same in response to neighboring osteocytes, undergoing apoptosis (Prideaux et al., 2016), that can be due to invasion of lacunae by *S. aureus*). The RANK-L expression was upregulated by osteoblasts, while TNF receptor-1 binds with SpA through bacterial internalization (Shodo et al., 2017), PMN s are responsible for osteoclastogenesis and resorption of bone through osteoclast-mediated secretion of IL-8 (Yin et al., 2019). Thus osteoclastogenesis and osteoclast activity predispose continuous inflammatory environment itself, through the secretion of osteoblastic proresorptive cytokines IL-6, TNF-a, and IL-1b (Michel et al., 2018). Again development of hypoxia itself causes persistent inflammation (Chlubnová et al., 2011).

Osteocytes plays another important role osteocytic osteolysis or perilacunar remodeling, where it matures and maintain the mineralized matrix, through the expression of several enzymes, capable to work reversibly in removing mineral as well as remodeling the organic phase of bone matrix (Stephanie et al., 2010). Thus *S. aureus* infected human osteocytes exhibits matrix metalloproteinase expression during osteomyelitis (Khan et al., 2022), indicating osteolysis of osteocytes occurred due to *S. aureus* infection, thus overall possessing inhibition of bone formation, leading to bone loss (Li et al., 2020).

10.2.4 REACTIONS IN HOST CELLS

Existence of bacteria within the bone tissues triggers the host response, exhibits innate immune response, driven by PMNs and macrophages. The adaptive immune responses also triggered and mediated by B cells, T cells and pathogen-specific antibodies. During bacterial recognition the osteocytes (Yang et al., 2018), osteoblasts (Josse et al., 2015) and resident macrophages (Chang et al., 2008; Josse et al., 2015) secrete chemoattractants to activate an influx of immune cells to the site of infection. PMNs influx during acute osteomyelitis was common both in human as well as in rodent osteomyelitis models (Horst et al., 2012). The macrophages responsible for inflammation (myeloid related protein 8 (MRP8)/MRP14 positive) and T cells (CD4) remains activated in human (Klosterhalfen et al., 1996). The activated PMNs are responsible for killing planktonic *S. aureus* via phagocytosis, production of antimicrobial peptides and using oxidative bursts. Meanwhile the proinflammatory cytokines and chemokines like IL-1b, tumor necrosis factor (TNF)-a, CXCL2, CXCL3 and other related factors activates & recruits PMNs, leading to pathogen clearance (Rigby et al., 2012). Also there is direct host defense response by PMNs and macrophages through forming neutrophils, mediated extracellular traps, to trap bacteria, which are ultimately cleared by immune cells. Again while infection persist and osteomyelitis becomes chronic, the bacteria tends synthesize biofilm phenotype and the influx of viable PMNs decreases significantly (Klosterhalfen et al., 1996; Horst et al., 2012). In osteomyelitis a chronic infection sites mostly the wound-healing M2 macrophages (CD163 positive) resides along with small number of CD8 T cells and plasma cells (Tiemann et al., 2014; Klosterhalfen et al., 1996; Wagner et al., 2008). Due to biofilm surrounding the M2 macrophages were unable to phagocytosing bacteria, tends to promote profibrotic environments as well as wound healing responses and thereby generating abscesses during chronic osteomyelitis infections (Scherr et al., 2014; Scherr et al., 2013). *S. aureus* evolves many protective mechanisms against adaptive immune responses in bone infection, thus resulting in chronic osteomyelitis. In porcine osteomyelitis the antibody responses against *S. aureus biofilms* are most commonly skewed to type 1 and type 17 helper T cell biased immune response, as a result they fails in bacterial clearance (Jensen et al., 2015). Furthermore *S. aureus* efficiently alter B cells, through secreting staphylococcal protein A (SpA), responsible for deteriorating survival and functionality of B cells also it is associated with Fcg and Fab domains of certain anti- bodies (Moks et al., 1986; Schneewind et al., 1992), and thereby ultimately inhibit antibody-mediated phagocytosis and resulting B cell apoptosis (Graille et al., 2000; Goodyear et al., 2003).

10.2.5 OSTEOMYELITIS AND OXIDATIVE STRESS

Study represents many bacterial osteomyelitis are associated with significant disability and poor clinical outcomes. Despite modern treatments around 30% of bone infection become chronic, affecting more in males than females (Borsiczky et al., 2003). Considering first line of defense the neutrophils create protections against many pathogens bacteria and plays crucial role in osteomyelitis development. During bone infection the circulating neutrophils count increased many folds, focusing on bone infection site. During that condition patient will also have high level of inflammatory cytokines in blood parameters (Frei et al., 1989; Sandukji et al., 2011). IL-1, the pro-inflammatory cytokines and the tumor necrosis factor (TNF-α) initiate activation of the inducible isoform of nitric oxide synthase (iNOS or iNOS-2), hence the nitric oxide (NO) derived from such pathway promotes the bone loss (Bonhoeffer et al., 2001, Miric et al., 2010). Increased NO production took place due to persistence of bacteria at the infection site as high activation of endothelial nitric oxide synthase (eNOS or eNOS-3) (McCormic et al., 1994). As a result, this accumulates high reactive oxygen species generation (ROS), leads to the damage of lipids, proteins as well as cell wall integrity and ultimately leads to high oxidative stress in osteomyelitis (McCormic et al., 1994). Furthermore, increased inducible and endothelial NO generation causes bone loss through sharp rise in bone resorption and accounts for osteolysis, typical characteristics of osteomyelitis. Again during the neutrophil phagolysosomes, Myeloperoxidase (MPO) is one oxidative enzyme that possesses antibacterial activity, releases ample amounts of H_2O_2, to synthesize hypochloric acid and other toxic materials. Such phenomenon again responsible for high ROS generation in cells (Koracevic et al., 2001, Liberatore et al., 2007).

10.3 DIAGNOSIS AND TREATMENT CURRENT APPROACH

Acute osteomyelitis is primarily a clinical diagnosis in children based on the localization and rapid onset of symptoms. Fever, lethargy, and irritability are some of the symptoms that can be present. Erythema, characterized as swelling of soft tissue, bony tenderness and decreased motion of joint range is one of the common findings that should be focused in physical examination.

The identification of a microorganism infection is also difficult as a result of blood cultures are positive in barely regarding half of cases (Chen et al., 2012). Identification of this disease is difficult among adults. While considering high index of clinical suspicion is required, otherwise recognition of symptoms, imaging and laboratory reports are beneficial in this case. Patient history and general symptoms are crucial in this case considering fever, extremity or back pain, lethargy, malaise, etc., Again understanding the status of preexisting diseases like diabetes, peripheral vascular disease, intravenous drug use and history of trauma. The physical examination ought to specialize in locating a possible nidus of infection, assessing sensory and peripheral vascular performance and also helps to identify ulcers in bone. If a contiguous infection with lesions is present, like diabetic foot infections, the use of a sterile steel probe is helpful to confirm the presence of osteomyelitis. In a recent study lower prevalence of osteomyelitis is found with a positive predictive value of 57% (Lavery et al., 2007). In general there is no specific laboratory test for identification of this disease, like all common infections the C-reactive protein levels, leukocytosis and rise in erythrocyte sedimentation rate considered for proper monitoring. Microbial culture of osteomyelitis sample is one integral part of treating this patients. Again complete histopathology including necrosis and positive microbial culture from bone biopsy are critical diagnostic criteria (Lipsky et al., 2006).

10.3.1 IMAGING AS IMPORTANT TOOL

For bone infections proper imaging is crucial for better identifications, as it helps in proper characterization of the disease and rule out the other possibilities of diseases. The common radiography techniques, magnetic resofaggot imaging (MRI) and technetium-99 bone scintigraphy are most commonly used tool for proper imaging. Whereas plain radiography never gives correct picture, especially when osteomyelitis is at initial stage, only it is detectable while 50% of the bone mineral content is lost due to infection (Pineda et al., 2009). Thus understanding metastases or factures plain radiography is beneficial. Additional advanced imaging techniques now a days playing crucial role in early identification (Schweitzer et al., 2008). MRI signifies higher efficiency in detection and diagnosis of osteomyelitis (Pineda et al., 2009). Considering diabetic foot ulcer and identification of osteomyelitis, proper imaging plays a crucial role in sleuthing osteo redness enclosed patients. Hence imaging helps in achieving more than 90% accuracy in this disease cases. Even proper imaging helps in finding dead bone, abscesses or sinus tracts. Furthermore, implementation of modern technology brings more advancement in this field especially in bone scintigraphy and becoming revolutionary in diagnosis of osteomyelitis at different stages.

10.3.2 POSSIBLE TREATMENTS AVAILABLE

Proper antibiotic therapy is the primary treatment of this bone infection. Although the surgical removal of infected and dead tissues, applicable antibiotic medical care are important steps in this case. In this case choice of antibiotics is based on results of microbial culture and susceptibility results (Roblot et al., 2007a). False-negative results sometimes misinterpreted the therapy.

Considering acute stages of osteomyelitis need less durations of antibiotic therapy, where chronic needs longer durations and highly depends on microbial culture and drug sensitivity results. Continuous use of antibiotic therapy, surgical debridement increases the recurrence rate of osteomyelitis infections by about 30% within 12 months of incidence (Tice et al., 2003). Parenteral antibiotic therapy for two to six weeks are always recommended in common osteomyelitis treatment. Of Late oral antibiotic therapy for four to eight weeks also common in this case (Cierny et al., 2003). Rules of applying empiric antibiotic treatment of acute osteomyelitis, especially in case of children, should incorporate an agent directed against *S. aureus* (Tice et al., 2003). Overall still there is no specific drug available for the treatment of osteomyelitis. MDR is one booming concern too in this case. As the pool of available antibiotics against MRSA or MDR strains are shrinking day by day hence a great amount of interest in growing regarding application of nutraceuticals or plant-based medicinal compounds in treating serious illness like osteomyelitis.

10.4 NUTRACEUTICALS: REMEDIAL ACTIONS IN OSTEOMYELITIS

The word 'Nutraceutical' combines two words "Nutrient" and "Pharmaceutical." A supplement regarded as a product procured from food products which resulted into a new outcome, it can be capsule, liquid, powdered or pill form. This are mainly used as dietary supplements which have beneficial effect on health (Keservani et al., 2010a, 2010b, 2017, 2020; Sasi et al., 2017). Nutraceuticals can be isolated nutrients, dietary supplements, herbal products (Sharma et al., 2020) and processed products like cereals, soups and even beverages (Andlauer et al., 2002). It is widely used to combat major health problems like obesity, cancer, Diabetes Mellitus, osteoporosis, Cholesterol and many more. Nutraceuticals is new area between medicine and health and food industry plays an important role in it (Das et al., 2011).

Human immune responses are composed of the first line of defense or innate immunity and second line of defense or adaptive immunity. The first line of defense is characterized by biochemical and physical barriers, along with non-specific cell-mediated response, including macrophages, granulocytes, humoral elements and natural killer cells, leading to proper counteract pathogenic infection and malignant transformation (Moser et al., 2010). However, the second line of defense gets activated only after macrophage-mediated presentation of antigens, directed to B lymphocytes, through T lymphocytes. After which the B cell mediated humoral immunity started enacting across the production of high-affinity antibodies and implant immunological memory (Kumar et al., 2020). Here helper T cells release cytokines after activation and then T lymphocytes can mediate cellular immunity (Moser et al., 2010). Importantly the immunomodulatory agents can alter immune responses in positive or negative manner and thus can be categorized as immunosuppressant or stimulants (Wen et al., 2012). Activation of immune responses are inhibited by immunosuppressors and thereby decreases hyper-activity of the immune cells, required to maintain body homeostasis. In the modern health care system these immunosuppressors are required during organ transplantations or in autoimmune disorders (Wen et al., 2012). Nutraceuticals plays important role in such cases, for example vitamin D act as a potent immunomodulator in human body and there are several reports that proofs its capability to counteract the aberrant immune responses, without compromising the physiological defenses (Untersmayr et al., 2020). Immunostimulants stimulate or boost the immune responses and thus again allow to restore body homeostasis (Wen et al., 2012). There is another group of immunomodulators capable of acting as immunoadjuvants, enhancing immune response to vaccines without producing specific antigenic effects (DiPasquale et al., 2015). Immune associated disorders or diseases like bacterial or fungal infections (predominant cause of osteomyelitis), usually correlated with acute inflammation, the key component for immune activation. Also in case of chronic inflammatory responses, inflammation pursue negative influences in immune functions, affects both B and T lymphocytes, hence implementation of immunomodulatory agent as part of therapeutic approach remain beneficial in such cases (Pawelec et al.,

2019). As previously discussed the relationship between oxidative stress and osteomyelitis, thus immunomodulatory agents with anti-inflammatory and antioxidant capacity have important remedial role as possible chemopreventive agents, due to its ability to counteract chronic inflammation (Dougan et al., 2012). Growing evidences also suggest the gut microbial influences in maintenance of proper immune function, thus disruption of microbiome seems to have crucial role in disease development and prognosis, especially when responsible pathogens are MDR (Zheng et al., 2020). Thus boosting residential microbiome through pre/probiotic supplements may reinforce proper immune defenses, as one alternative immunomodulatory strategy (Michel et al., 2018). In this scenario the symptomatic treatments and immunity boost are the most acceptable methods nowadays as the pool of efficient antibiotic therapy is shrinking day by day due to significant rise in MDR. Even though modern technologies are available and reduce the fatality rate in osteomyelitis, still there are several constraints associated with current antimicrobial treatments such as safety, efficacy and high cost. Since ancient times several medicinal plants and phytochemicals have been serving mankind for their ability to modulate the immune function (Mukherjee et al., 2012). As one major immunomodulator these nutraceuticals specifically boost the immune system, through inducement of both innate and adaptive humoral as well as cellular immunity (Details explained at 3.4.4). Furthermore, such phytonutrients have the capability to interfere in proinflammatory pathways and the modulation of the gut microbiome (Wang et al., 2017). Implementation of nutraceuticals as therapeutic approach in osteomyelitis treatment possess many advantages such as: 1) the nonconventional therapy by using vitamins, functional foods, multi-nutrients, nutraceuticals, and prebiotics can play a significant role to combat this critical bone infection. These natural components are not only bactericidal in action but can also boost the host immunity and sharp rise in physiological condition of the human body by reducing oxidative stress (such as replenishing glutathione (GSH) and modulating the free radicals inside the cells). Thus, it increases microbial clearance and it becomes difficult for microbes to replicate inside the host body and also the severity of the diseases reduced significantly, leads to better recovery rate) Nutraceuticals are effective immunomodulator by modulating immune responses mediated by macrophages, neutrophils, dendritic cells, etc. Also through generation of cytotoxic chemicals (antimicrobial) and cells like nitric oxide-(NO-)free radicals, T helper cells from CD4+T lymphocytes, cytotoxic T lymphocytes from CD8+T lymphocytes, activation of phagocytosis, proliferation of B lymphocytes, etc. Furthermore with an increase in age, the human immune system starts to produce a lesser number of T cells due to atrophy in the thymus gland, thus it is predisposed to be susceptible to lethal infections, simultaneously poor bone health makes them vulnerable to bone infections as well. Hence adequate nutrition support plays a significant role in supporting the immune system through optimizing cell functions especially involved in the immune system. Besides, nutraceuticals play a crucial role in such cases as to functionalize food and boost daily nourishment (Jayawardena et al., 2020).

10.4.1 NUTRACEUTICALS AGAINST PATHOGENS INCLUDING MDR, RESPONSIBLE FOR OSTEOMYELITIS AND OTHER INFECTIONS

Many studies have proves that phytochemicals exhibits antimicrobial activities against various pathogen through variety of mechanisms like doing damage to the bacterial cell membrane, promoting oxidative stress to the bacterial cell, suppressing virulence activity among pathogens, inhibition of many enzymes and toxins produced by pathogen, inhibition of biofilm formation, etc. According to the current available literature the nutraceuticals predisposing antibacterial activity belongs to many different chemical classes; including sulfur-containing phytochemicals, alkaloids, terpenoids, polyphenols, etc. Few of them also capable to have synergistic effects while tested in combination with antibiotics (conventional), also capable to modify the development of resistance as well. Phytochemicals are now becoming possible source of cheap, effective and safe antimicrobial agents, though still there is lack of more details work to be carried in *in vivo* conditions especially to ensure its efficiency but still of late it is well recognized therapeutic approach in treating many infection with low side and adverse effects. Furthermore the microbial drug resistance or antibiotic resistance is emerging globally as a serious human threat and the MDR strains are exhibiting major hurdles to any fatality related to infections. Thus the newer development of antibiotic -resistance mechanisms contributes to inefficiency to the existing drugs with prolonged illness and increased expenditure. Again imprudent application of conventional and commonly available antibiotics in human health, hygiene, agricultural and veterinary practices is enforcing to be a major drive for persistent, spread and evolution of antibiotic-resistance at a frightening rate.

TABLE 10.1 List of Major Nutraceuticals (Plant Based), Their Chemical Class, Source and Mode of Action Against Pathogenic Strains Responsible for Several Infections Including Osteomyelitis

Source	Chemical Class	Phytochemical	Mode of action	Target pathogen	Reference
Berberis species	Alkaloid	Berberine	Cell division/Protein/DNA synthesis	*Escherichia coli*	(Boberek et al., 2010)
Holarrhena	Alkaloid	Conessine	inhibitor	*Pseudomonas aeruginosa*	(Siriyong et al., 2017)
			Efflux pump inhibitor		
Anti-dysenterica	Alkaloid	Lysergol	Efflux pump inhibitor	*Escherichia coli*	(Maurya et al., 2013)
Ipomoea muricata					
Solanaceous	Alkaloid	Tomatidine	ATP synthase inhibitor	*Listeria, Bacillus and*	(Boulet et al., 2018)
Rauwolfia	Alkaloid	Reserpine	Efflux pump inhibitor	*Staphylococcus sp., Streptococcus sp.*	(Sridevi et al., 2017)
Dioscorea Bulbifera	Alkaloid	8-episodios bulbin	Plasmid curing	*Escherichia coli, Enterococcus faecalis,*	(Shriram et al., 2008)
Scutellaria	Phenolic compound	Baicalein	inhibitor	*Methicillin-resistant Staphylococcus aureus*	(Chan et al., 2011)
			Efflux pump inhibitor		
baicalensis	Phenolic compound	Kaempferol	Efflux pump inhibitor	*Methicillin-resistant Staphylococcus*	(Randhawa et al., 2016)
Alpinia calcarata					
Vitis vinifera	Phenolic compound	Resveratrol	Efflux pump inhibitor	*Campylobacter jejuni*	(Klancˇ nik et al., 2017)
Allium cepa	Phenolic compound	Taxifolin	Beta-Ketoacyl acyl carrier protein	*Enterococcus faecalis*	(Jeong et al., 2009)
Allium sativum	Organosulfur	Allicin	Sulfhydryl-dependent enzyme inhibitor,	*Staphylococcus epidermidis, Pseudomonas aeruginosa,*	(Reiter et al., 2017)

TABLE 10.1 *(Continued)*

Source	Chemical Class	Phytochemical	Mode of action	Target pathogen	Reference
Allium sativum	Organosulfur	Ajoene	DNA/protein synthesis inhibitor, Sulfhydryl-dependent enzyme inhibitor	*Campylobacter jejuni, Staphylococcus,*	(Rehman and Mairaj et al., 2013)
Diplotaxisharra	Organosulfur	Sulforaphane	Membrane destruction, ATP synthase inhibitor, DNA/ protein synthesis	*Escherichia coli*	(Wu et al., 2012)
Thymus capitatus	Terpene	Carvacrol	Cell membrane disruption	*Escherichia coli, Enterobacter aerogenes,*	(Althunibat et al., 2016)
Essential oils	Terpene	Cinnamaldehyde	Cell membrane disruption	*E. coli and S. aureus*	(Zhang et al., 2015)
Essential oils	Terpene	Eugenol	Cell membrane disruption	Methicillin-resistant *Staphylococcus aureus*	(Yadav et al., 2015)
Essential oils	Terpene	Farnesol	Cell membrane disruption	*Staphylococcus aureus*	(Togashi et al., 2010)
Ferulago	Coumarin	Aegelinol	DNA gyrase inhibitor	*Enterobacter aerogenes, Salmonella*	(Basile et al., 2009)
Asphodelus	Coumarin	Asphodelin A	DNA gyrase inhibitor	*Staphylococcus aureus, Escherichia coli,*	(El-Seedi et al., 2007)
Microcarpus	Coumarin			*Pseudomonas aeruginosa*	
Ferulaszowitsiana		Galbanic acid	Efflux pump inhibitor	*Staphylococcus aureus*	(Bazzaz et al., 2010)
Prangoshulusii	Coumarin	Osthole	DNA gyrase inhibitor	*Bacillus subtilis, Staphylococcus aureus, Klebsiella pneumoniae,*	(Tan et al., 2017)

These medicinal plants while applied in form of extracts have proven to impose inhibitory effects against clinical isolates. According to Dahiya and Purkayastha (2012) observed that different solvent-based extracts including Neem, Lemongrass, Aloe vera Oregano, Thyme, Rosemary, Bryophyllum and Tulsi exhibits inhibitory effects versus ten MDR clinical isolates (including Gram-positive and Gram-negative bacteria). This study correlated the two important nutraceuticals tannins and flavonoids, present in methanolic extract with antimicrobial activities against methicillin-resistant *S. aureus*, responsible for common osteomyelitis. *Lawsonia inermis* with its four different extract exhibits significant bactericidal activity against *S. typhi, E. coli, Klebsiella* spp., *S. sonnei S. aureus, B. subtilis, S. epidermidis* clinical isolates and interestingly never predispose any sign of toxidrome (Gull et al., 2013). As already discussed *S. aureus* is still the most frequent pathogen that causes osteomyelitis (De Mesy et al., 2018). Effective treatment for such infection is still inadequate, as it is capable of producing biofilm and adapts many preventive approaches against antibiotics (see Section 3.2.3). Therefore it is very common that *S. aureus* survives the drug therapy and develops resistance. Besides, conventional antibiotic therapy has several adverse effects on the patient's health, resulting in poor immune health development. According research report, there are some drugs which can temporarily be able to suppress the osteomyelitis (in animal model) developed by methicillin-resistant *S. aureus* (MRSA), but they develops severe side effects like diarrhea, poor digestion and absorption of food leading to malnutrition and sometimes death of the animal model (Vergidis et al., 2015). Thus development of new, effective and safe therapeutic strategies is urgently needed to overcome MDR strains and the adverse effects. Eventually, a common strategy is observed to be effective such as combination of plant derived nutraceuticals with different antibiotics (Bezerra Dos Santos et al., 2015).

Extracts isolated from *Zingiber officinale, Curcuma longa* and *Tinospora cordifolia* exhibits significant killing of osteomyelitis causing bacteria *S. aureus*, and others like *K. pneumoniae, P. aeruginosa, E. coli, P. mirabilis* and *B. subtilis* (Chakraborty et al., 2014). One summarized list of potential medicinal plant resources possessing effective antibacterial activity against a series of pathogens is given in Table 10.1. Other than that, in case of nutraceuticals, in one single formulation may represents complex mixture of many phytochemicals poses potential role against MDR pathogens with distinct mechanism of action. Again pure compounds like Berberine obtained from roots/ stem of Berberis species, possess significant antibacterial activity through RNA polymerase, DNA intercalation, gyrase /topoisomerase, also the inhibition of cellular division (Domadia et al., 2008). The alkaloid compound piperine extracted from *Piper longum* and *Piper nigrum* exhibited inhibitory actions against methicillin-resistant *S. aureus* while co-administered with antibiotics like gentamicin or ciprofloxacin. Piperine used to inhibit the efflux pump against NorA efflux pumps in the consideration of MDR pathogens (Khameneh et al., 2015). Tomatidine the steroidal alkaloid, extracted from *Solanaceous* plants (tomato, potato, eggplant, etc.), exhibits antibacterial activity against drug-resistant *S. aureus*, in synergistic manner along with aminoglycosides (Mitchell et al., 2012). This phytochemical also possess synergistic role while applied along with some common antibiotics like ampicillin, ciprofloxacin, cefepime and gentamicin, tested against microorganisms named *Enterococcus faecalis, S. aureus* and *P. aeruginosa* (Soltani et al., 2017). The organosulfur compound allicin, (belongs to the family of *Alliaceae*) observed to possess antibacterial activity against variety of pathogens such as methicillin resistant *S. aureus, S. epidermidis, P. aeruginosa* and human lung pathogens as well as the periodontitis causing oral pathogens (Reiter et al., 2017). The common mechanism of action of this phytochemical is observed as it inhibits the sulfhydryl-dependent enzyme activity like thioredoxin reductase, RNA polymerase and alcohol dehydrogenase (Lanzotti et al., 2014). Another volatile organosulfur compound, isothiocyanates, isolated from Armoracia rusticana exhibits strong antimicrobial activity against oral pathogens (Fahey et al., 2013).

The flavonoid named kaempferol and its natural derivatives (glycoside), obtained from *Persea lingue,* observed to increase the antimicrobial potentials of ciprofloxacin against NorA overexpressed *S. aureus* strain (Holler et al., 2012). Potential efflux pump inhibitory activities were also observed among Chalcones and catechin gallates (Belofsky et al., 2004). The frequent location of the genes or plasmid is the common characteristics of antibiotic resistance microbes. It also has the capacity to self-replicate; it can circulate the DNAs coding for various gene groups with various functionality. Phytochemicals are also capable to target such plasmid as well (Buckner et al., 2018). For example 8-epidiosbulbin-E-acetate obtained from *Dioscorea bulbifera* exhibits cure to the antibiotic-resistant R-plasmids of clinical isolates of *E. coli*, *E. faecalis*, *P. aeruginosa* and *Shigella sonnei* with 12–48% curing efficiency (Shriram et al., 2008). Table 10.2 summarizes different nutraceuticals having potent antibacterial activities with specific doses effective against *S. aureus,* the common pathogen of osteomyelitis. There are many more phytocompounds and its successful application against various MDR pathogens observed lately, although very few are still now implemented in therapeutic treatment of infections and the commercial application of such phytochemicals is still low and urgently needs one drastic improvement regarding the correct application of this natural compounds against treating infection.

TABLE 10.2 Nutraceuticals and its Effective Dose Against *S. aureus* the Common Pathogen in Osteomyelitis

Plant Name (Family) Parts	Name of the phytocompound	Bacterial species	Antibacterial Dosage	Reference
Rauwolfia serpentina	Reserpine	*Staphylococcus* sp.		(Sridevi et al. 2017)
Bay laurel	*Laurus nobilis*	*S. aureus*	3.91 to 15.62 mg/ml	(Merghni et al, 2012)
Brassicaceae family (broccoli, cauliflowers &cabbage)	isothiocyanates	*S. aureus*	0.84 mg/ml	(Fahey et al., 2013; Moon et al., 2010)
Cruciferous vegetables, especially watercress	2-phenylethyl-isothiocyanate	*S. aureus*	100 µg/ml	(Abreu et al., 2013)
Shation Pummelo		*S. aureus*	35 µg/ml	(Neng Guo et al. 2010)
Coffee	Caffeic acid	*S. aureus*	1600 µg/ml	(Lim et al., 2016)
the leaves of bearberry, in pomegranate root bark	Gallic acid	*S. aureus*	1750 µg/ml	(Sarjit et al., 2015)
whole grains, spinach, parsley, grapes	Farulic acid	*S. aureus*	1100µg/ml	(Borges et al., 2013).
Essential oils	Eugenol	Methicillin-resistant *Staphylococcus aureus*		(Yadav et al., 2015)
Asphodelus	Asphodelin A	*Staphylococcus aureus*,		(El-Seedi et al., 2007)

10.4.2 NUTRACEUTICALS IMPROVING BONE HEALTH

Development of bone depends on various factors and one of the primary contributing factors is Nutrition. In the last few decades many researches have heightened the relationship between bone health and immunity. Evidences revealed immune cells including regulatory T cells, natural killer T cells, B cells, conventional T cells, dendritic cells, neutrophils, mesenchymal stem cells all are detected in bone marrow (Zhao et al., 2011). In Rheumatoid arthritis, osteoporosis and in even bone fractures, immune cells are present to release a variety of effectors to regulate bone health (Guder et al., 2020). For instance, macrophages developed from the same precursor cells as osteoclast, even many immune cells influence the activities of osteoblast and osteoclast (Mori et al., 2013). The pathway of osteoclast requires the presence of a receptor activator for NFkappaB ligand (RANKL) and macrophage colony stimulating factors (M-CSF). Replacement for RANKL constitutes LIGHT chain enhancer of activated beta cells, TNFalpha and interleukins 6,11 and 8 and in case of M-CSF substitute includes vascular endothelial growth factor (VEGF), hepatocyte growth factor (HGF) and placental growth factor (PGF) (Knowles & Athanasou, 2009). The immune system are regulated in Bone marrow reservoir, in research it has been found that the memory lymphocyte have an important place there (Walsh et al., 2006).

Osteoblasts (OBs) known as the bone forming cells, and osteoclasts (OCs) the bone resorbing cells tightly maintained the bone remodeling which takes place continuously, (Rachner et al., 2011; Roodman, 1999). In osteogenesis and bone remodeling the transition of mesenchymal stem cells (MSCs) and monocyte/macrophage precursors into differentiating OCs are the main events (Boyle et al., 2003; Teitelbaum, 2000). Some signaling pathways are there which play a critical role in modulating the expression of genes and differentiation of lineages (Khanna et al., 2007). Some of the bone diseases like osteopetrosis, periodontal problems, and bone cancer metastases are caused be causes of impair function of these two cell types (Sethi & Aggarwal, 2007).

95% of our bone cells are made with osteocytes, it is located within the bone matrix. Ratio of osteocytes increases with age and size of the bones. It sends their dendritic process through tiny tunnels canaliculi to form a network which connects to the cell and to the vasculature (Dallas et al., 2013). In a mice model it has seen ablation of osteocytes leads to severe lymphopenia and the effects were reversed once osteocytes were replenished (Sato et al., 2013). Myelopoiesis (non-lymphoid leukocytes) and osteopenia is regulated by osteocyte by GS α signaling which affects G-CSF production (Azab et al., 2020; Fulzele et al., 2013). In another mice study it has seen elevation of Tumor necrosis factor-α, Interleukin-6, osteoclastogenesis regulators (RANKL and OPG) increases bone resorption which decreases bone formation and effects osteocytes synthesis (Metzger et al., 2017). Thus, osteocytes play an important role in bone formation. Osteocyte functioning is affected by pro-inflammatory signals which affects bone turnover as well (Metzger et al, 2019).

Osteoblast are cells which synthesize bones and mineralization of skeletons, required in initial formation of bone and even in bone remodeling. Though primary function of osteoblast is synthesis of bone matrix but it has seen it plays an important role in osteomyelitis. The homeostasis is disrupted during infection of *S aureus* results in production of cytokines, chemokines which affects innate and adaptive immunity (Josse et al., 2015b). On a recent study it has been mentioned osteoblast helps in trading leukocytes to sites of bacterial challenge, initiate development of cell mediated host responses to intracellular pathogen of bone tissue (Marriott, 2004). Osteoblast ablation leads to lymphopenia as it reduces common lymphoid progenitors (CLPs) even activation of osteoblast pharmacologically boost condition of sepsis induced lymphopenia (Terashima et al., 2016). Infection by *Staphylococcus aureus* is one of the primary factors to develop osteomyelitis and interlink between *S. auereus* and osteoblast plays a pivotal role in bone homeostasis. *S. aureus* interacts between extracellular spaces and intracellular space for internalization (Hudson et al., 1995). *S aureus* decreases the activity of osteoblast by apoptosis and independent mechanism.

10.4.3 SIGNALING PATHWAYS PLAYS CRUCIAL ROLE IN BONE HEALTH AND ROLE OF NUTRACEUTICALS IN IT

Signaling pathways are vital in commencement and isolation of OBs and OCs.

Various studies have linked the crucial players in the process of osteogenesis (Iniguez-Ariza & Clarke, 2015).

10.4.3.1 THE TGFβ SIGNALING

The TGFβ superfamily members bind and activates through binary type I and II transmembrane receptors, consists of *serine/threonine kinase* disciplines.

Smad proteins play an important part in transmitting signals from the receptor to the nexus (Miyazono et al., 2000; Derynck & Zhang, 2003). One such members of TGFβ superfamily, bone morphogenetic proteins (BMPs) are critical in the osteogenesis process (Chen et al., 2012; Wozney et al., 1992; Zhang et al., 2014). Being in colorful isoforms, BMPs intervene in a variety of conditioning in shell apkins (Rahman et al., 2015). BMP-2,-6, and-7, and-9 promotes bone conformation (Kochanowska et al., 2007). In this case BMP-3 inhibits bone conformation (Kokabu et al., 2012; Wang et al., 2014a).

BMP intermediated signaling pathways are both Smad-dependent as well as independent (Derynck & Zhang, 2003). In case of BMP stimulation recap factor Runx2 and Smads physically interacts and regulate the recap of genes which facilitates isolation of mesenchymal stem cells (MSCs) to osteoblast (Phimphilai et al., 2006; Javed et al., 2008).

It is not yet clear how BMP modulates the expression of Runx2. Recent studies have shown that BMP doesn't regulate directly Runx2 expression in mesenchymal cells (Lee et al., 2003), but rather it modulates the expression of distal lower homeobox 5(DLX5) in osteoblasts (Ryoo et al., 1997; Lee et al., 2003), which induces Runx2 in osteo-ancestor cells (Lee et al., 2003).

Also, BMP signaling also regulates genes critical in osteoblastic isolation similar as Hairy/enhancer of split related with YRPW motif 1 (Hey1; also called HesR1 and Herp2), Tcf7, ITF-2 (Tcf4), and interferon nonsupervisory factor 8 (ICSBP) (Franceschi et al., 2003; Chen et al., 2012). The initiation of MAPKs including ERK, p38 and JNK regulated by BMP-2 in osteoblastic cells. The BM signaling convinced MAPKs have distinct places in regulation of osteocalcin (Aubin et al., 2004; Broege et al., 2013; Rahman et al., 2015). Likewise, BMP stimulated p38 MAPK and Smads controls the MSCs isolation by Runx2 (Aubin et al., 2004). Also, studies using mouse ancestor cells and chondrocytes have demonstrated that BMP-2 induces Osterix (Osx) Matsubara et al., 2008).

Several studies have demonstrated that BMP signaling is intertwined in pathogenesis of variety of bone diseases similar as bone metastasis, brachydactyly type A2, and osteoarthritis (Dathe et al., 2009; Katz et al., 2013; Shen et al., 2014).

The group of Phimphilai et al. (2006), showed that repression of BMP signaling is responsible for the disruption of the osteoblast isolation (Phimphilai et al., 2006). According to Tsuji et al. (2006) it stoked the part of BMP-2 fracture form (Tsuji et al., 2006). Overall, reports on it suggest that BMP-2 signaling is needed for bone redoing.

Several studies have illustrated that natural agents induce osteoblast specific isolation through TGFβ-Smad signaling pathways. Along these lines it's shown that quercetin, celastrol, andrographolide, silymarin, and honokiol shown reduce the TGFβ-Smad signaling pathways and therefore induces the osteoblast specific isolation (Kim et al., 2006; Choi, 2011; Kim et al., 2012a; Casado Diaz et al., 2016; Zou et al., 2016).

10.4.3.2 FIBROBLAST GROWTH FACTOR MEDIATED SIGNALING

The fibroblast growth factors (FGFs) belong to the buried polypeptide family. FGFs bind to FGF tyrosine kinase receptors (FGFRs) and modulate several important natural events critical in endochondral and intramembranous ossification (Ornitz, 2005; Su et al., 2014). During embryonic development, FGFRs are expressed in the condensing mesenchyme, perichondrium and periosteum, which evolve in cartilage, osteoblasts, and cortical bone, independently (Ornitz & Marie, 2015; Teven et al., 2014). Mutations in FGFRs are linked with pathological conditions in humans including craniosynostosis and chondrodysplasia runs (Johnson and Wilkie, 2011; Ko, 2016). Still, the part of FGFR 1 signaling in OBs isolation is antithetical, as it's also reported that this signaling suppresses isolation of OBs. Presumably, FGFR1 signaling does act in a stage-specific manner (Jacob et al., 2006). Nevertheless, it has been demonstrated that FGF activates the recap factor Runx2 by MAPK pathways, and therefore regulates bone conformation (Park et al., 2010). Studies have shown that apigenin, wedelolactone, berberine, curcumin, oleanolic acid, emodin, genistein, quercetin, silymarin, resveratrol and withanolide spark the recap factor Runx2 through MAPK pathways, thus regulate the bone conformation (Lee et al., 2008; Gu et al., 2012; Kim et al., 2012a; Khedgikar et al., 2013; Goto et al., 2015; Zhang et al., 2015).

10.4.3.3 WNT SIGNALING PATHWAY

For bone cell signaling the Wnt/β-catenin pathway is important (Baron and Kneissel, 2013; Wang et al., 2014c). The requirement of Wnt signaling for that embryonic skeletal development and adult skeletal remodeling is unequivocally established (Wang et al., 2014c). Report of Krishnan et al. in 2006, first demonstrated that Wnt signaling plays a critical part in bone development. The main cause of alternation in bone mass is mutation in LRP5, a co-receptor of Wnt (Krishnan et al., 2006). In LRP5 eight missense mutations have been marked as causing high bone mass (Balemans et al., 2007; Niziolek et al., 2015), while several heterozygous and homozygous frameshift, nonsense, and missense mutations have been identified in osteoporosis-pseudo glioma patients responsible for the occurrence of low bone density (Cheung et al., 2006; Laine et al., 2011). By upregulating the development of OBs, canonical Wnt/β-catenin signaling increases bone mass (Yavropoulou & Yovos, 2007; Regard et al., 2012;). Higher levels of β-catenin signaling overexpress the expression of genes implicated in various OBs (Zhang et al., 2013).

10.4.3.4 PARATHYROID HORMONE SIGNALING

To build bone Parathyroid hormone (PTH) can be targeted therapeutically (Morley et al., 2001). PTH has been shown effective treatment for osteoporosis because depending on the method of administration PTH exerts either a catabolic or anabolic effect, (Morley et al., 2001; Thomas, 2006). Recent studies explaining the structure of PTH, PTH/PTHrP receptor and parathyroid hormone-related protein (PTHrP), have further enriched the understanding of its function in calcium and bone biology (Mundy & Edwards, 2008). It is well accepted that PTH/PTHrP axis play a crucial role in bone metabolism, many more studies are required in order to get proper explanation whether agents derived from natural agents modulate PTH/PTHrP axis for bone metabolism. It has been also identified that Phytoestrogens slow down osteoporosis in menopausal group (Kotecha & Lockwood, 2005). Overall, studies strongly proposed that natural agents have the ability to induce osteogenesis by modulating different signaling pathways integral in bone metabolism. Anyhow, further clinical trials and various models are needed to fully acknowledge the potential of natural agents.

10.4.3.5 EPHRINS SIGNALING

Bidirectional signaling mediates by ephrins (Pasquale, 2008). Ephrins are classified into two, class A and B. In B class molecules (ephrin B1 to B3) ligands for EphB tyrosine kinase receptors are available there (B1 to B6), furthermore the class A ephrins (A1 to A5) are ligands for glycosyl phosphatidylinositol (GPI) – anchored EphA receptors (A1 to A10) (Lisabeth et al., 2013). Pattern of the developing skeleton controlled by the axis of ephrin B/EphB receptors. The uncontrolled ephrin signaling is correlated with frontonasal syndrome (Wieland et al., 2004; Edwards and Mundy, 2008;). In communication between OCs and OBs the ephrin signaling is critical (Zhao et al., 2006). EphrinB2 ligand in OCs and EphB4, a tyrosine kinase receptor mediated bidirectional communication (Zhao et al., 2006). Also the reverse signaling from EphB4 in osteoblasts to ephrinB2 in OC progenitors predispose to the inhibition of osteoclast differentiation (Takyar et al., 2013). Some of the osteogenic regulatory factors, such as Osx, Dlx5, and Runx2, in calvarial osteoblasts, suggesting that EphB4 is responsible for rising up in the regulation of osteoblast differentiation induced by EphB4 (Zhao et al., 2006). Such studies show the concept that ephrin-Eph signaling is needed in maintaining bone homeostasis. Many laboratories have shown that curcumin, resveratrol, quercetin, wedelocatone, oleanolic acid, emodin, celastrol, fisetin, genistein, and berberine influences osteoblast differentiation, and thereby helps in preventing bone decay (Lee et al., 2008; Bian et al., 2012; Dai et al., 2013; Mobasheri & Shakibaei, 2013; Zhou & Lin, 2014; Leotoing et al., 2014; Liu et al., 2016a; Casado-Diaz et al., 2016; Zou et al., 2016).

10.4.3.6 HEDGEHOG SIGNALING

The role of hedgehog (Hh) in skeletal signaling pathways has been implicated in many studies (Yang et al., 2015). In patterning the axial, appendicular, and facial skeleton, Indian hedgehog (Ihh) and sonic hedgehog (Shh) are involved during skeletogenesis (St-Jacques et al., 1999; Pan et al., 2013). It has been demonstrated that Ihh is expressed by pre hypertrophic chondrocytes and coordinates growth and distinction of chondrocytes (Chung et al., 2001). Mutation in the Hh pathway is responsible for many skeletal deformities such as brachydactyly type A1, and Gorlin syndrome in humans (Onodera et al., 2017). The effect of homozygous mutation of Ihh is responsible for deregulation of chondrocyte homeostasis (Cai & Liu, 2016). Report by Sreekumar V et al, demonstrated that resveratrol induces osteogenic isolation (Sreekumar et al., 2017), through the inhibition of hedgehog signaling. Many natural agents have been explored for their osteogenic eventuality through hedgehog signaling pathways. Further studies are required to confirm whether natural agents induce bone conformation by modulating the Hh signaling.

10.4.3.7 SIGNALING PATHWAYS IN OSTEOCLASTS

In bone remodeling the role of osteocytes is enlightening in some recent studies. A number of signaling pathways attached with osteoclastogenesis have been discovered. Thus understanding the procedures in the differentiation of OC is crucial in order to evolve novel agents to prevent bone loss. In the TNF superfamily RANKL is an important member. RANKL is also named as TNF-related activation-induced cytokine (TRANCE), OC differentiation factor (ODF) osteoprotegerin ligand (OPGL) (Darnay et al., 1998; Boyce

& Xing, 2007). Demonstrated Mice with profound osteopetrosis resulting from lack of RANK and also shows OC differentiation, also proposing that RANK is essential for OC differentiation (Dougall et al1999). The RANKL mediated signaling is negatively regulated by osteoprotegrin (OPG), which is encoded by TNF Receptor Superfamily Member 11b (TNFRSF11b). The pre-OCs and OCs convey OPG and compete with RANKL for binding to RANK. Low numbers of osteoclasts result because autosomal recessive osteopetrosis influences mutations in the RANKL gene in humans (Pangrazio et al., 2012). The role of RANKL in rheumatoid arthritis has recently been elucidated at sites of pannus invasion into bone. By modulating RANKL signaling variety of nutraceuticals have been shown to inhibit bone loss, some of these are quercetin, resveratrol, curcumin and plumbagin (Bharti et al., 2004; Wattel et al., 2004; Sung et al., 2012; Zhao et al., 2015). Using animal models and *in vitro* studies have revealed that these natural agents inhibit osteoclastogenesis induced by RANKL, indicating that natural agents have the potential to inhibit bone loss.

10.4.3.8 MACROPHAGE-COLONY STIMULATING FACTOR

The macrophage-colony stimulating factor (M-CSF) responsible for adherence of MSCs to pre-OCs (Udagawa et al., 1990; Qiao et al., 1997), which is one of the principal events in OCs differentiation. Different research reports have suggested that proliferation, differentiation, and survival of hematopoietic cells are all based on M-CSF signaling (Feng & Teitelbaum, 2013). Moreover, M-CSF also manages cytoskeletal alterations (Boyle et al., 2003). Reports exhibited that M-CSF deficient osteopetrotic (op/op) mutant mouse develop osteopetrotic phenotype, which firmly indicates that M-CSF is required for OC differentiation (Umeda et al., 1996; Takatsuka et al., 1998). As these deficiencies can be altered by injecting M-CSF (Umeda et al., 1996). Elaborated studies were required in this case to understand full function of MCSF. Likewise, the studies of Cappellen et al. demonstrate that M-CSF induces RANK, TRAF2A, PI3- kinase, and MEKK3 which are critical in regulation of OCs differentiation (Cappellen et al., 2002). Furthermore, M-CSF influences the continuation of OCs precursors through activation of survival protein Bcl-xL (Boyce, 2013). As discussed above, the role of M-CSF on osteoclastogenesis is not fully understandable till date. Nonetheless, studies on natural agents suggest that quercetin (Casado-Diaz et al., 2016), syringetin (Tsai et al., 2015), caffeic acid (Wu et al.,

2012) may prohibit the activation of MSCs to pre-OCs by modulating PI3K kinase pathways. Anyhow, more studies are required to fully map out the mechanism.

10.4.3.9 NF-κB SIGNALING

The inactive dimers of NF-κB are sequestered in cytoplasm by combining with inhibitory protein IκB (Aggarwal, 2004; Hayden & Ghosh, 2008). The upstream kinase, IκB kinase (IKK), phosphorylates IκB, which activates the degradation of inhibitory protein (Aggarwal, 2004; Hayden & Ghosh, 2008). As inhibitory protein is degraded, dimer is liberated from the complex and translocated to the nucleus (Karin and Lin, 2002; Aggarwal, 2004). Patients having X-linked osteopetrosis, anhidrotic ectodermal dysplasia, lymphedema, harbor a X420W point mutation in IKKγ, thus constitutively activated NF-κB, leads to severe osteopetrosis (Roux et al., 2002; Walsh and Choi, 2014; Boyce et al., 2015). Several reports using genetically engineered mouse models suggest that NF-κB pathway is key in RANKL which is responsible for osteoblast development and function ability. RankL can help in both traditional and alternative NF-kB activation in osteoclast and its precursor (Novack, 2011). Removal of IKK2 in mice cause faulty osteoclastogenesis in OC precursor in response to RANKL, TNF- which cause osteoporosis (excessive bone formation) and resistance to inflammatory-associated bone loss *in vivo* (Ruocco et al., 2005). Osteopetrosis and faulty osteoclastogenesis were seen in mice with p50/p52 expunge Franzoso et al., 1997; Iotsova et al., 1997). The proinflammatory cytokines fastens osteoclastogenesis via activation of precursor cells (McLean, 2009; Schett, 2011). However debate is still going on regarding how RANKL signaling responsible for the osteoclast activation. Antibodies against the TNF -αreceptors efficiently diminish the process. For that reason TNF -αmediated OC activation is independent of RANKL signaling While. Inhibition of RANKL by OPG during an earlier stage of OC differentiation (exposed to M-CSF alone) augmented TNF-α-mediated OC activation. Surprisingly, bone marrow myeloid cells deficiency in RANK can differentiate into OCs by TNF-α stimulation. It demonstrated the existence of RANKL/RANK-independent pathway of OC activation (Kim et al., 2005a). It is evident from the above studies the envoy which inhibits NF-kB signaling will have positive effects on bone mass. Many nutraceuticals like example resveratrol, curcumin, quercetin, withanolide, silibinin, rosmarinic acid, obovatol, wedelolactone, gambogic acid, embelin, zerumbone, ursolic acid, silymarin, butein, fisetin, plumbagin, honokiol, and berberine modulate NF-κB signaling pathways thus control the losses of bones (Kim et al., 2009; Leotoing et al., 2013; Kavitha et al., 2014; Kim et al., 2014a; Pandey et al., 2014; Jiang et al., 2015a; Omori et al., 2015; Liu et al., 2016b).

10.4.3.10 SRC SIGNALING

Src is the member of non-receptor tyrosine kinase (NRTK) family (Boggon & Eck, 2004). In genetically engineered mouse models the critical role of Src is suggested and role in osteoclastogenesis was demonstrated. These mice have deficient functional Src and produce an osteopetrotic skeletal phenotype (Marzia et al., 2000). The presence of Src affects the bone-resorbing activity of mature OCs, but remains ineffective in OC formation (Miyazaki et al., 2004). A key function of Src in OCs is to promote the rapid accumulation and non accumulation of the podosomes, the specialized integrin-based attachment structures of OCs and other extremely motile cells (Destaing et al., 2008). All together, such studies have proposed that Src is crucial for bone resorption. It has been reported that wedelolactone (Liu et al., 2016a), and fisetin (Choi et al., 2012) decrease bone loss through the inhibition of Src (Figure 10.3).

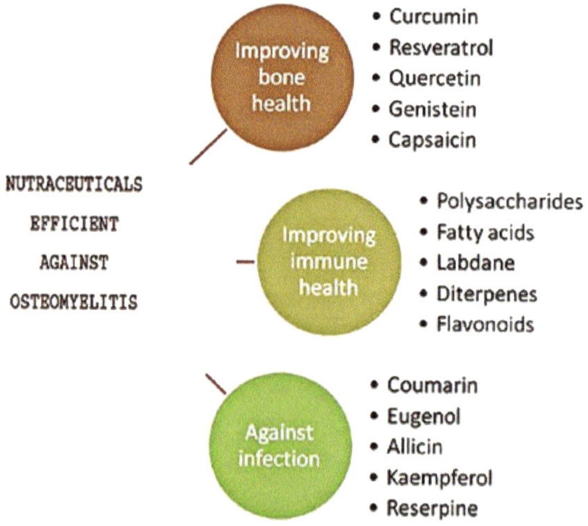

FIGURE 10.3 Beneficial effect of nutraceuticals in treating osteomyelitis.

10.4.4 NUTRACEUTICALS AS IMMUNOMODULATOR

As previously mentioned nutraceuticals have proofs to act as potent immunomodulators to human health. A number of research demonstrated such beneficial effects of nutraceuticals. While fighting against any infection one needs a strong immune response in the host defense. Such benefits are lacking while antibiotic therapy were solely continued to treat chronic infections like osteomyelitis. Hence applications of nutraceutical in treating any infection become beneficial to the host immune system as well, while antibiotic therapy alters the same and made the host vulnerable to have other diseases including poor immune responses. Several clinical and preclinical studies highlighted the capability of *Curcuma longa* L. as an immune booster through initiating the cellular immunity as well as capable to modulate inflammation (Catanzaro et al., 2018). Interestingly, Astragalus *membranaceus* (Fisch.) Bge., *Panax ginseng* C.A. Meyer, and *Withania Somnifera* (L.) Dunal exhibits immune stimulant properties, through improving the resistance to stress (Kaur et al., 2017; Wagner et al., 1994). Ginseng, reported to be able to modulate immune system and gut microbiome and thus remains an efficient immune booster as dietary supplement (Riaz et al., 2019). The details regarding several medicinal plants and its specific role in immune modulation is explained in table 10.3.

Several phytochemicals especially alkaloids, phenols, saponins and glycoproteins shown to possess immunomodulatory and anti-inflammatory properties (Cundell et al., 2014). Polysaccharides, fatty acids and terpenoids are also important immunomodulator and shown promising activity in this function (Cundell et al., 2014). Alkaloids isolated from *Aristolochiaclematitis L.* exhibit potent immunostimulatory properties through enhancing phagocytic activity of macrophages and leukocytes, although it consists of carcinogenic risks. Similarly staurosporine and vincristine exhibits immunostimulant activity at its low dose, while exhibits immunosuppressive activity at higher doses (Jantan et al., 2015). Resveratols, the polyphenols stimulated both humoral and cellular immunity in preclinical models and also it prevents pathogen replication and inflammation. It is responsible for promoting antitumor immune response too (Meng et al., 2020). Again *echinacea* extracted cichoric acid initiates phagocytic activity in both *in vivo* and *in vitro* conditions (Kuban-Jankowska et al., 2016).

Examples of Some important Nutraceuticals act as potent immunomodulator in human body are as follows:

TABLE 10.3 Example of Various Bioactive Constituents Obtained from Plants, and Their Effect on Immune Function for Better Immune Health to Fight Against Infections Like Osteomyelitis

References	Phytochemicals	Effect on Immune Function	Plant source
(Stohs et al, 2015)	Flavonoids, phenolic acids, catechins	Anti Inflammatory activity/Immunoadjuvant	*Acacia catechu* Willd. heartwood
(Zhang et al., 2006)	Acemannan, dihydrocoumarin	Adaptive immunity activation/Immunostimulant	*Aloe vera* (L.) Burm.f
(W. Wang et al., 2010)	Diterpene lactones (andrographolide)	Modulation of innate and adaptive immunity/ Immune Suppressor	*Andrographis paniculate* (Burm.f.) Nees
(Adorisio et al., 2019)	Auronol glycosides (maesopsin 4-O-glucoside and alphitonin-4-O-glucoside)	Inhibition of humoral and cellular adaptive immunity/Immune Suppressor	*Artocarpus tonkinensis A.Chev.* Ex Gagnep
(Zheng et al., 2020)	Polysaccharides	Activation of cellular immunity/ Immunostimulant	*Astragalus membranaceus* (Fisch.) Bge.
(Thomson et al., 2002)	Phenolics (gingerols)	Activation of adaptive humoral	*Zingiber officianalis* Roscoe
(Chandran et al., 2017)	Steroidal lactones (Withaferin A)	Adaptogen effects, stimulation of adaptive humoral and cellular immunity/Immunostimulant	*Withania somnifera* (L.) Dunal
(Haque et al., 2017)	Arabinogalactan	Activation of adaptive immunity/Immunostimulant	*Tinospora cordifolia* (Thunb.) Miers
(Wael et al., 2018)	Monoterpenes (eugenol), sesquiterpenes (β-caryophyllene)	Activation of adaptive humoral immunity/Immunostimulant	*Syzigium aromaticum (L.)* Merr. and L.M. Perry

TABLE 10.3 *(Continued)*

References	Phytochemicals	Effect on Immune Function	Plant source
(Wagner et al., 1994)	Triterpene saponins (ginsenosides)	Adaptogen effects, stimulation of immune systems via cytokine activation, modulation of gut microbiome	*Panax ginseng* C.A. Meyer
(Di Sotto et al., 2020)	Monoterpenes (eugenol and methyleugenol), sesquiterpenes (β-caryophyllene)	Activation of innate and adaptive immunity/Immunostimulant	*Ocimum sanctum* L.
(Matyanga et al., 2020a)	Phenolic glycosides (hypoxoside)	Anti-Inflammatory properties/Immunoadjuvant	*Hypoxis rooperi* T. Moore
(Matyanga et al., 2020b)	Triterpene saponins (glycyrrhizin)	anti-inflammatory activity/Immunostimulant	*Glycyrrhiza glabra* L.
(Khanizadeh et al., 2017)	Alkylamides, polysaccharides, glycoproteins,	Activation of cellular immunity and modulation of gut microbiome	*Echinacea purpurea (L.)* Moench
(Mollazadeh et al., 2017)	Curcuminoids	IL-10-mediated anti-inflammatory and immuno-suppressive activity/Immunosuppression	*Curcuma longa* L.
(Gohil et al., 2010)	Madecassoside	Regulation the abnormal humoral and cellular immunity/Immune Suppressor	*Centella asiatica* (L.) Urban

10.4.5 FATTY ACIDS AS IMMUNOMODULATOR

Fatty acids (FA) are the lipids, characterized by several numbers of carbon, structured in a linear skeleton of carbon with variable length having a terminal carboxylic group (Fahy et al., 2009). FA occurs widely in nature. Human diet consists both from plant origin and animal origin as well. Among them lauric acid and myristic acid the medium chain fatty acids (MCFA) are the principal component of oils obtained from some *Lauraceae* and *Myristice* especies (Arkcoll et al., 1988). The common saturated fatty acid (SFA), the palmitic acid present in vegetable oil like palm oil (Fattore et al., 2013). Among the FAs the most potent immunomodulators are from long-chain classes; punicic acid one conjugated triene, obtained as a unique component of pomegranate seed oil (Aruna, 2016). Table 10.4 summarizes common sources of polyunsaturated fatty acids (PUFA) and long-chain monounsaturated (MUFA) associated with immune system modulating activities and relative amounts.

Plant based edible oil like sunflower oils, hemp seed, rapeseed oils consist of high linolenic acid (LA), essential ω6 fatty acids. FA is the principal component of phospholipids, monoglycerides, diglycerides, and triglycerides. The long chain FA s are capable to modulate both the innate and adaptive immunity (Yaqoob et al., 2002). One potent immunomodulating nutrient is oleic acid, as it diminishes NK (Natural Killer) cell activity (Jeffery et al., 1997), capable to initiate expression of the leukocyte adhesion molecules (Jeffery et al., 1997), capable to enhance neutrophil aggression, phagocytic and candidacidal capacities, and enhancement of neutrophil-endothelial cell attachment (Mastrangelo et al., 1998). Considering adaptive response oleic acid capable to inhibit proliferation of some specific immune cells like Jurkat T cells and lymphocytes and thus capable to regulate cell cycle (Yaqoob et al., 2002). Again minerval and oleic acid predispose proapoptotic effects in B lymphocyte cells and Jurkat (T lymphocyte), through mitochondrial depolarization and ROS production (Llado et al., 2010; Cury-Boaventura et al., 2004; Cury-Boaventura et al., 2005). Oleate exhibits protective role among macrophages from palmitic induced lipotoxicity, also thee is association with the rise regulatory phenotype of the myeloid MSC-2 suppressor cells and as a result suppression of activated T cells took place respectively (Kim et al., 2017; Wu et al., 2017). The oleic acid altogether with other unsaturated FA, exhibits important role in the lipid moiety of *Staphylococcus aureus* Lpp, thus provides one protective role against pathogen (Nguyen et al., 2015). Punicic acid, the conjugated PUFA helps in improving immune system development, through stimulating $CD8^+$ and $CD4^+$ lymphocyte-mediated

TABLE 10.4 Major Natural Sources of Long-chain Monounsaturated (MUFA) and Polyunsaturated Fatty Acids (PUFA) Associated with Immune System Modulating Activities and Relative Amounts

Natural sources	Species (Byproduct)	Amount%	References
Conjugated PUFA Punicic acid	*(Pomegranate seed oil) (bitter Punica granatum L gourd seed oil)*	70–85%	(Aruna et al., 2016)
	Momordica charantia L	1.5–16%	(Yoshime et al., 2016)
Essential ω6-PUFA	*Borago sp. (seed oil)*	26.8–37.9%	(Guil-Guerrero et al., 2018a)
Linoleic acid (LA)			
	OenotherabiennisL.	70–74%	(Timoszuk et al., 2018a)
	CannabissativaL.	27–58%	(Farinon et al., 2020; Saini et al, 2018)
	Glycine max (L.)Merr.	50.4%	(Farinon et al., 2020; Saini et al., 2018)
	Bertholletia excelsa	23.8%	(Farinon et al., 2020; Saini et al., 2018)
	JuglansregiaL.	33.8%	(Farinon et al., 2020; Saini et al., 2018)
	Zeamays L.	53.2%	(Farinon et al., 2020; Saini et al., 2018)
γ-Linolenicacid (GLA)	*Borago sp. (seed oil)*	9.6–39.8%	(Guil-Guerrero et al., 2018b)
	OenotherabiennisL.	8–10%	(Timoszuk et al., 2018b)
	Ribesnigrum (seed oil)	15–19%[a]	(Timoszuk et al., 2018b)
	Echium spp.	31.2–47.1%	(Guil-Guerrero et al., 2018b)
	Ranunculus spp.	37.9–39.7%	(Guil-Guerrero et al., 2018b)
Essential ω3-PUFA			
Linoleic acid (LA)	*Linumusitatissimum* L.	55–57%	(Saini & Keum et al., 2018b; Parikh et al., 2018)
	*Salviahispanica*L.	17.8%	(Saini & Keum
	*Perilla frutescens var. frutescens*L.	54–64%	(Asif et al., 2011)
EPA, DHA	Fish Oils	4–13%	(Saini & Keum et al.,
MUFA			
Oleic acid	*Olea europaea L*	56–84%	(Servili et al., 2013)
	*Vitisvinifera*L (rapeseed oil)	9–17%	(Lachman et al., 2015)
	Brassica Spp.	60–80%	(Nath et al., 2016)

immunity, also increases immune responses against pathogens (Viladomiu et al., 2013). Such immune boosting by punicic acid done by the nuclear peroxisome proliferator-activated receptor (PPAR)γ- and δ-dependent mechanisms, as punicic acid act as agonists to these receptors, thus the loss of PPARγin immune cells impaired the effects (Yuan et al., 2015). Interestingly TNF-α-induced priming ROS production was inhibited by punicic acid through Ser345-p47phox phosphorylation and also upstreaming kinase p38 MAPK. Thereby TNF-α-induced release of myeloperoxidase from neutrophils was blocked and reduced the ROS/MPO-mediated tissue damage and neutrophil-activation *in vivo* (Boussetta et al., 2009). Furthermore, activation of PPARγ has relationship with anti-inflammatory response as it suppressed the expressions of inflammatory genes (chemokines, encoding cytokines, cyclooxygenase, NO synthase, and metalloproteinases) (Aruna et al., 2016).

ω-3PUFA and ω-6 PUFA both has important immunomodulatory properties exhibits strong potential in modulating different inflammatory processes (Lorente-Cebrián et al., 2013). All of them share common biosynthetic enzymes, commonly mediate the synthesis of different series of eicosanoids, including arachidonic acid (AA), dihomo-γ-linoleic acid (DGLA), and eicosapentaenoic acid (EPA). Importantly the three prostaglandins (PG), the PG1, PG2 and PG3 also produced through the same pathways. The PG1 is responsible for lowering inflammation, thus being recognized as anti-inflammatory prostanoid; whereas PG2 induce inflammation, vasoconstriction and blood clotting. PG3 poses different approach as mixture of function, though capable to reduce PG2 mediated inflammation (Lorente-Cebrián et al., 2013). In the overall process the activity of PG1,2 and 3 relied on the Δ5-desaturase and Δ6-desaturase enzymes activity, often compromised during infections like osteomyelitis. Study proofs that diet rich with ω-3 fatty acids are capable to initiate the conversion of DGLA into PG1, whereas reduced ω-3 intake increases the conversion in AA along with the synthesis of proinflammatory prostanoids (Caldar et al., 2005). During cell injuries or during changes in the biomembrane composition AA can be released through the action of phospholipase A2, critical trigger of inflammation as one defense response. During the process of PG synthesis eicosapentaenoic acid (EPA) and AA compete for the different PG synthesis, mediated by cyclooxygenase, while 5-lipoxygenase (LOX) is responsible in their conversion into leukotrienes and thromboxanes. AA is responsible for production of leukotriene B4 and thromboxane A2, while EPA produces thromboxane A3 and leukotriene B5. The most importantly DHA and EPA are responsible for producing protectins, resolvins and lipoxins, responsible for regulating vascular tone, anti-inflammatory effect and blood pressure (Calder et al., 2005). Like punicic acid the EPA and DHA mediated anti-inflammatory effects are predisposed through PPARα/γ (Echeverria et al., 2016). ω-3 and ω-6-based diets have benefits as anti-inflammatory agents but still AA increases the plasmatic levels of proinflammatory eicosanoids, responsible for an increased rate of allergic and inflammatory disorders along with excessive cell proliferation. However, the importance of FA as incorporated in the cellular membrane seems to represent key mechanisms in exhibiting immunomodulating properties. Importantly the immune cells like neutrophils and T cells incorporate exogenous fatty acids into the membrane with alteration in the efficiency of cell surface pattern recognition receptors (Kumar et al., 2019). Dietary ω-3 PUFA activates G protein coupled receptors (GPR) and induced shift to an anti-inflammatory phenotype, thus it is capable of modulating macrophage function (Radzikowska et al., 2019). ω-3 PUFA capable to modulate signaling via GPRs receptor activation, promote leukocyte function. For example ω-3 PUFA capable to inhibit proinflammatory phenotype of T cells and dendritic cells responses (Radzikowska et al., 2019). They are also responsible to inhibit monocyte and neutrophil adhesion, depending on the activation of PPAR-α (Radzikowska et al., 2019). ω-6 PUFAs also promote inflammation associated with raised ROS levels in neutrophils (Radzikowska et al., 2019).

Linoleic acid responsible for raising up the neutrophils in tissues through induction of adhesion molecules, complexed with 1-antitrypsin, the anti-inflammatory molecule and thus helps in reduction of LPS-induced IL-1 secretion in neutrophils, thereby promoting innate immunity to the human immune system (Radzikowska et al., 2019). Considering adaptive immunity, ω-3 PUFA plays important role by improving the mitogen-mediated immune cell activation, helps in promoting development of aTH2-type immune response (Kumar et al., 2019). Additionally it is responsible for increase in anti-inflammatory cytokine IL-4 and reduction in pro-inflammatory TNF-α (Kumar et al., 2019). Similar immunomodulatory effect was observed in both fish oil enriched diet and the purified DHA and EPA (Radzikowska et al., 2019). ω-3 PUFA is also responsible to modulate inflammation of epithelial cells, capable to restore impaired barrier function and thus responsible for the reduction of proinflammatory mediators (Radzikowska et al., 2019). Considering gut microbiota a strong relationship observed between body immune response and omega-3 fatty acids, and seems to have essential component to maintain intestinal wall integrity. Omega-3 FA responsible for symbiosis effect in gut microbiota and thus resulting sharp rise in the production of anti-inflammatory compounds, like short-chain fatty acids (Costantini et al., 2017). ω-3and/orω-6PUFA from plant source also exhibiting promising immunomodulatory activities recently, the major focuses on *Linumusitatis-simun*L., *Boragoofficinalis* L and *Oenotherabiennis* L (Asadi-Samani et al., 2014). The flaxseed oil (*L.usitatissimum*), known to exhibit suppressing cell mediated immunity, without interfering with humoral immunity. It is also a rich source of ALA (Kasote et al., 2012). Also known to be effective in reduction of skin inflammation (Park et al., 2011). Oils extracted from *O.biennis* are *known* to enhance PGE1 synthesis in peritoneal macrophages (Hamburger et al., 2002).

10.4.6 PLANT POLYSACCHARIDES AS IMMUNOMODULATOR

Polysaccharides are naturally present in every living cells playing different role according to its requirements. It is present in body fluids, cell walls, bacteria, fungi and yeast, extracellular fluids, plant seeds, leaves, stemsetc. Plant polysaccharides are potent Immunomodulator in this case, constitute an ideal alternative for immune modulation. Several plant polysaccharides with its function as immunomodulators. Some of the most studied polysaccharides include glucomannan from *Amorphallus konjac* Koch, acemannan and aloeride from *Aloe vera* L., panaxanes from *Panax ginseng* C. A. Meyer and fucogalactoxyloglucan from *Echinacea purpurea* L. (Moench) etc. (Ferreira et al., 2015; Schepetkin et al., 2006). Plant polysaccharides capable to modulate innate and acquired intestinal immunity through direct or indirect mechanisms, including activation of macrophages, dendritic cells, Tcell, B cell and natural killer cells also short chain fatty acid formation (SCFA). Plant polysaccharides capable to modulate ROS & NOS generation in macrophages and stimulates cytokines secretion, proliferation of cell and macrophages phagocytic activity (Ramesh et al., 2002). Membranaceus polysaccharides promoted NO synthesis in macrophages through inducing iNOS (inducible Nitric Oxide Synthase) gene expression through induction of nuclear factor kappa-B(NF-κB)/Rel (Yin et al., 2019). Furthermore, it increases macrophage mediated phagocytic activity through enhancing their secretion of release factor as well as increasing the intracellular Ca^{2+} concentration (Lee et al., 2005; Eman et al., 2014). From *Citrus unshiu* Marc the pectic polysaccharides were extracted, exhibits pro-inflammatory cytokine regulations through increasing the production of tumor necrosis factor (TNF)-α, interleukin (IL)-6 and anti-inflammatory cytokine IL-12 in macrophage RAW264.7 (Qin et al., 2012). From *E. purpurea* arabinogalactan was extracted and exhibits macrophage activation as well as increased production of IL-1, TNF-α and interferon (IFN)-β (Zha et al., 2007). Plant polysaccharides basically interact with specific receptors present on their surface, initiates immune response and exerts immunomodulatory response. These pattern recognition molecules include: CD14, complement receptor 3 (CR3), Toll-like receptor 4 (TLR4), mannose receptor (MR), Dectin-1 and scavenger receptor (SR). These specific receptor mediated activities determines a series of intracellular signaling cascades, resulting transcriptional activation and regulation of inflammation-related cytokines (Ramesh et al., 2002). Pectin exhibits reduced IL-6 and IL-10 release. Inulin, pectin β-glucan and arabinoxylan observed to elevate IL-10/IL-12 ratio. Also they are responsible to reduce the release of IL-6, IL-12, IFN-γ, IL-1, IL-8, TNF-α and monocyte chemoattractant protein (MCP)-1 by dendritic cells, macrophage inflammatory proteins (MIP)-1α, and RANTES (Sahasrabudhe et al., 2018). *E. purpurea* extracted polysaccharides exhibits phenotypic promotion and maturation of dendritic cells by modulating *c-Jun N-terminal kinase* (JNK), and NF-κB pathways, p38 *mitogen-activated protein kinase* (MAPK) (Cai et al., 2020). Plant polysaccharides capable to initiate natural killer (NK) cells activation (Li et al., 2017). It is capable of rising CD3-CD4-CD8+ NKs in peripheral blood lymphocytes (Li et al., 2009). Adaptive immunity also modulated by plant polysaccharides (*A. membranaceus*) through significant upregulations of B lymphocyte proliferations via interrelation with immunoglobulin on the surface of B cells (Ghoneum et al., 2004; Fan et al., 2012). It also capable to enhance $CD4^+/CD8^+$ T cell ratio (Hong et al., 2018). Plant based nutraceuticals prove to be potent immunomodulatory agents and represent one interesting tool for treatment and prevention of immune system disorders, as they exhibit multiple bioactivities, well tolerability and good patient compliances.

10.4.7 SPECIFIC NUTRACEUTICALS AND ITS ROLE IN OSTEOMYELITIS AND BONE HEALTH

For healthy bones it is always desirable to maintain homeostasis between osteoblasts (OBs) and osteoclasts (OCs). Thus, other than common dietary constituents there are many plant-based components, can modulate the expression of key modulators involved in bone health. And these are becoming focused areas nowadays as an alternative approach towards better bone health. Table 10.5 summarized the different nutraceuticals, and exhibits a key role in regulating proper bone health.

Brief discussion of such nutraceuticals are as follows:

10.4.7.1 RESVERATROL (VITIS VINIFERA)

Resveratrol extracted from white hellebore roots (*Veratrum grandiflorum O. Loes*. Also, it is extracted from various plants including berries, grapes and peanuts. Resveratrol exhibits suppression of the NF-κB activation as key regulator of bone loss (Manna et al., 2000). It also induces SIRT1, and capable to regulates RUNX2 the by inducing the differentiation of OBs (Zainabadi et al., 2017). Resveratrol inhibits osteoclastogenesis through deacetylating Fox Osin mouse model (Kim e tal., 2015a). Again *in vivo* model inhibits methotrexate induced bone loss (Lee et al., 2017). Resveratrol supplementation inhibits bone loss (Dosier et al., 2012; Tou, 2015; Chen et al., 2015); Resveratrol attenuates liposaccharide (RPL) inhibits osteoblastic differentiation which plays an important role in the pathogenesis of osteomyelitis. RPL reduces osteogenic differentiation as indicated by the significant decreased levels ALP, OCN, OPN thus inhibits osteogenic differentiation in MC3T3-E1 cells and protect against osteomyelitis (Ma et al., 2018). It is a polyphenolic phytoestrogen with osteogenic and osteoinductive properties (Mobasheri et al, 2013). Due to its anti-oxidant properties Resveratrol preserves the microstructure and strength of the bone as it has seen in mice model (Durbin et al., 2013). Even resveratrol has positive effect on post-menopausal osteoporosis prevention (Lin et al., 2005). With Type 2 Diabetes it exhibits positive effects on bone density if 500 mg Resveratrol supplemented (Bo et al., 2018)

10.4.7.2 CURCUMIN (CURCUMA LONGA)

Curcumin is extracted from turmeric, traditionally used as important spice in Indian Subcontinent Forman Years (Aggarwal et al., 2006). Several related studies conducted *in vitro* and *invivo* models indicate the protective role of curcumin against bone loss, through downregulating the NF-κB activation, -Wnt/β-catenin, TNF-α and RANKL all are interconnected with bone loss (Kim et al., 2012; von Metzler et al., 2008). It is also capable to inhibit the osteoclastogenesis through HMGB1 deliver in a P38-MAPK dependent mechanism (Fujii et al., 2012). Curcumin also inhibits RANKL induced osteoclastogenesis (Bharti et al., 2004). Again curcumin possesses potential bone loss (Kim et al., 2012).

It reduces human osteoclastogenesis through inhibition of the signal some-associated IκB kinase (Von Metzler et al., 2009). Considering ovariectomy-induced bone loss curcumin plays protective role (Kim et al., 2012). It also protective against breast cancer induced bone loss (Yamaguchi et al., 2015). Curcumin exhibits osteoblast differentiation of human adipose-derived MSCs through inhibiting Wnt/β-catenin signaling (Wang et al., 2009). Erythromycin and curcumin can subdue S. *aureus* growth. On a study it has been seen on methicillin resistant *S.aureus* rats after treating with erythromycin and curcumin reduces pro-inflammatory cytokines tumor necrosis factor (TNF) and Interleukin -6, it was also evident curcumin suppressed bacterial growth (Zhou et al., 2017). Curcumin does have inhibitory effect on NorA pumps and reduces antibiotic resistance in *Staphylococcus aureus* this helps in the treatment of osteomyelitis (Jaberi et al., 2018). In rat model it has seen curcumin increases mesenchymal stem cells osteoblast differentiation but inhibits adipocytes formation as it correlates with HO-1 expression (Gu et al., 2012).

10.4.7.3 GENISTEIN (GLYCINE MAX)

Mostly found in soy and a variety of food products. This isoflavone demonstrated to prevent bone loss, observed in many *in vitro* and *in vivo* studies (King et al., 2015; Zhang et al., 2016; An et al., 2016). Isoflavone induces osteoblast differentiation through upregulating PPARγ (Zhang et al., 2016), Runx2, Smad5 (Dai et al., 2013), Wnt/β-catenin BMP2, (Kushwaha et al., 2014), estrogen and estrogen receptors (ERs) (Liao et al., 2014). It also inhibits osteoclastogenesis through inhibiting number of transcription factors, critical in bone loss such as NF-κB, AP-1(Liao et al., 2014), also through bone loss associated cytokines such as IGF-1 and TGFβ (Joo et al., 2004)). Thus dietary supplementation of soy product benefits in prevention of bone loss. Vitamin k2 and Genistein enhance stimulatory effect on Type 1 collagen and osteocalcin expression thus improves bone formation (Karieb et al., 2016).

10.4.7.4 QUERCETIN (ALLIUM CEPA)

Quercetin the flavonoids (Keservani & Sharma, 2014; Keservani et al., 2016a, 2016b) extracted from variety of fruits and vegetables, especially found in onion. Study shows it capable to initiate osteoblast or inhibit the OC differentiation and thereby helps in regulating bone metabolism (Casado-Díaz et al., 2016a). OB differentiation was caused by Quercetin through activating Smad as well as P38 MAPK (Fan et al., 2017; Yamaguchi & Weitzmann, 2011), and inhibiting WNT/β-Catenin pathways (Casado-Díaz et al., 2016b). Studies also demonstrated that quercetin reduces RANKL induced osteoclastogenesis by inhibition of NF-κB activation (Yamaguchi & Weitzmann, 2011). Supplementation of quercetin found to inhibit bone loss in ovariectomy mouse model (Tsuji et al., 2009)). Thus, this flavonoid poses good potential to be used as bone health supplements. Quercetin has a wide spectrum of antioxidant activities; the effect of quercetin and magnesium together helps in bone regeneration (Preethi et al, 2021) (Figure 10.4).

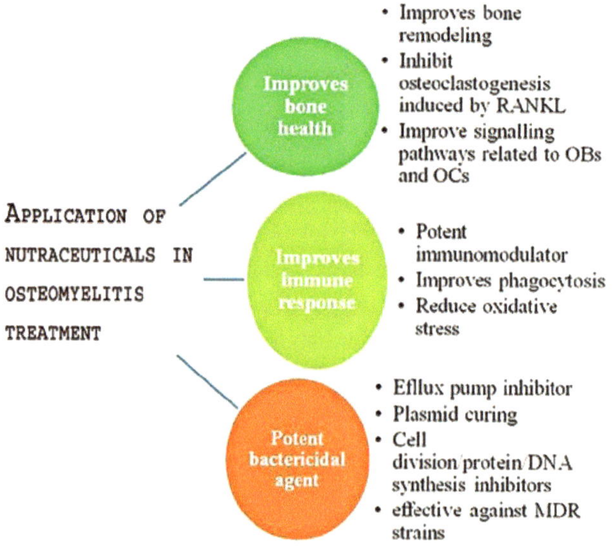

FIGURE 10.4 Name of some promising nutraceuticals in treating osteomyelitis and its complications.

10.5 CONCLUSION

The osteomyelitis treatment is most challenging, especially while the causative agents become MDR, and bacterial biofilm already established abscess and bone deformities. Furthermore, the microbial drug resistance or antibiotic resistance is emerging globally as a serious human threat and the MDR strains are exhibiting major hurdles to any

TABLE 10.5 Different Nutraceuticals and its Role in Better Bone Health

Source	Natural agent	Models	Mechanism	Reference
Petroselinum crispum (parsley), *Thymus vulgaris* (thyme)	Apigenin	*In vitro*	↓IL-6,↓RANTES,↓MCP-1,↓MCP-3,↓MCP-1,↓RANKL,↓c-Fms,↑BMP-6,↑ Osteopontin,↑Runx2,↑p-JNK,↑p-p38	(Bandyopadhyay et al., 2006), Goto et al. (2015), Zhang et al. (2015)
Berberis vulgaris (barberry)	Berberine	*In vitro*	↓NF-κB,↓Akt,↑osteopontin,↑osteocalcin,↑Runx 2,↑p38 MAPK,↑COX-2,↑ Wnt/β-catenin	Hu et al. (2008), Lee et al. (2008)/ Tao et al. (2016), Xu et al. (2010)
Betula Pubescens (downy birch)	Betulinic Acid	*In vitro & Invivo*	↓MMP-2,↓MMP-9,↓cathepsin-K	Park et al. (2014)
Eucalyptus Globulus (bluegum)	Caffeic acid	*In vitro & In vivo*	↓ERK1/2,↓p38,↓JNK,↓AP-1,↓c-Fos,↓p-Akt.↓NFATc1,↓TRAP,↓ cathepsin-K,↓c-Src,↑Runx2	Duan et al. (2014), Tolba et al. (2017), Wu et al. (2012), Zawawi et al. (2015)
Juglans Regia (Walnut) and many other fruits and vegetables	Ellagic Acid	*In vitro*	↓p38 MAPK	Rantlha et al. (2017)
Rheum Emodi (Himalayan rhubarb	Emodin	*In vitro &Invivo*	↓RANKL,↓PPARγ,↓C/EBPα,↑Runx2,↑osterix,↑ COL1,↑osteocalcin,↑ALP	Chen et al. (2017), Kang et al. (2014), Yang et al. (2014)
Acacia Greggii, Acacia Berlandieri, Quebracho colorado (Catclaw)	Fisetin	*In vitro & Invivo*	↓p38 MAPK,↓c-Fos,↓NFATc1,↓DC-STAMP,↓cathepsin-K,↓ERK,↓Akt,↓ JNK,↓NF-κB,↓TRAF6,↓c-Src-2,↓osteocalcin,↑HO-1,↑MKP-1,↑Runx-1, ↑Col1A1	Choi et al. (2012), Kim et al. (2014b), Leotoing et al. (2014), Leotoing et al. (2013), Sakai et al. (2013)
Genista Tinctoria (dyer's broom)	Genistein	*In vitro & Invivo*	↓osteocalcin,↓TNF-α,↓PPARγ,↓adipsin,↓NF-κB,↑ALP,↑ACP,↑ osteocalcin,↑ALP,↑TGFβ1,↑ Estrogen Receptor,↑p38 MAPK-Runx2,↑NO/ cGMP,↑OPG,↑SMAD5,↑BMP2	Dai et al. (2013), Heim et al. (2004), Li & Yu et al. (2003), Ming et al. (2013), Zhang et al. (2016)
Found In Many Cruciferous Vegetables, fruits and herbs	Luteolin	*In vitro*	↓ATF2,↓p38 MAPK,↓NFATc1,↓TNF-α,↓CTX,↑Dnajb1,↑Hsp90b1	Crasto et al. (2013), Kwon et al. (2016), Lee et al. (2009), Shin et al. (2012)

TABLE 10.5 *(Continued)*

Source	Natural agent	Models	Mechanism	Reference
Solanumly copersicum (tomato	Lycopene	*Invivo*	↓ALP,↓IL-6	Ardawi et al. (2016), Iimura et al. (2014), Liang et al. (2012)
Vitis Vinifera (red grapes)	Resveratrol	*In vitro & Invivo*	↓IL1,↓IL-6,↓(COX-2),↓RANKL,↓Wnt,↓PPAR-γ,↓TNF-α,↓VEGF,↓lamin,↓NF-κB,↑FOXO1,↑HO-1,↑Sirt1,↑Runx2,↑osterix,↑p-ERK1/2,↑AMPK,↑ALP,↑osteocalcin	Ke et al. (2015), Lee et al. (2017), Matsuda et al. (2018), Mobasheri and Shakibaei et al. (2013), Tou et al. (2015), Zhao et al. (2015)
Curcuma longa (turmeric)	Curcumin	*In vitro & Invivo*	↓NF-κB,↓RANKL,↓MMP-13,↓CTX,↓ROSlevels,↓PPAR-γ,↓C/EBPalpha, ↓ERK,↓JNK,↓p38,↓NFAT2,↓NF-κB,↑osteocalcin,↑Wnt-signaling pathway ,↑Runx2,↑Osterix,↑Col1A1,↑Osteonectin ,↑HO-1,↑ALP	Bharti et al. (2004), Chen et al. (2016), Gu et al. (2012), Hou et al. (2016), Kim et al. (2011b), Oh et al. (2008), TenBroek et al. (2016), von Metzler et al. (2009), Wang et al. (2016), Xin et al. (2015)
Acnistusar borescens (hollow heart)	Withanolide (Withaferin A)	*In vitro & Invivo*	↓RANKL,↓Smurf2,↓NF-κB,↑Runx2	Khedgikar et al. (2013)
Alliumcepa (onions)	Quercetin	*In vitro & Invivo*	↓p-AKT,↓RANKL,↓PGE2,↓TRAF6,↓COX-2,↓Bcl-2,↓NF-κB,↓Smad,↓AP1,↑ER-α,↑ALP,↑ERK,↑p38,↑MAPK,↑Osx,↑Runx2,↑BMP-2,↑Col-1,↑ OPN,↑OCN,↑Bax,↑Bcl-2	Casado-Diaz et al. (2016), Derakhshanian et al. (2013), Gomez-Florit et al. (2015), Guo et al. (2012), Kim et al. (2006), Masuhara et al. (2016), Wang et al. (2014b), Wattel et al. (2004), Yamaguchi and Weitzmann et al. (2011), Zhou and Lin (2014), Zhou et al. (2015)
Rosmari nusofficinalis (rosemary)	Rosmarinic Acid	*In vitro*	↓NFATc1,↓MMP-9,↓TRAP,↓cathepsin-K,↓NF-κBInvitro	Hsu et al. (2011), Omori et al. (2015)

TABLE 10.5 *(Continued)*

Source	Natural agent	Models	Mechanism	Reference
Lysimachiac ongestiflora (Creeping Jenny)	Syringetin	*In vitro*	↓AKT,↓mTOR,↓RANKL	Im et al. (2016)
PhytolaccaAmericana, olive oil	Oleanolic Acid	*In vivo*	↑osteocalcin, and↑Runx2	Tsai et al. (2015)
cruciferous vegetables	Indole-3-carbinol	*In vivo*	↓iNOS,↓IL-6,↓NF-κB,↓TNF-α,↓IL-1,↓NO,↓PGE2	Bian et al. (2012)
*Magnolia Obovata (*Japanese Big Leaf)	Obovatol	*In vitro & Invivo*	↓NF-κB,↓JNK,↓ERK,↓c-Fos,↓NFATc1	Dong et al. (2010), Yu et al. (2015)
*Garciniahanburyi (*false mango steen)	Gambogic Acid	*In vitro*	↓NF-κB,↓CXCR4,↓p-Akt,↓p38,↓Erk1/2,↓IL-6	Kim et al. (2014a)
*Harpagophytumprocumbens (*wood spider)	Harpagoside	*In vitro and in vivo*	↓ERK,↓JNK,↓Syk-Btk-PLCγ2-Ca(2+),↓c-Fos, and↓NFATc1	Pand et al. (2014)
Eclipta Alba (false daisy)	Wedelolactone	*In vitro and in vivo*	↓PLCγ2,↓GSK3βactivity,↓NFATc1,↓NF-κB, ↓c-Src,↓c-Fos,↓Cathepsin-K, ↓NF- κB/c-Fos/NFATc1,↑β-catenin,↑Runx2, ↑Wnt/GSK3β/β-catenin	Chung et al. (2017), Kim et al. (2015b)

fatality related to infections. Thus the newer development of antibiotic-resistance mechanisms contributes to inefficiency to the existing drugs with prolonged illness and increased expenditure. Again imprudent application of conventional and commonly available antibiotics in human health, hygiene, agricultural and veterinary practices is enforcing to be a major drive for persistent, spread and evolution of antibiotic-resistance at a frightening rate. Besides, conventional antibiotic therapy has several adverse effects on the patient's health, resulting in poor immune health development. According to research reports, there are some drugs which can temporarily be able to suppress osteomyelitis (in animal model), but they develop severe side effects like diarrhea, poor digestion and absorption of food leading to malnutrition and sometimes death of the animal model. Thus development of new, effective and safe therapeutic strategies is urgently needed to overcome MDR strains and the adverse side effects. Of late one common strategy is observed to be effective such as combination of plant derived nutraceuticals with different antibiotics. There are many more phytocompounds and its successful application against various MDR pathogens observed lately, along with better bone health and improved immune functions. Thus application of nutraceuticals in osteomyelitis treatment may not only help the host to kill the pathogenic causative bacteria efficiently, but also it will improve the bone health and immune health of the host, most required criteria to recover from any bone infection and its associated health complications. Although very few are still now implemented in therapeutic treatment of infections, the commercial application of such phytochemicals is still low and urgently needs one drastic improvement regarding the correct application of these natural compounds against treating infection. Thus there is need for fast-tracking more research, clinical trials and applications of these nutraceuticals to battle against osteomyelitis and its complications.

ACKNOWLEDGMENTS

The authors acknowledge Dr Subhasis Maity, Director NSHM-Knowledge Campus Kolkata Group of Institute, West Bengal for the facilities provided. The authors would like to acknowledge Dr Dilip Ghosh, Director, Nutriconnect, Sydney, Australia for his guidance and support.

KEYWORDS

- **bone**
- **immunity**
- **infection**
- **multidrug resistance**
- **nutraceuticals**
- **osteomyelitis**

REFERENCES

Abdulrazak, A., Ibrahim, B. Z., Al-Shamali, A. A., & Ahmed, M. L., (2005). Bacteriological study of diabetic foot infections. *Journal of Diabetes and Its Complications, 19*(3), 138–141.

Abreu, A. C., Borges, A., Simões, L. C., Saavedra, M. J., & Simões, M., (2013). Antibacterial activity of phenyl isothiocyanate on *Escherichia coli* and *Staphylococcus aureus*. *Med. Chem., 9*(5), 756–761.

Adorisio, S., Fierabracci, A., Muscari, I., Liberati, A. M., Calvitti, M., Cossignani, L., Blasi, F., et al., (2019). *Artocarpus tonkinensis* protects mice against collagen-induced arthritis and decreases Th17 cell function. *Frontiers in Pharmacology, 10*, 503.

Aggarwal, B. B., (2004). Nuclear factor-kappaB: The enemy within. *Cancer Cell, 6*(3), 203–208.

Aggarwal, S., Ichikawa, H., Takada, Y., Sandur, S. K., Shishodia, S., & Aggarwal, B. B., (2006). Curcumin (diferuloylmethane) down-regulates expression of cell proliferation and antiapoptotic and metastatic gene products through suppression of IκBα kinase and Akt activation. *Molecular Pharmacology, 69*(1), 195–206.

Althunibat, O. Y., Qaralleh, H., Al-Dalin, S. Y. A., Abboud, M., Khleifat, K., Majali, I. S., et al., (2016). Effect of thymol and carvacrol, the major components of *Thymus capitatus* on the growth of *Pseudomonas aeruginosa*. *J. Pure Appl. Microbiol., 10*, 367–374.

Andlauer, W., & Fürst, P., (2002). Nutraceuticals: A piece of history, present status and outlook. *Food Research International, 35*(2, 3), 171–176.

Aragón-Sánchez, J., Lázaro-Martínez, J. L., Quintana-Marrero, Y., Hernández-Herrero, M. J., García-Morales, E., Cabrera-Galván, J. J., & Benefit-Montesinos, J. V., (2009). Are diabetic foot ulcers complicated by MRSA osteomyelitis associated with worse prognosis? Outcomes of a surgical series. *Diabetic Medicine: A Journal of the British Diabetic Association, 26*(5), 552–555.

Ardawi, M. S., Badawood, M. H., Hassan, S. M., Rouzi, A. A., Ardawi, J. M., AlNosani, N. M., Qari, M. H., & Mousa, S. A., (2016). Lycopene treatment against loss of bone mass, microarchitecture and strength in relation to regulatory mechanisms in a post- menopausal osteoporosis model. *Bone, 83*, 127–140.

Aruna, P., Venkataramanamma, D., Singh, A. K., & Singh, R. P., (2016). Health benefits of punicic acid: A review. *Comprehensive Reviews in Food Science and Food Safety, 15*(1), 16–27.

Asif, M., (2011). Health effects of omega-3,6,9 fatty acids: *Perilla frutescens* is a good example of plant oils. *Oriental Pharmacy & Experimental Medicine, 11*(1), 51–59.

Aubin, J., Davy, A., & Soriano, P., (2004). *In vivo* convergence of BMP and MAPK signaling pathways: Impact of differential Smad1 phosphorylation on development and homeostasis. *Genes Dev., 18*(12), 1482–1494.

Auh, J. S., Binns, H. J., & Katz, B. Z., (2004). Retrospective assessment of subacute or chronic osteomyelitis in children and young adults. *Clinical Pediatrics, 43*(6), 549–555.

Azab, E., Chandler, K. B., Uda, Y., Sun, N., Hussein, A., Shuwaikan, R., Lu, V., et al., (2020). Osteocytes control myeloid cell proliferation and differentiation through Gsα-dependent and -independent mechanisms. *The FASEB Journal, 34*(8), 10191–10211.

Bachur, R., & Pagon, Z., (2007). Success of short-course parenteral antibiotic therapy for acute osteomyelitis of childhood. *Clinical Pediatrics, 46*(1), 30–35.

Balemans, W., Devogelaer, J. P., Cleiren, E., Piters, E., Caussin, E., & Van, H. W., (2007). Novel LRP5 missense mutation in a patient with a high bone mass phenotype results in decreased DKK1-mediated inhibition of Wnt signaling. *J. Bone Miner. Res., 22*(5), 708–716.

Baron, R., & Kneissel, M., (2013). WNT signaling in bone homeostasis and disease: From human mutations to treatments. *Nat. Med., 19*(2), 179–192.

Barrett, A. M., Lucero, M. A., Le, T., Robinson, R. L., Dworkin, R. H., & Chappell, A. S., (2007). Epidemiology, public health burden, and treatment of diabetic peripheral neuropathic pain: A review. *Pain Medicine (Malden, Mass.), 8*(suppl. 2).

Basile, A., Sorbo, S., Spadaro, V., Bruno, M., Maggio, A., Faraone, N., et al., (2009). *Antimicrobial and Antioxidant Activities of Coumarins from the Roots of Ferulago Campestris (Apiaceae), 14*, 939–952.

Bazzaz, B. S. F., Memariani, Z., Khashiarmanesh, Z., Iranshahi, M., & Naderinasab, M., (2010). Effect of galbanic acid, a sesquiterpene coumarin from *Ferula szowitsiana*, as an inhibitor of efflux mechanism in resistant clinical isolates of *Staphylococcus aureus*. *Braz. J. Microbiol., 41*, 574–580.

BBorsiczky, B., Szabo, Z., Jaberansari, M. T., Mack, P. P. O., & Roth, E., (2003). Activated PMNs lead to oxidative stress on chondrocytes: A study of swine knees. *Acta Orthopaedica Scandinavica, 74*(2), 190–195.

Belofsky, G., Percivill, D., Lewis, K., Tegos, G. P., & Ekart, J., (2004). Phenolic metabolites of Dalea versicolor that enhance antibiotic activity against model pathogenic bacteria. *Journal of Natural Products, 67*(3), 481–484.

Bernard, L., Assal, M., Garzoni, C., & Uçkay, I., (2010). Predicting the pathogen of diabetic toe osteomyelitis by two consecutive ulcer cultures with bone contact. *European Journal of Clinical Microbiology & Infectious Diseases, 30*(2), 279–281.

Bezerra, D. S. A. T., Araújo, T. F., Nascimento Da, S. L. C., Da Silva, C. B., De Oliveira, A. F., Araújo, J. M., Correia, M. T., & Lima, V. L., (2015). Organic extracts from *Indigofera suffruticosa* leaves have antimicrobial and synergic actions with erythromycin against *Staphylococcus aureus*. *Frontiers in Microbiology, 6*, 13.

Bharti, A. C., Takada, Y., & Aggarwal, B. B., (2004). Curcumin (diferuloylmethane) inhibits receptor activator of NF-kappa B ligand-induced NF-kappa B activation in osteoclast precursors and suppresses osteoclastogenesis. *J. Immunol., 172*(10), 5940–5947.

Bian, Q., Liu, S. F., Huang, J. H., Yang, Z., Tang, D. Z., Zhou, Q., Ning, Y., et al., (2012). Oleanolic acid exerts an osteoprotective effect in ovariectomy-induced osteoporotic rats and stimulates the osteoblastic differentiation of bone mesenchymal stem cells *in vitro. Menopause, 19*(2), 225–233.

Bo, S., Gambino, R., Ponzo, V., Cioffi, I., Goitre, I., Evangelista, A., Ciccone, G., Cassader, M., & Procopio, M., (2018). Effects of resveratrol on bone health in type 2 diabetic patients. A double-blind randomized-controlled trial. *Nutrition & Diabetes, 8*(1), 1–10.

Boberek, J. M., Stach, J., & Good, L., (2010). Genetic evidence for inhibition of bacterial division protein FtsZ by berberine. *PLoS One, 5*, 0013745.

Boggon, T. J., & Eck, M. J., (2004). Structure and regulation of Src family kinases. *Oncogene, 23*(48), 7918–7927.

Bonhoeffer, J., Haeberle, B., Schaad, U. B., & Heininger, U., (2001). Diagnosis of acute haematogenous osteomyelitis and septic arthritis: 20 years experience at the university children's hospital basel. *Swiss Medical Weekly, 131*(39, 40), 575–581.

Borges, A., Ferreira, C., Saavedra, M. J., & Simões, M., (2013). Antibacterial activity and mode of action of ferulic and gallic acids against pathogenic bacteria. *Microb. Drug Resist., 19*, 256–265.

Boulet, M. L., Isabelle, C., Guay, I., Brouillette, E., Langlois, J. P., JacquesÉ, P., et al., (2018). Tomatidine is a lead antibiotic molecule that targets *Staphylococcus aureus* ATP synthase subunit C. *Antimicrob. Agents Chemother., 62*, e02197–2117.

Boyce, B. F., & Xing, L., (2007). Biology of RANK, RANKL, and osteoprotegerin. *Arthritis Res. Ther., 9*(Suppl. 1), S1.

Boyce, B. F., (2013). Advances in the regulation of osteoclasts and osteoclast functions. *J. Dent. Res., 92*(10), 860–867.

Boyce, B. F., Xiu, Y., Li, J., Xing, L., & Yao, Z., (2015). NF-kappaB-mediated regulation of osteoclastogenesis. *Endocrinol. Metab. (Seoul), 30*(1), 35–44.

Boyle, W. J., Simonet, W. S., & Lacey, D. L., (2003). Osteoclast differentiation and activation. *Nature, 423*, 337–342.

Brandt, S. L., Putnam, N. E., Cassat, J. E., & Serezani, C. H., (2018). Innate immunity to *Staphylococcus aureus*: Evolving paradigms in soft tissue and invasive infections. *The Journal of Immunology, 200*(12), 3871–3880.

Broege, A., Pham, L., Jensen, E. D., Emery, A., Huang, T. H., Stemig, M., Beppu, H., et al., (2013). Bone morphogenetic proteins signal via SMAD and mitogen-activated protein (MAP) kinase pathways at distinct times during osteoclastogenesis. *J. Biol. Chem., 288*(52), 37230–37240.

Cai, Y., Folkerts, J., Folkerts, G., Maurer, M., & Braber, S., (2020). Microbiota-dependent and -independent effects of dietary fiber on human health. *Br. J. Pharmacol., 177*: 1363–1381.

Calhoun, J. H., Manring, M. M., & Shirtliff, M., (2009). Osteomyelitis of the long bones. *Seminars in Plastic Surgery, 23*(2), 59.

Cappellen, D., Luong-Nguyen, N. H., Bongiovanni, S., Grenet, O., Wanke, C., & Susa, M., (2002). Transcriptional program of mouse osteoclast differentiation governed by the macrophage colony-stimulating factor and the ligand for the receptor activator of NFkappa B. *J. Biol. Chem., 277*(24), 21971–21982.

Casado-Diaz, A., Anter, J., Dorado, G., & Quesada-Gomez, J. M., (2016). Effects of quercetin, a natural phenolic compound, in the differentiation of human mesenchymal stem cells (MSC) into adipocytes and osteoblasts. *J. Nutr. Biochem., 32*, 151–162.

Casado-Díaz, A., Anter, J., Dorado, G., & Quesada-Gómez, J. M., (2016a). Effects of quercetin, a natural phenolic compound, in the differentiation of human mesenchymal stem cells (MSC) into adipocytes and osteoblasts. *The Journal of Nutritional Biochemistry, 32*, 151–162.

Casado-Díaz, A., Anter, J., Dorado, G., & Quesada-Gómez, J. M., (2016b). Effects of quercetin, a natural phenolic compound, in the differentiation of human mesenchymal stem cells (MSC) into adipocytes and osteoblasts. *The Journal of Nutritional Biochemistry, 32*, 151–162.

Chakraborty, B., Nath, A., Saikia, H., & Sengupta, M., (2014). Bactericidal activity of selected medicinal plants against multidrug resistant bacterial strains from clinical isolates. *Asian Pacific Journal of Tropical Medicine, 7S1*, S435–S441.

Chambers, T. J., (2000). Regulation of the differentiation and function of osteoclasts. *J. Pathol., 192*(1), 4–13.

Chan, B. C. L., Ip, M., Lau, C. B. S., Lui, S. L., Jolivalt, C., Ganem-Elbaz, C., et al., (2011). Synergistic effects of baicalein with ciprofloxacin against NorA overexpressed methicillin-resistant *Staphylococcus aureus* (MRSA) and inhibition of MRSA pyruvate kinase. *J. Ethnopharmacol., 137*, 767–773.

Chandran, U., & Patwardhan, B., (2017). Network ethnopharmacological evaluation of the immunomodulatory activity of *Withania somnifera. Journal of Ethnopharmacology, 197*, 250–256.

Chattopadhyay, N., (2006). Attenuation of osteoclastogenesis and osteoclast function by apigenin. *Biochem. Pharmacol., 72*(2), 184–197.

Chen, G., Deng, C., & Li, Y. P., (2012). TGF-beta and BMP signaling in osteoblast differentiation and bone formation. *Int. J. Biol. Sci., 8*(2), 272–288.

Chen, S., Zhao, X., Ran, L., Wan, J., Wang, X., Qin, Y., Shu, F., et al., (2015). Resveratrol improves insulin resistance, glucose and lipid metabolism in patients with non-alcoholic fatty liver disease: A randomized controlled trial. *Digestive and Liver Disease, 47*(3), 226–232.

Chen, X., Ren, S., Zhu, G., Wang, Z., & Wen, X., (2017). Emodin suppresses cadmium-induced osteoporosis by inhibiting osteoclast formation. *Environ. Toxicol. Pharmacol., 54*, 162–168.

Chen, Z., Xue, J., Shen, T., Mu, S., & Fu, Q., (2016). Curcumin alleviates glucocorticoid-induced osteoporosis through the regulation of the Wnt signaling pathway. *Int. J. Mol. Med., 37*(2), 329–338.

Cheung, W. M., Jin, L. Y., Smith, D. K., Cheung, P. T., Kwan, E. Y., Low, L., & Kung, A. W., (2006). A family with osteoporosis pseudoglioma syndrome due to compound heterozygosity of two novel mutations in the LRP5 gene. *Bone, 39*(3), 470–476.

Chlubnová, I., Sylla, B., Nugier-Chauvin, C., Daniellou, R., Legentil, L., Kralová, B., & Ferrières, V., (2011). Natural glycans and glycoconjugates as immunomodulating agents. *Natural Product Reports, 28*(5), 937–952.

Choi, S. W., Son, Y. J., Yun, J. M., & Kim, S. H., (2012). Fisetin inhibits osteoclast Differentiation via downregulation of p38 and c-Fos-NFATc1 signaling pathways. *Evid. Based Complement. Alternate. Med., 810563*.

Chung, H. J., Kim, W. K., Oh, J., Kim, M. R., Shin, J. S., Lee, J., Ha, I. H., & Lee, S. K., (2017). Anti-Osteoporotic activity of harpagoside by upregulation of the BMP2 and Wnt signaling pathways in osteoblasts and suppression of differentiation in osteoclasts. *J. Nat. Prod., 80*(2), 434–442.

Cierny, G., Mader, J. T., & Penninck, J. J., (2003). A clinical staging system for adult osteomyelitis. *Clinical Orthopedics and Related Research, 414*(414), 7–24.
Claro, T., Widaa, A., McDonnell, C., Foster, T. J., O'Brien, F. J., & Kerrigan, S. W., (2013). Staphylococcus aureus protein A binding to osteoblast tumor necrosis factor receptor 1 results in activation of nuclear factor kappa B and release of interleukin-6 in bone infection. *Microbiology (Reading, England), 159*(1), 147–154.
Claro, T., Widaa, A., O'Seaghdha, M., Miajlovic, H., Foster, T. J., O'Brien, F. J., & Kerrigan, S. W., (2011). *Staphylococcus aureus* protein A binds to osteoblasts and triggers signals that weaken bone in osteomyelitis. *PLOS One*, *6*(4), e18748.
Conterno, L. O., & Turchi, M. D., (2013). Antibiotics for treating chronic osteomyelitis in adults. *Cochrane Database of Systematic Reviews, 2013*(9).
Crasto, G. J., Kartner, N., Yao, Y., Li, K., Bullock, L., Datti, A., & Manolson, M. F., (2013). Luteolin inhibition of V-ATPase a3-d2 interaction decreases osteoclast resorptive activity. *J. Cell. Biochem., 114*(4), 929–941.
Dahiya, P., & Purkayastha, S., (2012). Phytochemical screening and antimicrobial activity of some medicinal plants against multi-drug resistant bacteria from clinical isolates. *Indian Journal of Pharmaceutical Sciences, 74*(5), 443–450.
Dai, J., Li, Y., Zhou, H., Chen, J., Chen, M., & Xiao, Z., (2013). Genistein promotion of osteogenic differentiation through BMP2/SMAD5/RUNX2 signaling. *Int. J. Biol. Sci., 9*(10), 1089–1098.
Dallas, S. L., Prideaux, M., & Bonewald, L. F., (2013). The osteocyte: An endocrine cell and more. *Endocrine Reviews, 34*(5), 658–690.
Darnay, B. G., Haridas, V., Ni, J., Moore, P. A., & Aggarwal, B. B., (1998). Characterization of the intracellular domain of receptor activator of NF-kappaB (RANK). Interaction with tumor necrosis factor receptor-associated factors and activation of NF-kappab and c- Jun N-terminal kinase. *J. Biol. Chem., 273*(32), 20551–20555.
Das, L., Bhaumik, E., Raychaudhuri, U., & Chakraborty, R., (2011). Role of nutraceuticals in human health. *Journal of Food Science and Technology, 49*(2), 173–183.
Dathe, K., Kjaer, K. W., Brehm, A., Meinecke, P., Nurnberg, P., Neto, J. C., Brunoni, D., et al., (2009). Duplications involving a conserved regulatory element downstream of BMP2 are associated with brachydactyly type A2. *Am. J. Hum. Genet., 84*(4), 483–492.
Davis, J. S., (2005). Management of bone and joint infections due to *Staphylococcus aureus*. *Internal Medicine Journal, 35*(suppl. 2), S79–96.
De Barros, A. P. D. N., Takiya, C. M., Garzoni, L. R., Leal-Ferreira, M. L., Dutra, H. S., Chiarini, L. B., Meirelles, M. N., et al., (2010). Osteoblasts and bone marrow mesenchymal stromal cells control hematopoietic stem cell migration and proliferation in 3D *in vitro* model. *PLOS One*, *5*(2), e9093.
De Mesy, B. K. L., MacDonald, A., Schwarz, E. M., & Oh, I., (2018). Chronic osteomyelitis with *Staphylococcus aureus* deformation in submicron canaliculi of osteocytes: A case report. *JBJS Case Connector, 8*(1), e8.
De Oliveira, E. M., De Arruda, J. A. A., Nascimento, P. A. M., Neiva, I. M., Mesquita, R. A., & Souza, L. N., (2018). An update on osteomyelitis treatment in a pycnodysostosis patient. *Journal of Oral and Maxillofacial Surgery, 76*(10), 2136.e1–2136.e10.
Derakhshanian, H., Djalali, M., Djazayery, A., Nourijelyani, K., Ghadbeigi, S., Pishva, H., Saedisomeolia, A., et al., (2013). Quercetin prevents experimental glucocorticoid-induced osteoporosis: A comparative study with alendronate. *Can. J. Physiol. Pharmacol., 91*(5), 380–385.

Derynck, R., & Zhang, Y. E., (2003). Smad-dependent and smad-independent pathways in TGF-beta family signaling. *Nature, 425*(6958), 577–584.

Destaing, O., Sanjay, A., Itzstein, C., Horne, W. C., Toomre, D., De Camilli, P., & Baron, R., (2008). The tyrosine kinase activity of c-Src regulates actin dynamics and organization of podosomes in osteoclasts. *Mol. Biol. Cell., 19*(1), 394–404.

Di Sotto, A., Vitalone, A., & Di Giacomo, S., (2020). Plant-derived nutraceuticals and immune system modulation: An evidence-based overview. *Vaccines, 8*(3), 468.

Dohin, B., Gillet, Y., Kohler, R., Lina, G., Vandenesch, F., Vanhems, P., Floret, D., & Etienne, J., (2007). Pediatric bone and joint infections caused by Panton-valentine leukocidin-positive *Staphylococcus aureus. Pediatric Infectious Disease Journal, 26*(11), 1042–1048.

Domadia, P. N., Bhunia, A., Sivaraman, J., Swarup, S., & Dasgupta, D., (2008). Berberine targets assembly of *Escherichia coli* cell division protein FtsZ. *Biochemistry, 47*(10), 3225–3234.

Dong, L., Xia, S., Gao, F., Zhang, D., & Chen, J., (2010). Zhang, J. 3,3'-diindolylmethane attenuates experimental arthritis and osteoclastogenesis. *Biochem. Pharmacol., 79*(5), 715–721.

Dosier, C. R., Erdman, C. P., Park, J. H., Schwartz, Z., Boyan, B. D., & Guldberg, R. E., (2012). Resveratrol effect on osteogenic differentiation of rat and human adipose derived stem cells in a 3-D culture environment. *Journal of the Mechanical Behavior of Biomedical Materials, 11*, 112–122.

Dougall, W. C., Glaccum, M., Charrier, K., Rohrbach, K., Brasel, K., De Smedt, T., Daro, E., et al., (1999). RANK is essential for osteoclast and lymph node development. *Genes Dev., 13*(18), 2412–2424.

Duan, W., Wang, Q., Li, F., Xiang, C., Zhou, L., Xu, J., Feng, H., & Wei, X., (2014). Anti-catabolic effect of caffeic acid phenethyl ester, an active component of honeybee propolis on bone loss in ovariectomized mice: A micro-computed tomography study and histological analysis. *Chin. Med., 127*(22), 3932–3936.

Durbin, S. M., Jackson, J. R., Ryan, M. J., Gigliotti, J. C., Alway, S. E., & Tou, J. C., (2013). Resveratrol supplementation preserves long bone mass, microstructure, and strength in hindlimb-suspended old male rats. *Journal of Bone and Mineral Metabolism, 32*(1), 38–47.

Edwards, C. M., & Mundy, G. R., (2008). Eph receptors and ephrin signaling pathways: A role in bone homeostasis. *Int. J. Med. Sci., 5*(5), 263–272.

Ellington, J. K., Harris, M., Hudson, M. C., Vishin, S., Webb, L. X., & Sherertz, R., (2006). Intracellular *Staphylococcus aureus* and antibiotic resistance: Implications for treatment of staphylococcal osteomyelitis. *Journal of Orthopadic Research, 24*(1), 87–93.

El-Seedi, H. R., (2007). Antimicrobial aryl coumarins from *Asphodelus microcarpus. J. Nat. Prod., 70*, 118–120.

Eman, Z., Engy, R., Fatma, A., Hebata, A. M., & Tarek, I., (2014). Effects of dietary astragalus polysaccharides (APS) on growth performance, immunological parameters, digestive enzymes, and intestinal morphology of Niletilapia (*Oreochromis niloticus*). *Fish Shellfish Immunol., 38*, 149–157.

Fahey, J. W., Stephenson, K. K., Wade, K. L., & Talalay, P., (2013). Urease from *Helicobacter pylori* is inactivated by sulforaphane and other isothiocyanates. *Biochem. Biophys. Res. Commun., 435*(1), 1–7.

Fan, A. Q., Feng, J. Q., Liu, W., Zhang, M. J., Liu, T., Zhou, Y. L., & Xu, Y. J., (2017). Antagonistic effect of quercetin on PM 2.5 toxicity in the rat's embryonic development *in*

vitro. Beijing Da Xue Xue Bao. Yi Xue Ban = Journal of Peking University. Health Science., 49(3), 388–393.
Fan, Y., Hu, Y., Wang, D., Liu, J., Zhang, J., Zhao, X., Liu, X., et al., (2012). Effects of astragalus polysaccharide liposome on lymphocyte proliferation *in vitro* and adjuvanticity *in vivo*. *Carbohydr. Polym., 88*, 68–74.
Farinon, B., Molinari, R., Costantini, L., & Merendino, N., (2020). The seed of industrial hemp (*Cannabis sativa* L.): Nutritional quality and potential functionality for human health and nutrition. *Nutrients, 12*(7), 1937.
Feng, X., & Teitelbaum, S. L., (2013). Osteoclasts: New insights. *Bone Res., 1*(1), 11–26.
Ferreira, S. S., Passos, C. P., Madureira, P., Vilanova, M., & Coimbra, M. A., (2015). Structure-function relationships of immunostimulatory polysaccharides: A review. *Carbohydr. Polym., 132*: 378–396.
Franceschi, R. T., Xiao, G., Jiang, D., Gopalakrishnan, R., Yang, S., & Reith, E., (2003). Multiple signaling pathways converge on the Cbfa1/Runx2 transcription factor to regulate osteoblast differentiation. *Connect. Tissue Res., 44* (Suppl. 1), 109–116.
Franzoso, G., Carlson, L., Xing, L., Poljak, L., Shores, E. W., Brown, K. D., Leonardi, A., et al., (1997). Requirement for NF-kappaB in osteoclast and B-cell development. *Genes Dev., 11*(24), 3482–3496.
Frei, B., England, L., & Ames, B. N., (1989). Ascorbate is an outstanding antioxidant in human blood plasma. *Proceedings of the National Academy of Sciences of the United States of America, 86*(16), 6377–6381.
Fujii, T., Minagawa, T., Shimizu, T., Takeguchi, N., & Sakai, H., (2012). Inhibition of ecto-ATPase activity by curcumin in hepatocellular carcinoma HepG2 cells. *Journal of Physiological Sciences, 62*(1), 53–58.
Fulzele, K., Krause, D. S., Panaroni, C., Saini, V., Barry, K. J., Liu, X., Lotinun, S., et al., (2013). Myelopoiesis is regulated by osteocytes through Gsα-dependent signaling. *Blood, 121*(6), 930–939.
Ghoneum, M., & Abedi, S., (2004). Enhancement of natural killer cell activity of aged mice by modified arabinoxylan rice bran (MGN-3/Biobran). *J. Pharm. Pharmacol., 56*: 1581–1588.
Gohil, K. J., Patel, J. A., & Gajjar, A. K., (2010). Pharmacological review on *Centella asiatica*: A potential herbal cure-all. *Indian Journal of Pharmaceutical Sciences., 72*(5), 546.
Gomez-Florit, M., Monjo, M., & Ramis, J. M., (2015). Quercitrin for periodontal regeneration: Effects on human gingival fibroblasts and mesenchymal stem cells. *Sci. Rep., 5*, 16593.
González-Martín, M., Silva, V., Poeta, P., Corbera, J. A., & Tejedor-Junco, M. T., (2022). *Microbiological Aspects of Osteomyelitis in Veterinary Medicine: Drawing Parallels to the Infection in Human Medicine*, *42*(1), 1–11.
Gornitzky, A. L., Kim, A. E., O'Donnell, J. M., & Swarup, I., (2020). Diagnosis and management of osteomyelitis in children: A critical analysis review. *JBJS Reviews*, *8*(6).
Goto, T., Hagiwara, K., Shirai, N., Yoshida, K., & Hagiwara, H., (2015). Apigenin inhibits osteoblastogenesis and osteoclastogenesis and prevents bone loss in ovariectomized mice. *Cytotechnology*, *67*(2), 357–365.
Govaert, G. A. M., Kuehl, R., Atkins, B. L., Trampuz, A., Morgenstern, M., Obremskey, W. T., Verhofstad, M. H. J., et al., (2020). Diagnosing fracture-related infection: Current concepts and recommendations. *Journal of Orthopedic Trauma, 34*(1), 8–17.
Gu, Q., Cai, Y., Huang, C., Shi, Q., & Yang, H., (2012). Curcumin increases rat mesenchymal stem cell osteoblast differentiation but inhibits adipocyte differentiation. *Pharmacognosy Magazine, 8*(31), 202.

Guay, I., Boulanger, S., Isabelle, C., Brouillette, E., Chagnon, F., Bouarab, K., et al., (2018). Tomatidine and analog FC04-100 possess bactericidal activities against *Listeria, Bacillus* and *Staphylococcus* Spp. *BMC Pharmacol. Toxicol., 19*: 7.

Guder, C., Gravius, S., Burger, C., Wirtz, D. C., & Schildberg, F. A., (2020). Osteoimmunology: A current update of the interplay between bone and the immune system. *Frontiers in Immunology, 11*, 58.

Guil-Guerrero, J. L., Gómez-Mercado, F., Ramos-Bueno, R. P., González-Fernández, M. J., Urrestarazu, M., Jiménez-Becker, S., & De Bélair, G., (2018a). Fatty acid profiles and sn-2 fatty acid distribution of γ-linolenic acid-rich Borago species. *Journal of Food Composition and Analysis, 66*, 74–80.

Guil-Guerrero, J. L., Gómez-Mercado, F., Ramos-Bueno, R. P., González-Fernández, M. J., Urrestarazu, M., Jiménez-Becker, S., & De Bélair, G., (2018b). Fatty acid profiles and sn-2 fatty acid distribution of γ-linolenic acid-rich Borago species. *Journal of Food Composition and Analysis, 66*, 74–80.

Gull, I., Sohail, M., Aslam, M. S., & Amin, A. M., (2013). Phytochemical, toxicological and antimicrobial evaluation of *Lawsonia inermis* extracts against clinical isolates of pathogenic bacteria. *Annals of Clinical Microbiology and Antimicrobials, 12*, 36.

Guo, C., Hou, G. Q., Li, X. D., Xia, X., Liu, D. X., Huang, D. Y., & Du, S. X., (2012). Quercetin triggers apoptosis of lipopolysaccharide (LPS)-induced osteoclasts and inhibits bone resorption in RAW264.7 cells. *Cell. Physiol. Biochem., 30*(1), 123–136.

Gutierrez, K., (2005). Bone and joint infections in children. *Pediatric Clinics of North America, 52*(3), 779–794.

Haque, M. A., Jantan, I., & Abbas, B. S. N., (2017). *Tinospora* species: An overview of their modulating effects on the immune system. *Journal of Ethnopharmacology, 207*, 67–85.

Hayden, M. S., & Ghosh, S., (2008). Shared principles in NF-kappaB signaling. *Cell, 132*(3), 344–362.

Heim, M., Frank, O., Kampmann, G., Sochocky, N., Pennimpede, T., Fuchs, P., Hunziker, W., et al., (2004). The phytoestrogen genistein enhances osteogenesis and represses adipogenic differentiation of human primary bone marrow stromal cells. *Endocrinology, 145*(2), 848–859.

Holler, J. G., Christensen, S. B., Slotved, H. C., Rasmussen, H. B., Guzman, A., Olsen, C. E., Petersen, B., & Mølgaard, P., (2012). Novel inhibitory activity of the *Staphylococcus aureus* NorA efflux pump by a kaempferol rhamnoside isolated from *Persea lingue* nees. *Journal of Antimicrobial Chemotherapy, 67*(5), 1138–1144.

Hong, H., Kim, J., Lim, T., Song, Y., Cho, C., & Jang, M., (2018). Mixing ratio optimization for functional complex extracts of *Rhodiola crenulata, Panax quinquefolius*, and *Astragalus membranaceus* using mixture design and verification of immune functional efficacy in animal models. *J. Funct. Foods, 40*, 447–454.

Hou, M., Song, Y., Li, Z., Luo, C., Ou, J. S., Yu, H., Yan, J., & Lu, L., (2016). Curcumin attenuates osteogenic differentiation and calcification of rat vascular smooth muscle cells. *Mol. Cell. Biochem., 420*(1, 2), 151–160.

Hsu, Y. C., Cheng, C. P., & Chang, D. M., (2011). *Plectranthus amboinicus* attenuates inflammatory bone erosion in mice with collagen-induced arthritis by downregulation of RANKL-induced NFATc1 expression. *J. Rheumatol., 38*(9), 1844–1857.

Hu, J. P., Nishishita, K., Sakai, E., Yoshida, H., Kato, Y., Tsukuba, T., & Okamoto, K., (2008). Berberine inhibits RANKL-induced osteoclast formation and survival through suppressing the NF-kappaB and Akt pathways. *Eur. J. Pharmacol., 580*(1, 2), 70–79.

Hudson, M. C., Ramp, W. K., Nicholson, N. C., Williams, A. S., & Nousiainen, M. T., (1995). Internalization of *Staphylococcus aureus* by cultured osteoblasts. *Microbial Pathogenesis., 19*(6), 409–419.

Ichikawa, H., Murakami, A., & Aggarwal, B. B., (2006a). 1'-acetoxychavicol acetate inhibits RANKL-induced osteoclast differentiation of RAW 264.7 monocytic cells by sup- pressing nuclear factor-kappaB activation. *Mol. Cancer Res., 4*(4), 275–281.

Im, N. K., Lee, D. S., Lee, S. R., & Jeong, G. S., (2006c). Lupeol isolated from *Sorbus commixta* suppresses 1alpha,25-(OH)2D3-mediated osteoclast differentiation and bone loss *in vitro* and *in vivo*. *J. Nat. Prod.,79*(2), 412–420.

Iniguez-Ariza, N. M., & Clarke, B. L., (2015). Bone biology, signaling pathways, and therapeutic targets for osteoporosis. *Maturitas, 82*(2), 245–255.

Inzana, J. A., Trombetta, R. P., Schwarz, E. M., Kates, S. L., & Awad, H. A., (2015). 3D printed bioceramics for dual antibiotic delivery to treat implant-associated bone infection. *European Cells & Materials, 30*, 232.

Jaberi, S., Fallah, F., Hashemi, A., Karimi, A. M., & Azimi, L., (2018). Inhibitory effects of curcumin on the expression of NorA efflux pump and reduce antibiotic resistance in staphylococcus aureus. *Journal of Pure and Applied Microbiology, 12*(1), 95–102.

Jacob, A. L., Smith, C., Partanen, J., & Ornitz, D. M., (2006). Fibroblast growth factor receptor 1 signaling in the osteo-chondrogenic cell lineage regulates sequential steps of osteoblast maturation. *Dev. Biol., 296*(2), 315–328.

Javed, A., Bae, J. S., Afzal, F., Gutierrez, S., Pratap, J., Zaidi, S. K., Lou, Y., et al., (2008). Structural coupling of Smad and Runx2 for execution of the BMP2 osteogenic signal. *J. Biol. Chem., 283*(13), 8412–8422.

Jeong, K. W., Lee, J. Y., Kang, D. I., Lee, J. U., Shin, S. Y., & Kim, Y., (2009). Screening of flavonoids as candidate antibiotics against *Enterococcus faecalis*. *J. Nat. Prod., 72*, 719–724.

Johnson, D., & Wilkie, A. O., (2011). Craniosynostosis. *Eur. J. Hum. Genet., 19*(4), 369–376.

Joo, S. S., Won, T. J., Kang, H. C., & Lee, D. I., (2004). Isoflavones extracted from *Sophorae fructus* upregulate IGF-I and TGF-beta and inhibit osteoclastogenesis in rat bone marrow cells. *Archives of Pharmacal Research, 27*(1), 99–105.

Josse, J., Velard, F., & Gangloff, S. C., (2015a). Staphylococcus aureus vs. osteoblast: Relationship and consequences in osteomyelitis. *Frontiers in Cellular and Infection Microbiology, 5*.

Josse, J., Velard, F., & Gangloff, S. C., (2015b). *Staphylococcus aureus* vs. osteoblast: Relationship and consequences in osteomyelitis. *Frontiers in Cellular and Infection Microbiology, 5,* 85.

Kalinka, J., Hachmeister, M., Geraci, J., Sordelli, D., Hansen, U., Niemann, S., Oetermann, S., et al., (2014). Staphylococcus aureus isolates from chronic osteomyelitis are characterized by high host cell invasion and intracellular adaptation, but still induce inflammation. *International Journal of Medical Microbiology, 304*(8), 1038–1049.

Kang, D. M., Yoon, K. H., Kim, J. Y., Oh, J. M., Lee, M., Jung, S. T., Juhng, S. K., & Lee, Y. H., (2014). CT imaging biomarker for evaluation of emodin as a potential drug on LPS-mediated osteoporosis mice. *Acad. Radiol., 21*(4), 457–462.

Kaplan, S. L., (2005). Osteomyelitis in children. *Infectious Disease Clinics of North America, 19*(4), 787–797.

Karamanis, E. M., Matthaiou, D. K., Moraitis, L. I., & Falagas, M. E., (2008). Fluoroquinolones versus beta-lactam based regimens for the treatment of osteomyelitis: A meta-analysis of randomized controlled trials. *Spine, 33*(10).

Karieb, S. S., Jawad, M. M., Al-Shmgani, H. S., & Kadri, Z. H. M., (2016). The effect of the combination of vitamin K2 and genistein, coumestrol and daidzein on the osteoblast differentiation and bone matrix formation. *Journal of Biotechnology Research Center, 10*(2), 12–19.

Karin, M., & Lin, A., (2002). NF-kappaB at the crossroads of life and death. *Nat. Immunol., 3*(3), 221–227.

Kavanagh, N., Ryan, E. J., Widaa, A., Sexton, G., Fennell, J., O'Rourke, S., Cahill, K. C., et al., (2018). Staphylococcal osteomyelitis: Disease progression, treatment challenges, and future directions. *Clinical Microbiology Reviews, 31*(2).

Keservani, R. K., & Sharma, A. K., (2014). Flavonoids: Emerging trends and potential health benefits. *Journal of Chinese Pharmaceutical Sciences, 23*(12), 815.

Keservani, R. K., Kesharwani, R. K., Sharma, A. K., Vyas, N., & Chadoker, A., (2010b). Nutritional supplements: An overview. *International Journal of Current Pharmaceutical Review and Research, 1*(1), 59–75.

Keservani, R. K., Kesharwani, R. K., Vyas, N., Jain, S., Raghuvanshi, R., & Sharma, A. K., (2010a). Nutraceutical and functional food as future food: A review. *Der Pharmacia Lettre, 2*(1), 106–116.

Keservani, R. K., Sharma, A. K., & Kesharwani, R. K., (2016a). Nutraceutical and functional foods for cardiovascular health. In: *Food Process Engineering* (pp. 291–312). Apple Academic Press, CRC Press, (pp. 257–278). ISBN: 9781771884020.

Keservani, R. K., Sharma, A. K., & Kesharwani, R. K., (2016b). Medicinal effect of nutraceutical fruits for the cognition and brain health. *Scientifica*, 10. Article ID 3109254. doi: 1 0.1155/2016/3109254.

Keservani, R. K., Sharma, A. K., & Kesharwani, R. K., (2017). An overview and therapeutic applications of nutraceutical and functional foods. *Recent Advances in Drug Delivery Technology*, 160–201. ISBN:9781522507543.

Keservani, R. K., Sharma, A. K., & Kesharwani, R. K., (2020). *Nutraceuticals and Dietary Supplements: Applications in Health Improvement and Disease Management.* CRC Press. ISBN: 9781771888738.

Khameneh, B., Iranshahy, M., Ghandadi, M., Ghoochi, A. D., Fazly, B. B. S., & Iranshahi, M., (2015). Investigation of the antibacterial activity and efflux pump inhibitory effect of co-loaded piperine and gentamicin nanoliposomes in methicillin-resistant *Staphylococcus aureus*. *Drug Development and Industrial Pharmacy, 41*(6), 989–994.

Khan, A., Hassan, S. N., Farhat, M. F., Riaz, A., Mukhtar, B., & Ali, A., (2022). External fixation as a primary and definitive treatment for tibial diaphyseal fractures: A retrospective longitudinal study. *Pakistan Journal of Medical and Health Sciences, 16*(1), 624–626.

Khanizadeh, S., Baninasab, B., Tabori, M., Yu, J., Zhang, Y., Wang, X., & Deschiffart, I., (2017). Low temperature storage and *in vitro* pollen germination of selected spring wheat accessions. *Journal of Agricultural Science, 9*(9), 12–237.

Khedgikar, V., Kushwaha, P., Gautam, J., Verma, A., Changkija, B., Kumar, A., Sharma, S., et al., (2013). Withaferin A: A proteasomal inhibitor promotes healing after injury and exerts anabolic effect on osteoporotic bone. *Cell Death Dis., 4*, e778.

Kim, D. C., Ku, S. K., & Bae, J. S., (2012). Anticoagulant activities of curcumin and its derivative. *BMB Reports, 45*(4), 221–226.

Kim, J. H., Gupta, S. C., Park, B., Yadav, V. R., & Aggarwal, B. B., (2012). Turmeric (*Curcuma longa*) inhibits inflammatory nuclear factor (NF)-kappaB and NF-kappaB-regulated gene products and induces death receptors leading to suppressed proliferation, induced chemosensitization, and suppressed osteoclastogenesis. *Mol. Nutr. Food Res., 56*(3), 454–465.

Kim, Y. H., Kim, J. L., Lee, E. J., Park, S. H., Han, S. Y., Kang, S. A., & Kang, Y. H., (2014b). Fisetin antagonizes cell fusion, cytoskeletal organization and bone resorption in RANKL-differentiated murine macrophages. *J. Nutr. Biochem., 25*(3), 295–303.

King, T. J., Shandala, T., Lee, A. M., Foster, B. K., Chen, K. M., Howe, P. R., & Xian, C. J., (2015). Potential effects of phytoestrogen genistein in modulating acute methotrexate chemotherapy-induced osteoclastogenesis and bone damage in rats. *International Journal of Molecular Sciences, 16*(8), 18293–18311.

Klanˇcnik, A., ˇSikiˊc, P. M., Troˇst, K., Tuˇsek, ˇZ. M., Mozetiˇc, V. B., & Smole, M. S., (2017). Anti-campylobacter activity of resveratrol and an extract from waste pinot noir grape skins and seeds, and resistance of camp. jejuni planktonic and biofilm cells, mediated via the CmeABC efflux pump. *J. Appl. Microbiol., 122*, 65–77.

Knowles, H. J., & Athanasou, N. A., (2009). Canonical and non-canonical pathways of osteoclast formation. *Histology and Histopathology, 24*(3), 337–346.

Ko, J. M., (2016). Genetic syndromes associated with craniosynostosis. *J. Korean Neurosurg Soc., 59*(3), 187–191.

Kobayashi, S., Takahashi, H. E., Ito, A., Saito, N., Nawata, M., Horiuchi, H., Ohta, H., et al., (2003). Trabecular mini modeling in human iliac bone. *Bone, 32*(2), 163–169.

Kochanowska, I., Chaberek, S., Wojtowicz, A., Marczynski, B., Wlodarski, K., Dytko, M., & Ostrowski, K., (2007). Expression of genes for bone morphogenetic proteins BMP-2, BMP-4 and BMP-6 in various parts of the human skeleton. *BMC Musculoskelet. Disord., 8*, 128.

Kohli, R., & Hadley, S., (2005). Fungal arthritis and osteomyelitis. *Infectious Disease Clinics of North America, 19*(4), 831–851.

Kokabu, S., Gamer, L., Cox, K., Lowery, J., Tsuji, K., Raz, R., Economides, A., Katagiri, T., & Rosen, V., (2012). BMP3 suppresses osteoblast differentiation of bone marrow stromal cells via interaction with Acvr2b. *Mol. Endocrinol., 26*(1), 87–94.

Koracevic, D., Koracevic, G., Djordjevic, V., Andrejevic, S., & Cosic, V., (2001). Method for the measurement of antioxidant activity in human fluids. *Journal of Clinical Pathology, 54*(5), 356–361.

Kotecha, N., & Lockwood, B., (2005). Soy-relieving the symptoms of menopause and fighting osteoporosis. *Pharm., 275*, 5.

Krauss, J. L., Roper, P. M., Ballard, A., Shih, C. C., Fitzpatrick, J. A. J., Cassat, J. E., Pei, Y. N., et al., (2019). Staphylococcus aureus infects osteoclasts and replicates intracellularly. *MBio, 10*(5).

Kubica, M., Guzik, K., Koziel, J., Zarebski, M., Richter, W., Gajkowska, B., Golda, A., et al., (2008). A potential new pathway for staphylococcus aureus dissemination: The silent survival of *S. aureus* phagocytosed by human monocyte-derived macrophages. *PLOS ONE., 3*(1), e1409.

Kwon, S. M., Kim, S., Song, N. J., Chang, S. H., Hwang, Y. J., Yang, D. K., Hong, J. W., et al., (2016). Antiadipogenic and pro osteogenic effects of luteolin, a major dietary flavone, are mediated by the induction of DnaJ (Hsp40) homolog, subfamily B, member 1. *J. Nutr. Biochem., 30*, 24–32.

Lachman, J., Hejtmánková, A., Táborský, J., Kotíková, Z., Pivec, V., Střalková, R., Vollmannová, A., et al., (2015). Evaluation of oil content and fatty acid composition in the seed of grapevine varieties. *LWT – Food Science and Technology, 63*(1), 620–625.

Lanzotti, V., Scala, F., & Bonanomi, G., (2014). Compounds from allium species with cytotoxic and antimicrobial activity. *Phytochem. Rev., 13*, 769–791.

Lee, H. W., Suh, J. H., Kim, H. N., Kim, A. Y., Park, S. Y., Shin, C. S., Choi, J. Y., & Kim, J. B., (2008). Berberine promotes osteoblast differentiation by Runx2 activation with p38 MAPK. *J. Bone Miner. Res., 23*(8), 1227–1237.

Lee, J. W., Ahn, J. Y., Hasegawa, S., Cha, B. Y., Yonezawa, T., Nagai, K., Seo, H. J., et al., (2009). Inhibitory effect of luteolin on osteoclast differentiation and function. *Cytotechnology, 61*(3), 125–134.

Lee, K. Y., & Jeon, Y. J., (2005). Macrophage activation by polysaccharide isolated from *Astragalus membranaceus. Int. Immunopharm., 5*, 1225–1233.

Lee, M. H., Kim, Y. J., Kim, H. J., Park, H. D., Kang, A. R., Kyung, H. M., Sung, J. H., et al., (2003). BMP-2-induced Runx2 expression is mediated by Dlx5, and TGF-beta 1 opposes the BMP-2-induced osteoblast differentiation by sup- pression of Dlx5 expression. *J. Biol. Chem., 278*(36), 34387–34394.

Leotoing, L., Davicco, M. J., Lebecque, P., Wittrant, Y., & Coxam, V., (2014). The flavonoid fisetin promotes osteoblasts differentiation through Runx2 transcriptional activity. *Mol. Nutr. Food Res., 58*(6), 1239–1248.

Leotoing, L., Wauquier, F., Guicheux, J., Miot-Noirault, E., Wittrant, Y., & Coxam, V., (2013). The polyphenol fisetin protects bone by repressing NF-kappaB and MKP-1-dependent signaling pathways in osteoclasts. *PLoS One., 8*(7), e68388.

Lew, D. P., & Waldvogel, F. A., (1997). Osteomyelitis. *The New England Journal of Medicine, 336*(14), 999–1007.

Lew, P. D. P., & Waldvogel, P. F. A., (2004). Osteomyelitis. *Lancet (London, England), 364*(9431), 369–379.

Li, F., Zhao, F., Li, M., Pan, M., Shi, F., Xu, H., Zheng, D., et al., (2020). Decreasing New York esophageal squamous cell carcinoma 1 expression inhibits multiple myeloma growth and osteolytic lesions. *Journal of Cellular Physiology, 235*(3), 2183–2194.

Li, R., Chen, W., Wang, W., Tian, W., & Zhang, X., (2009). Extraction, characterization of astragalus polysaccharides and its immune modulating activities in rats with gastric cancer. *Carbohydr. Polym., 78*, 738–742.

Liberatore, M., Calandri, E., Pavoni, G. L., et al., (2007). Reliability of white blood cell scan in the follow-up of osteomyelitis. *Biomedicine and Pharmacotherapy, 61*(5), 272–276.

Lima, L. M., Babakhani, B., Boldaji, S. A. H., Asadi, M., & Boldaji, R. M., (2013). Essential oils composition and antibacterial activities of *Eucalyptus camaldulensis* Dehn. *Med. Plants—Int. J. Phytomed. Relat. Indust., 5*(4), 214–218.

Lin, Q., Huang, Y., Xiao, B., & Ren, G. F., (2005). Effects of resveratrol on bone mineral density in ovarectomized rats. *International Journal of Biomedical Science, 1*(1).

Lipsky, B. A., Berendt, A. R., Deery, H. G., et al., (2006). Infectious diseases society of America. Diagnosis and treatment of diabetic foot infections. *Plast Reconstr. Surg., 117*(7), 212S–238S.

Lisabeth, E. M., Falivelli, G., & Pasquale, E. B., (2013). Eph receptor signaling and ephrins. *Cold Spring Harb. Perspect. Biol., 5*(9).

Liu, P., Yang, D. Q., Xie, F., Zhou, B., & Liu, M., (2014). Effect of calcitonin on anastrozole-induced bone pain during aromatase inhibitor therapy for breast cancer. *Genet. Mol. Res., 13*(3), 5285–5291.

Liu, Y. Q., Han, X. F., Bo, J. X., & Ma, H. P., (2016a). Wedelolactone enhances osteoblastogenesis but inhibits osteoclastogenesis through Sema3A/NRP1/PlexinA1 pathway. *Front. Pharmacol., 7*, 375.

Love, C., Patel, M., Lonner, B. S., Tomas, M. B., & Palestro, C. J., (2000). Diagnosing spinal osteomyelitis: A comparison of bone and Ga-67 scintigraphy and magnetic resonance imaging. *Clin. Nucl. Med., 25*(12), 963–977.

Ma, J., Wang, Z., Zhao, J., Miao, W., Ye, T., & Chen, A., (2018). Resveratrol attenuates lipopolysaccharides (LPS)-induced inhibition of osteoblast differentiation in MC3T3-E1 cells. *Medical Science Monitor: International Medical Journal of Experimental and Clinical Research, 24,* 2045.

Manna, S. K., Mukhopadhyay, A., & Aggarwal, B. B., (2000). Resveratrol suppresses TNF-induced activation of nuclear transcription factors NF-κB, activator protein-1, and apoptosis: Potential role of reactive oxygen intermediates and lipid peroxidation. *The Journal of Immunology, 164*(12), 6509–6519.

Marriott, I., (2004). Osteoblast responses to bacterial pathogens. *Immunologic Research, 30*(3), 291–308.

Marzia, M., Sims, N. A., Voit, S., Migliaccio, S., Taranta, A., Bernardini, S., Faraggiana, T., et al. (2000). Decreased c-Src ex- pression enhances osteoblast differentiation and bone formation. *J. Cell Biol., 151*(2), 311–320.

Masters, E. A., Muthukrishnan, G., Ho, L., Gill, A. L., De Mesy, B. K. L., Galloway, C. A., McGrath, J. L., et al., (2021). *Staphylococcus aureus* cell wall biosynthesis modulates bone invasion and osteomyelitis pathogenesis. *Frontiers in Microbiology, 12,* 2231.

Matsubara, T., Kida, K., Yamaguchi, A., Hata, K., Ichida, F., Meguro, H., Aburatani, H., et al., (2008). BMP2 regulates osterix through Msx2 and Runx2 during osteoblast differentiation. *J. Biol. Chem., 283*(43), 29119–29125.

Matsuda, Y., Minagawa, T., Okui, T., & Yamazaki, K., (2018). Resveratrol suppresses the alveolar bone resorption induced by artificial trauma from occlusion in mice. *Oral Dis., 24*(3), 412–421.

Matyanga, C. M. J., Morse, G. D., Gundidza, M., & Nhachi, C. F. B., (2020a). African potato (*Hypoxis hemerocallidea*): A systematic review of its chemistry, pharmacology and ethno medicinal properties. *BMC Complementary Medicine and Therapies, 20*(1), 182.

Matyanga, C. M. J., Morse, G. D., Gundidza, M., & Nhachi, C. F. B., (2020b). African potato (*Hypoxis hemerocallidea*): A systematic review of its chemistry, pharmacology and ethno medicinal properties. *BMC Complementary Medicine and Therapies, 20*(1), 182.

Maurya, A., Dwivedi, G. R., Darokar, M. P., & Srivastava, S. K., (2013). Antibacterial and synergy of clavine alkaloid lysergol and its derivatives against nalidixic acid-resistant *Escherichia coli. Chem. Biol. Drug Des., 81*, 484–490.

McCormic, D. B., & Greene, H. L., (1994). Methods for the determination of ascorbic acid. In: Burtis, C. A., & Ashwood, E. R., (eds.), *Tietz Textbook of Clinical Chemistry* (2nd edn., pp. 1313, 1314). WB Saunders, Philadelphia, Pa, USA.

McLean, R. R., (2009). Proinflammatory cytokines and osteoporosis. *Curr. Osteoporos. Rep., 7*(4), 134–139.

McNally, M., & Nagarajah, K., (2010). (iv) osteomyelitis. *Orthopaedics and Trauma, 24*(6), 416–429.

Merghni, A., Marzouki, H., Hentati, H., Aouni, M., & Mastouri, M., (2015). Antibacterial and antibiofilm activities of *Laurus nobilis* L. essential oil against *Staphylococcus aureus* strains associated with oral infections. *Pathol. Biol.*

Metsemakers, W. J., Morgenstern, M., McNally, M. A., Moriarty, T. F., McFadyen, I., Scarborough, M., Athanasou, N. A et al., (2018). Fracture-related infection: A consensus on definition from an international expert group. *Injury, 49*(3), 505–510.

Metzger, C. E., & Anand, N. S., (2019). The role of osteocytes in inflammatory bone loss. *Frontiers in Endocrinology, 10*, 285.

Metzger, C. E., Narayanan, A., Zawieja, D. C., & Bloomfield, S. A., (2017). Inflammatory bowel disease in a rodent model alters osteocyte protein levels controlling bone turnover. *Journal of Bone and Mineral Research, 32*(4), 802–813.

Ming, L. G., Chen, K. M., & Xian, C. J., (2013). Functions and action mechanisms of flavonoids genistein and icariin in regulating bone remodeling. *J. Cell. Physiol., 228*(3), 513–521.

Miric, D., Katanic, R., Kisic, B., et al., (2010). Oxidative stress and myeloperoxidase activity during bacterial meningitis: Effects of febrile episodes and the BBB permeability. *Clinical Biochemistry, 43*(3), 246–252.

Mitchell, G., Lafrance, M., Boulanger, S., Séguin, D. L., Guay, I., Gattuso, M., Marsault, E., et al., (2012). Tomatidine acts in synergy with aminoglycoside antibiotics against multiresistant *Staphylococcus aureus* and prevents virulence gene expression. *The Journal of Antimicrobial Chemotherapy, 67*(3), 559–568.

Miyazaki, T., Sanjay, A., Neff, L., Tanaka, S., Horne, W. C., & Baron, R., (2004). Src kinase activity is essential for osteoclast function. *J. Biol. Chem., 279*(17), 17660–17666.

Miyazono, K., Ten, D. P., & Heldin, C. H., (2000). TGF-beta signaling by Smad proteins. *Adv. Immuno., 75*, 115–157.

Mobasheri, A., & Shakibaei, M., (2013). Osteogenic effects of resveratrol *in vitro*: Potential for the prevention and treatment of osteoporosis. *Annals of the New York Academy of Sciences, 1290*(1), 59–66.

Mobasheri, A., & Shakibaei, M., (2013). Osteogenic effects of resveratrol *in vitro*: Potential for the prevention and treatment of osteoporosis. *Ann. N. Y. Acad. Sci., 1290*, 59–66.

Mollazadeh, H., Cicero, A. F. G., Blesso, C. N., Pirro, M., Majeed, M., & Sahebkar, A., (2017). *Immune Modulation by Curcumin: The Role of Interleukin-10, 59*(1), 89–101.

Monaco, C., Stranix, J. T., Avraham, T., Brecht, L., Saadeh, P. B., Hirsch, D., & Levine, J. P., (2016). Evolution of surgical techniques for mandibular reconstruction using free fibula flaps: The next generation. *Head & Neck, 38*(S1), E2066–E2073.

Moon, H., & Rhee, M. S., (2016). Synergism between carvacrol or thymol increases the antimicrobial efficacy of soy sauce with no sensory impact. *Int. J. Food Microbiol., 217*, 35–41.

Mori, G., D'Amelio, P., Faccio, R., & Brunetti, G., (2013). The interplay between the bone and the immune system. *Clinical and Developmental Immunology, 16.*

Morley, P., Whitfield, J. F., & Willick, G. E., (2001). Parathyroid hormone: An anabolic treatment for osteoporosis. *Curr. Pharm. Des., 7*(8), 671–687.

Mylona, E., Samarkos, M., Kakalou, E., Fanourgiakis, P., & Skoutelis, A., (2009). Pyogenic vertebral osteomyelitis: A systematic review of clinical characteristics. *Seminars in Arthritis and Rheumatism, 39*(1), 10–17.

Nagar, G. K., Singh, D., Trivedi, P. K., Sangwan, N. S., Mishra, P. R., & Trivedi, R., (2013). Withaferin A: A proteasomal inhibitor promotes healing after injury and exerts anabolic effect on osteoporotic bone. *Cell Death Dis., 4*, e778.

Nath, U. K., Kim, H. T., Khatun, K., Park, J. I., Kang, K. K., & Nou, I. S., (2016). Modification of fatty acid profiles of rapeseed (*Brassica napus* L.) oil for using as food, industrial feed-stock and biodiesel. *Plant Breeding and Biotechnology, 4*(2), 123–134.

Niziolek, P. J., Bullock, W., Warman, M. L., & Robling, A. G., (2015). Missense mutations in LRP5 associated with high bone mass protect the mouse skeleton from disuse- and ovariectomy-induced osteopenia. *PLoS One, 10*(11), e0140775.

Novack, D. V., (2011). Role of NF-kappaB in the skeleton. *Cell Res., 21*(1), 169–182.

Omori, A., Yoshimura, Y., Deyama, Y., & Suzuki, K., (2015). Rosmarinic acid and arbutin suppress osteoclast differentiation by inhibiting superoxide and NFATc1 down- regulation in RAW 264.7 cells. *Biomed Rep., 3*(4), 483–490.

Ornitz, D. M., & Marie, P. J., (2015). Fibroblast growth factor signaling in skeletal development and disease. *Genes Dev., 29*(14), 1463–1486.

Ornitz, D. M., (2005). FGF signaling in the developing endochondral skeleton. *Cytokine Growth Factor Rev., 16*(2), 205–213.

Pandey, M. K., Kale, V. P., Song, C., Sung, S. S., Sharma, A. K., Talamo, G., Dovat, S., & Amin, S. G., (2014). Gambogic acid inhibits multiple myeloma mediated osteoclastogenesis through suppression of chemokine receptor CXCR4 signaling pathways. *Exp. Hematol., 42*(10), 883–896.

Pangrazio, A., Cassani, B., Guerrini, M. M., Crockett, J. C., Marrella, V., Zammataro, L., Strina, D., et al., (2012). RANK-dependent autosomal recessive osteopetrosis: Characterization of five new cases with novel mutations. *J. Bone Miner., 27*(2), 342–351.

Parikh, M., Netticadan, T., & Pierce, G. N., (2018). Flaxseed: Its bioactive components and their cardiovascular benefits. *American Journal of Physiology – Heart and Circulatory Physiology, 314*(2), H146–H159.

Park, O. J., Kim, H. J., Woo, K. M., Baek, J. H., & Ryoo, H. M., (2010). FGF2-activated ERK mitogen-activated protein kinase enhances Runx2 acetylation and stabilization. *J. Biol. Chem., 285*(6), 3568–3574.

Park, S. Y., Kim, H. J., Kim, K. R., Lee, S. K., Lee, C. K., Park, K. K., & Chung, W. Y., (2014). Betulinic acid, a bioactive pentacyclic triterpenoid, inhibits skeletal-related events induced by breast cancer bone metastases and treatment. *Toxicol. Appl. Pharmacol., 275*(2), 152–162.

Pasquale, E. B., (2008). Eph-ephrin bidirectional signaling in physiology and disease. *Cell., 133*(1), 38–52.

Phimphilai, M., Zhao, Z., Boules, H., Roca, H., & Franceschi, R. T., (2006). BMP signaling is required for RUNX2-dependent induction of the osteoblast phenotype. *J. Bone Miner. Res., 21*(4), 637–646.

Pineda, C., Espinosa, R., & Pena, A., (2009). Radiographic imaging in osteomyelitis: The role of plain radiography, computed tomography, ultrasonography, magnetic resonance imaging, and scintigraphy. *Semin Plast Surg., 23*(2), 80–89.

Preethi, A. M., & Bellare, J. R., (2021). Concomitant effect of quercetin- and magnesium-doped calcium silicate on the osteogenic and antibacterial activity of scaffolds for bone regeneration. *Antibiotics, 10*(10), 1170.

Prideaux, M., Findlay, D. M., & Atkins, G. J., (*2016*). Osteocytes: The master cells in bone remodeling. *Current Opinion in Pharmacology, 28*, 24–30.

Qiao, J. H., Tripathi, J., Mishra, N. K., Cai, Y., Tripathi, S., Wang, X. P., Imes, S., et al., (1997). Role of macrophage colony-stimulating factor in atherosclerosis: Studies of osteopetrotic mice. Am. *J. Pathol., 150*(5), 1687–1699.

Qin, C. H., Zhou, C. H., Song, H. J., Cheng, G. Y., Zhang, H. A., Fang, J., & Tao, R., (2019). Infected bone resection plus adjuvant antibiotic-impregnated calcium sulfate versus infected bone resection alone in the treatment of diabetic forefoot osteomyelitis. *BMC Musculoskeletal Disorders, 20*(1), 1–8.

Qin, Q., Niu, J., Wang, Z., Xu, W., Qiao, Z., & Gu, Y., (2012). *Astragalus membranaceus* extract activates immune response in macrophages via heparanase. *Molecules, 17*, 7232–7240.

Rachner, T. D., Khosla, S., & Hofbauer, L. C., (2011). Osteoporosis: Now and the future. *Lancet, 377*, 1276–1287.

Rahman, M. S., Akhtar, N., Jamil, H. M., Banik, R. S., & Asaduzzaman, S. M., (2015). TGF-beta/ BMP signaling and other molecular events: Regulation of osteoblastogenesis and bone formation. *Bone Res., 3*, 15005.

Ramesh, H. P., Yamaki, K., & Tsushida, T., (2002). Effect of fenugreek (*Trigonella foenum-graecum* L.) galactomannan fractions on phagocytosis in rat macrophages and on proliferation and IgM secretion in HB4C5 cells. *Carbohydr. Polym., 50*, 79–83.

Randhawa, H. K., Hundal, K. K., Ahirrao, P. N., Jachak, S. M., & Nandanwar, H. S., (2016). Efflux pump inhibitory activity of flavonoids isolated from *Alpinia calcarata* against methicillin-resistant *Staphylococcus aureus*. *Biologia, 71*, 484–493.

Rantlha, M., Sagar, T., Kruger, M. C., Coetzee, M., & Deepak, V., (2017). Ellagic acid inhibits RANKL-induced osteoclast differentiation by suppressing the p38 MAP kinase pathway. *Arch. Pharm. Res., 40*(1), 79–87.

Regard, J. B., Zhong, Z., Williams, B. O., & Yang, Y., (2012). Wnt signaling in bone development and disease: Making stronger bone with Wnts. *Cold Spring Harb. Perspect. Biol., 4*(12).

Rehman, F., & Mairaj, S., (2013). Antimicrobial studies of allicin and ajoene. *Int. J. Pharm. Bio Sci., 4*, 1095–1105.

Reiter, J., Levina, N., Van, D. L. M., Gruhlke, M., Martin, C., & Slusarenko, A., (2017). Diallylthiosulfinate (allicin), a volatile antimicrobial from garlic (*Allium sativum*), kills human lung pathogenic bacteria, including MDR strains, as a vapor. *Molecules., 22*, 1711.

Reuter, S., Prasad, S., Phromnoi, K., Kannappan, R., Yadav, V. R., & Aggarwal, B. B., (2010). Embelin suppresses osteoclastogenesis induced by receptor activator of NF-kappaB ligand and tumor cells *in vitro* through inhibition of the NF-kappaB cell signaling pathway. *Mol. Cancer Res., 8*(10), 1425–1436.

Roblot, F., Besnier, J. M., Juhel, L., Vidal, C., Ragot, S., Bastides, F., Le Moal, G., et al., (2007a). Optimal duration of antibiotic therapy in vertebral osteomyelitis. *Seminars in Arthritis and Rheumatism, 36*(5), 69–277.

Roblot, F., Besnier, J. M., Juhel, L., Vidal, C., Ragot, S., Bastides, F., Le Moal, G., et al., (2007b). Optimal duration of antibiotic therapy in vertebral osteomyelitis. *Seminars in Arthritis and Rheumatism, 36*(5), 269–277.

Roodman, G. D., (1999). Cell biology of the osteoclast. *Exp. Hematol., 27*(8), 1229–1241.

Roux, S., Amazit, L., Meduri, G., Guiochon-Mantel, A., Milgrom, E., & Mariette, X., (2002). RANK (receptor activator of nuclear factor kappa B) and RANK ligand are expressed in giant cell tumors of bone. *Am. J. Clin. Pathol., 117*(2), 210–216.

Ruocco, M. G., Maeda, S., Park, J. M., Lawrence, T., Hsu, L. C., Cao, Y., Schett, G., et al., (2005). I{kappa}B kinase (IKK){beta}, but not IKK{alpha}, is a critical mediator of osteoclast survival and is required for inflammation-induced bone loss. *J. Exp.*, *201*(10), 1677–1687.

Ryoo, H. M., Hoffmann, H. M., Beumer, T., Frenkel, B., Towler, D. A., Stein, G. S., Stein, J. L., et al., (1997). Stage-specific expression of Dlx-5 during osteo- blast differentiation: Involvement in regulation of osteocalcin gene expression. *Mol. Endocrinol.*, *11*(11), 1681–1694.

Saavedra-Lozano, J., Mejías, A., Ahmad, N., Peromingo, E., Ardura, M. I., Guillen, S., Syed, A., et al., (2008). Changing trends in acute osteomyelitis in children: Impact of methicillin-resistant *Staphylococcus aureus* infections. *Journal of Pediatric Orthopedics, 28*(5), 569–575.

Sahasrabudhe, N. M., Beukema, M., Tian, L., Troost, B., Scholte, J., Bruininx, E., Bruggeman, G., Van, D. B. M., Scheurink, A., Schols, H. A., et al., (2018). Dietary fiber pectin directly blocks toll-like receptor 2-1 and prevents doxorubicin-induced ileitis. *Front. Immunol, 9*, 383.

Saini, R. K., & Keum, Y. S., (2018a). Omega-3 and omega-6 polyunsaturated fatty acids: Dietary sources, metabolism, and significance — A review. *Life Sciences, 203,* 255–267.

Sandukji, A., Al-Sawaf, H., Mohamadin, A., Alrashidi, Y., & Sheweita, S. A., (2011). Oxidative stress and bone markers in plasma of patients with long-bone fixative surgery: Role of antioxidants. *Human and Experimental Toxicology, 30*(6), 435–442.

Sarjit, A., Wang, Y., & Dykes, G. A., (2015). Antimicrobial activity of gallic acid against thermophilic campylobacter is strain specific and associated with a loss of calcium ions. *Food Microbiol., 46*, 227–233.

Sasi, S., (2017). Nutraceuticals – A review. *International Journal of Industrial Biotechnology and Biomaterials, 3*(1), 25–29.

Sato, M., Asada, N., Kawano, Y., Wakahashi, K., Minagawa, K., Kawano, H., Sada, A., et al., (2013). Osteocytes regulate primary lymphoid organs and fat metabolism. *Cell Metabolism, 18*(5), 749–758.

Schepetkin, I. A., & Quinn, M. T., (2006). Botanical polysaccharides: Macrophage immunomodulation and therapeutic potential. *Int. Immunopharmacol., 6*, 317–333.

Schett, G., (2011). Effects of inflammatory and anti-inflammatory cytokines on the bone. *Eur. J. Clin. Investig, 41*(12), 1361–1366.

Schmitt, S. K., (2017). Osteomyelitis. *Infectious Disease Clinics of North America., 31*(2), 325–338.

Schweitzer, M. E., Daffner, R. H., Weissman, B. N., et al., (2008). ACR appropriateness criteria on suspected osteomyelitis in patients with diabetes mellitus. *J. Am. Coll. Radiol.*, *5*(8), 881–886.

Servili, M., Sordini, B., Esposto, S., Urbani, S., Veneziani, G., Di Maio, I., Selvaggini, R., & Taticchi, A., (2014). Biological activities of phenolic compounds of extra virgin olive oil. *Antioxidants, 3*(1), 1–23.

Sethi, G., & Aggarwal, B. B., (2007). Mending the bones with natural products. *Chem. Biol, 14*(7), 738–740.

Sharma, A. K., Keservani, R. K., & Gautam, S. P., (2020). *Herbal Product Development.* Apple Academic Press, CRC Press, Taylor & Francis Group. ISBN: 9781771888776.

Shin, D. K., Kim, M. H., Lee, S. H., Kim, T. H., & Kim, S. Y., (2012). Inhibitory effects of luteolin on titanium particle-induced osteolysis in a mouse model. *Acta Biomater., 8*(9), 3524–3531.

Shishodia, S., & Aggarwal, B. B., (2005). *Resveratrol: A Polyphenol for All Seasons*. In: *Resveratrol in Health and Disease* (1st ed.). Aggarwal, B. B., & Shishodia, S. (Eds.). CRC Press, Boca Raton, Florida. pp. 1–15.

Shodo, R., Sato, Y., Ota, H., & Horii, A., (2017). Clavicle fracture with osteomyelitis after neck dissection and post-operative radiotherapy: Case report. *The Journal of Laryngology & Otology, 131*(11), 1026–1029.

Shriram, V., Jahagirdar, S., Latha, C., Kumar, V., Puranik, V., Rojatkar, S., et al., (2008). A potential plasmid-curing agent, 8-episodios bulbine acetate, from *Dioscorea bulbifera* L. against multidrug-resistant bacteria. *Int. J. Antimicrob. Agents, 32*, 405–410.

Siriyong, T., Srimanote, P., Chusri, S., Yingyongnarongkul, B. E., Suaisom, C., Tipmanee, V., et al., (2017). Conessine as a novel inhibitor of multidrug efflux pump systems *Pseudomonas aeruginosa. BMC Complement. Altern. Med, 17*, 405.

Smith, M. H., Harms, P. W., Newton, D. W., Lebar, B., Edwards, S. P., & Aronoff, D. M., (2011). Mandibular actinomyces osteomyelitis complicating florid cemento-osseous dysplasia: Case report. *BMC Oral Health, 11*(1), 1–6. h.

Soltani, R., Fazeli, H., Bahri, N. R., & Jelokhanian, A., (2017). Evaluation of the synergistic effect of tomatidine with several antibiotics against standard and clinical isolates of *Staphylococcus aureus*, *Enterococcus faecalis*, *Pseudomonas aeruginosa* and *Escherichia coli. Iranian Journal of Pharmaceutical Research, 16*(1), 290–296.

Sridevi, D., Shankar, C., Prakash, P., Park, J., & Thamaraiselvi, K., (2017). Inhibitory effects of reserpine against efflux pump activity of antibiotic resistance bacteria. *Chem. Biol. Lett., 4*, 69–72.

Stephanie, B., Eric, D., Sophie, F. M., Christian, B., & Yu, G., (2010). Carrageenan from *Solieria chordalis* (*Gigartinales*): Structural analysis and immunological activities of the low molecular weight fractions. *Carbohydrate Polymers, 81*(2), 448–460.

Stohs, S. J., & Bagchi, D., (2015). Antioxidant, anti-inflammatory, and chemoprotective properties of acacia catechu heartwood extracts. *Phytotherapy Research, 29*(6), 818–824.

Strobel, M., Pförtner, H., Tuchscherr, L., Völker, U., Schmidt, F., Kramko, N., Schnittler, H. J., et al., (2016). Post-invasion events after infection with *Staphylococcus aureus* are strongly dependent on both the host cell type and the infecting *S. aureus* strain. *Clinical Microbiology and Infection, 22*(9), 799–809.

Szafranska, A. K., Oxley, A. P. A., Chaves-Moreno, D., Horst, S. A., Roßlenbroich, S., Peters, G., Goldmann, O., et al., (2014a). *High-resolution Transcriptomic Analysis of the Adaptive Response of Staphylococcus Aureus During Acute and Chronic Phases of Osteomyelitis, 5*(6).

Takatsuka, H., Umezu, H., Hasegawa, G., Usuda, H., Ebe, Y., Naito, M., & Shultz, L. D., (1998). Bone remodeling and macrophage differentiation in osteopetrosis (op) mutant mice defective in the production of macrophage colony-stimulating factor. *J. Submicrosc. Cytol. Pathol., 30*(2), 239–247.

Takyar, F. M., Tonna, S., Ho, P. W., Crimeen-Irwin, B., Baker, E. K., Martin, T. J., & Sims, N., (2013). A EphrinB2/EphB4 inhibition in the osteoblast lineage modifies the anabolic response to parathyroid hormone. *J. Bone Miner. Res., 28*(4), 912–925.

Tan, N., Yazıcı-Tütünis, S., Bilgin, M., Tan, E., & Miski, M., (2017). Antibacterial activities of pyrenylated coumarins from the roots of *Prangos hulusii*. *Molecules., 22*, 1098.

Teitelbaum, S. L., (2000). Bone resorption by osteoclasts. *Science, 289*, 1504–1508.

Terashima, A., Okamoto, K., Nakashima, T., Akira, S., Ikuta, K., & Takayanagi, H., (2016). Sepsis-induced osteoblast ablation causes immunodeficiency. *Immunity., 44*(6), 1434–1443.

Teven, C. M., Farina, E. M., Rivas, J., & Reid, R. R., (2014). Fibroblast growth factor (FGF) signaling in development and skeletal diseases. *Genes Dis., 1*(2), 199–213.

Thomas, T., (2006). Intermittent parathyroid hormone therapy to increase bone formation. *Joint Bone Spine, 73*(3), 262–269.

Thompson, J., Mendoza, F., Tan, E., Bertol, J. W., Gaggar, A. S., Jun, G., Biguetti, C., & Fakhouri, W. D., (2019). A cleft lip and palate gene, Irf6, is involved in osteoblast differentiation of craniofacial bone. *Developmental Dynamics, 248*(3), 221–232.

Thomson, M., Al-Qattan, K. K., Al-Sawan, S. M., Alnaqeeb, M. A., Khan, I., & Ali, M., (2002). The use of ginger (*Zingiber officinale* Rosc.) as a potential anti-inflammatory and antithrombotic agent. *Prostaglandins, Leukotrienes and Essential Fatty Acids, 67*(6), 475–478.

Tice, A. D., Hoaglund, P. A., & Shoultz, D. A., (2003). Outcomes of osteomyelitis among patients treated with outpatient parenteral antimicrobial therapy. *The American Journal of Medicine, 114*(9), 723–728.

Timoszuk, M., Bielawska, K., & Skrzydlewska, E., (2018a). Evening primrose (*Oenothera biennis*) biological activity dependent on chemical composition. *Antioxidants, 7*(8), 108.

Timoszuk, M., Bielawska, K., & Skrzydlewska, E., (2018b). Evening primrose (*Oenothera biennis*) biological activity dependent on chemical composition. *Antioxidants, 7*(108), 108.

Togashi, N., Hamashima, H., Shiraishi, A., Inoue, Y., & Takano, A., (2010). Antibacterial activities against *Staphylococcus aureus* of terpene alcohols with aliphatic carbon chains. *J. Essent. Oil Res., 22*, 263–269.

Tolba, M. F., El-Serafi, A. T., & Omar, H. A., (2017). Caffeic acid phenethyl ester protects against glucocorticoid-induced osteoporosis *in vivo*: Impact on oxidative stress and RANKL/OPG signals. *Toxicol. Appl. Pharmacol., 324*, 26–35.

Tong, S. Y. C., Davis, J. S., Eichenberger, E., Holland, T. L., & Fowler, V. G., (2015). *Staphylococcus aureus* infections: Epidemiology, pathophysiology, clinical manifestations, and management. *Clinical Microbiology Reviews., 28*(3), 603–661.

Tou, J. C., (2015). Resveratrol supplementation affects bone acquisition and osteoporosis: Pre-clinical evidence toward translational diet therapy. *Biochimica et Biophysica Acta (BBA) – Molecular Basis of Disease, 1852*(6), 1186–1194.

Tsai, Y. M., Chong, I. W., Hung, J. Y., Chang, W. A., Kuo, P. L., Tsai, M. J., & Hsu, Y. L., (2015). Syringetin suppresses osteoclastogenesis mediated by osteoblasts in human lung adenocarcinoma. *Oncol. Rep., 34*(2), 617–626.

Tsuji, K., Bandyopadhyay, A., Harfe, B. D., Cox, K., Kakar, S., Gerstenfeld, L., Einhorn, T., et al., (2006). BMP2 activity, although dispensable for bone formation, is required for the initiation of fracture healing. *Nat. Genet., 38*(12), 1424–1429.

Tsuji, M., Yamamoto, H., Sato, T., Mizuha, Y., Kawai, Y., Taketani, Y., Kato, S., et al., (2009). Dietary quercetin inhibits bone loss without effect on the uterus in ovariectomized mice. *Journal of Bone and Mineral Metabolism, 27*(6), 673–681.

Tuchscherr, L., & Löffler, B., (2015). Staphylococcus aureus dynamically adapts global regulators and virulence factor expression in the course from acute to chronic infection. *Current Genetics, 62*(1), 15–17.

Tyagi, A. K., Prasad, S., Majeed, M., & Aggarwal, B. B., (2016). Calebin A down regulates osteoclastogenesis through suppression of RANKL signaling. *Arch. Biochem. Biophys., 593*, 80–89.

Udagawa, N., Takahashi, N., Akatsu, T., Tanaka, H., Sasaki, T., Nishihara, T., Koga, T., et al., (1990). Origin of osteoclasts: Mature monocytes and macro- phages are capable of differentiating into osteoclasts under a suitable micro-environment prepared by bone marrow-derived stromal cells. *Proc. Natl. Acad. Sci. U. S. A., 87*(18), 7260–7264.

Umeda, S., Takahashi, K., Naito, M., Shultz, L. D., & Takagi, K., (1996). Neonatal changes of osteoclasts in osteopetrosis (op/op) mice defective in production of functional macrophage colony-stimulating factor (M-CSF) protein and effects of M-CSF on osteoclast development and differentiation. *J. Submicrosc. Cytol. Pathol., 28*(1), 13–26.

Vergidis, P., Schmidt-Malan, S. M., Mandrekar, J. N., Steckelberg, J. M., & Patel, R., (2015). Comparative activities of vancomycin, tigecycline and rifampin in a rat model of methicillin-resistant *Staphylococcus aureus* osteomyelitis. *The Journal of Infection, 70*(6), 609–615.

Von, M. I., Krebbel, H., Kuckelkorn, U., Heider, U., Jakob, C., Kaiser, M., Fleissner, C., et al., (2008). Curcumin diminishes human osteoclastogenesis by inhibition of the signalosome-associated IκB kinase. *Journal of Cancer Research and Clinical Oncology, 135*(2), 173–179.

Wael, S., Wilhelmus, W. T., Arini, I., Smit, A., Matdoan, N., Retna, P. D., Bina, S. A., et al., (2018). Potential of *Syzygium aromaticum* (Clove) leaf extract on immune proliferation response in Balb/c mice infected with *Salmonella typhimurium. Case Reports in Clinical Medicine, 7*(12), 613–627.

Wagner, H., Nörr, H., & Winterhoff, H., (1994). Plant adaptogens. *Phytomedicine, 1*(1), 63–76.

Waldvogel, F. A., Medoff, G., & Swartz, M. N., (1970). Osteomyelitis: A review of clinical features, therapeutic considerations and unusual aspects. *The New England Journal of Medicine., 282*(4), 198–206.

Walsh, M. C., & Choi, Y., (2014). Biology of the RANKL-RANK-OPG system in immunity, bone, and beyond. *Front. Immunol., 5*, 11.

Walsh, M. C., Kim, N., Kadono, Y., Rho, J., Lee, S. Y., Lorenzo, J., & Choi, Y., (2006). Osteoimmunology: Interplay between the immune system and bone metabolism. *Annual Review of Immunology, 24*, 33–63.

Walter, G., Kemmerer, M., Kappler, C., & Hoffmann, R., (2014a). Treatment algorithms for chronic osteomyelitis. *Deutsches Arzteblatt International., 109*(14), 257–264.

Wang, R. N., Green, J., Wang, Z., Deng, Y., Qiao, M., Peabody, M., Zhang, Q., et al., (2014a). Bone morphogenetic protein (BMP) signaling in development and human diseases. *Genes Dis., 1*(1), 87–105.

Wang, W., Wang, J., Dong, S. F., Liu, C. H., Italiani, P., Sun, S. H., Xu, J., et al., (2010). Immunomodulatory activity of andrographolide on macrophage activation and specific antibody response. *Acta Pharmacologica Sinica, 31*(2), 191–201.

Wang, Y., Li, Y. P., Paulson, C., Shao, J. Z., Zhang, X., Wu, M., & Chen, W., (2014). Wnt and the Wnt signaling pathway in bone development and disease. *Front Biosci (Landmark Ed),19*, 379–407.

Wang, Y., Lu, Z., Wu, H., & Lv, F., (2009). Study on the antibiotic activity of microcapsule curcumin against foodborne pathogens. *International Journal of Food Microbiology, 136*(1), 71–74.

Wattel, A., Kamel, S., Prouillet, C., Petit, J. P., Lorget, F., Offord, E., & Brazier, M., (2004). Flavonoid quercetin decreases osteoclastic differentiation induced by RANKL via a mechanism involving NF kappa B and AP-1. *J. Cell. Biochem., 92*(2), 285–295.

Widaa, A., Claro, T., Foster, T. J., O'Brien, F. J., & Kerrigan, S. W., (2012). *Staphylococcus aureus* protein a plays a critical role in mediating bone destruction and bone loss in osteomyelitis. *PLOS One, 7*(7), e40586.

Wieland, I., Jakubiczka, S., Muschke, P., Cohen, M., Thiele, H., Gerlach, K. L., Adams, R. H., & Wieacker, P., (2004). Mutations of the ephrin-B1 gene cause craniofrontonasal syndrome. Am. *J. Hum. Genet., 74*(6), 1209–1215.

Wozney, J. M., (1992). The bone morphogenetic protein family and osteogenesis. *Mol. Reprod. Dev., 32*(2), 160–167.

Wu, H. Z., Fei, H. J., Zhao, Y. L., Liu, X. J., Huang, Y. J., & Wu, S. W., (2012). Antibacterial mechanism of sulforaphaneon *Escherichia coli. Sichuan Da Xue Xue Bao Yi Xue Ban., 43*,386–390.

Wu, X., Li, Z., Yang, Z., Zheng, C., Jing, J., Chen, Y., Ye, X., et al., (2012). Caffeic acid 3,4-dihydroxy-phenethyl ester suppresses receptor activator of NF-kappaB ligand-induced osteoclastogenesis and prevents ovariectomy-induced bone loss through inhibition of mitogen-activated protein kinase/activator protein 1 and Ca^{2+}-nuclear factor of activated T-cells cytoplasmic 1 signaling pathways. *J. Bone Miner. Res., 27*(6), 1298–1308.

Xin, M., Yang, Y., Zhang, D., Wang, J., Chen, S., & Zhou, D., (2015). Attenuation of hind-limb suspension-induced bone loss by curcumin is associated with reduced oxidative stress and increased vitamin D receptor expression. *Osteoporos. Int., 26*(11), 2665–2676.

Xu, D., Yang, W., Zhou, C., Liu, Y., & Xu, B., (2010). Preventive effects of berberine on glucocorticoid-induced osteoporosis in rats. *Planta Med., 76*(16), 1809–1813.

Yadav, M. K., Chae, S. W., Im, G. J., Chung, J. W., & Song, J. J., (2015). Eugenol: A phyto-compound effective against methicillin-resistant and methicillin sensitive *Staphylococcus aureus* clinical strain biofilms. *PLoS One, 10*, e0119564.

Yamaguchi, M., & Weitzmann, M. N., (2011). Quercetin, a potent suppressor of NF-κB and Smad activation in osteoblasts. *International Journal of Molecular Medicine, 28*(4), 521–525.

Yang, D., Wijenayaka, A. R., Solomon, L. B., Pederson, S. M., Findlay, D. M., Kidd, S. P., & Atkins, G. J., (2018). Novel insights into *Staphylococcus aureus* deep bone infections: The involvement of osteocytes. *MBio., 9*(2).

Yavropoulou, M. P., & Yovos, J. G., (2007). The role of the Wnt signaling pathway in osteoblast commitment and differentiation. *Hormones Athens., 6*(4), 279–294.

Yin, M., Zhang, Y., & Li, H., (2019). Advances in research on immunoregulation of macrophages by plant polysaccharides. *Front. Immunol., 10*, 145.

Yin, Z., Zhu, W., Wu, Q., Zhang, Q., Guo, S., Liu, T., Li, S., Chen, X., Peng, D., & Ouyang, Z., (2019). Glycyrrhizic acid suppresses osteoclast differentiation and postmenopausal osteoporosis by modulating the NF-κB, ERK, and JNK signaling pathways. *European Journal of Pharmacology, 859*, 172550.

Yoshime, L. T., De Melo, I. L. P., Sattler, J. A. G., De Carvalho, E. B. T., & Mancini-Filho, J., (2016). Bitter gourd (*Momordica charantia* L.) seed oil as a naturally rich source of bioactive compounds for nutraceutical purposes. *Nutrire, 41*(1), 1–7.

Yu, S., Jiang, B., Jia, C., Wu, H., Shen, J., Hu, X., & Xie, Z., (2020). Investigation of biofilm production and its association with genetic and phenotypic characteristics of

OM (osteomyelitis) and non-OM orthopedic *Staphylococcus aureus*. *Annals of Clinical Microbiology and Antimicrobials., 19*(1), 1–9.

Zawawi, M. S., Perilli, E., Stansborough, R. L., Marino, V., Cantley, M. D., Xu, J., Dharmapatni, A. A., et al., (2015). Caffeic acid phenethyl ester abrogates bone resorption in a murine calvarial model of polyethylene particle-induced osteolysis. *Calcif. Tissue Int., 96*(6), 565–574.

Zha, X. Q., Luo, J. P., Luo, S. Z., & Jiang, S. T., (2007). Structure identification of a new immunostimulating polysaccharide from the stems of *Dendrobium huoshanense*. *Carbohydr. Polym., 69*, 86–93.

Zhang, L. Y., Xue, H. G., Chen, J. Y., Chai, W., & Ni, M., (2016). Genistein induces adipogenic differentiation in human bone marrow mesenchymal stem cells and suppresses their osteogenic potential by upregulating PPARgamma. *Exp. Ther. Med., 11*(5), 1853–1858.

Zhang, X. F., Wang, H. M., Song, Y. L., Nie, L. H., Wang, L. F., Liu, B., Shen, P. P., & Liu, Y., (2006). Isolation, structure elucidation, antioxidative and immunomodulatory properties of two novel di hydrocoumarins from Aloe vera. *Bioorganic & Medicinal Chemistry Letters, 16*(4), 949–953.

Zhang, X., Guo, J., Zhou, Y., & Wu, G., (2014). The roles of bone morphogenetic proteins and their signaling in the osteogenesis of adipose-derived stem cells. *Tissue Eng. Part B Rev., 20*(1), 84–92.

Zhang, X., Zhou, C., Zha, X., Xu, Z., Li, L., Liu, Y., Xu, L., et al., (2015). Apigenin promotes osteogenic differentiation of human mesenchymal stem cells through JNK and p38 MAPK pathways. *Mol. Cell. Biochem., 407*(1, 2), 41–50.

Zhang, Y. B., Liu, X. Y., Jiang, P. P., Li, W. D., & Wang, Y. F., (2015). Mechanism and antibacterial activity of cinnamaldehyde against *Escherichia coli* and *Staphylococcus aureus*. *Mod. Food Sci. Techno*l., 31–35.

Zhao, C., Irie, N., Takada, Y., Shimoda, K., Miyamoto, T., Nishiwaki, T., Suda, T., & Matsuo, K., (2006). Bidirectional ephrinB2-EphB4 signaling controls bone homeostasis. *Cell Metab., 4*(2), 111–121.

Zhao, E., Xu, H., Wang, L., Kryczek, I., Wu, K., Hu, Y., Wang, G., & Zou, W., (2011). Bone marrow and the control of immunity. *Cellular & Molecular Immunology, 9*(1), 11–19.

Zhao, L., Wang, Y., Wang, Z., Xu, Z., Zhang, Q., & Yin, M., (2015). Effects of dietary resveratrol on excess-iron-induced bone loss via antioxidative character. *J. Nutr. Biochem., 26*(11), 1174–1182.

Zheng, Y., Ren, W., Zhang, L., Zhang, Y., Liu, D., & Liu, Y., (2020). A review of the pharmacological action of astragalus polysaccharide. *Frontiers in Pharmacology., 11*, 349.

Zhou, W., Spoto, M., Hardy, R., Guan, C., Fleming, E., Larson, P. J., Brown, J. S., & Oh, J., (2020). Host-specific evolutionary and transmission dynamics shape the functional diversification of *Staphylococcus* epidermidis in human skin. *Cell, 180*(3), 454–470.

Zhou, Y., Wu, Y., Jiang, X., Zhang, X., Xia, L., Lin, K., & Xu, Y., (2015). The effect of quercetin on the osteogenic differentiation and angiogenic factor expression of bone marrow-derived mesenchymal stem cells. *PLoS One, 10*(6), e0129605.

Zhou, Z., Pan, C., Lu, Y., Gao, Y., Liu, W., Yin, P., & Yu, X., (2017). Combination of erythromycin and curcumin alleviates *Staphylococcus aureus* induced osteomyelitis in rats. *Frontiers in Cellular and Infection Microbiology, 7*, 379.

Zimmerli, W., (2010). Clinical practice. Vertebral osteomyelitis. *The New England Journal of Medicine, 362*(11), 1022–1029.

Zou, Y. C., Yang, X. W., Yuan, S. G., Zhang, P., & Li, Y. K., (2016). Celastrol inhibits prostaglandin E2-induced proliferation and osteogenic differentiation of fibroblasts isolated from ankylosing spondylitis hip tissues *in vitro*. *Drug Des. Devel. Ther., 10*, 933–948.

CHAPTER 11

NUTRACEUTICALS: NEW PARADIGM FOR BETTER MANAGEMENT OF RICKETS

SHYAMLI THAKUR,[1] and LUXITA SHARMA[2]

[1]*School of Pharmaceutical and Populations Health Informatics, DIT University, Dehradun, Uttarakhand, India*

[2]*Faculty of Health and Allied Sciences, Amity University, Gurgaon, India*

ABSTRACT

Nutraceuticals offer a proactive healthcare approach with enormous therapeutic effects on human body. They are potent means which help towards maintaining and improving human health and acting against nutritionally caused acute and chronic diseases and thereby advocating a higher quality of life. Defective chondrocyte differentiation, mineralization of the growth plate and osteoid mineralization which are perpetually found in growing children results in nutritional rickets. Low dietary intake of calcium plays a critical role in developing rickets whereas in temperate countries it is the vitamin D deficiency which causes rickets. Prevention of rickets is achievable through either supplementation or food fortification with vitamin D, calcium and adequate sunlight exposure. In the year 1932, milk was fortified with vitamin D, as a step towards preventing the rickets in children in the community. To ensure that the integrity of the vitamins D is well-maintained and its bioavailability in the gastrointestinal tract is retained, microencapsulations have been found to be suitable to encapsulate both Vitamin D 2 and Vitamin

Nutraceuticals and Bone Health. Deepak Sharma, Madan Mohan Gupta, Anil K. Sharma, Raj K. Keservani, Rajesh K. Kesharwani (Eds.)

D 3 nutraceuticals to supplement and fortify foods. Statement of Global Consensus Recommendations of Nutritional Rickets have stated various recommendation for prevention of rickets for different population groups along with the route of administration and duration of therapy. Since colecalciferol-Vitamin D3, and ergocalciferol- Vitamin D2 and calcium is required as a highly significant health-care need of various population groups, they are considered to be of essential nature and mentioned under WHO Model List of Essential Medicines. Hence, they qualify to be available in adequate amounts with standard expected quality within the setting of operating health systems at a cost that the individual and the community can manage to pay for.

11.1 INTRODUCTION

11.1.1 WHAT ARE NUTRACEUTICALS?

In Stephen L. who was the founder of the Foundation of Innovation Medicine introduced the hybrid terminology in 1989 named "nutraceutical" as combination of terms, "nutrition" and "pharmaceutical," (Keservani et al., 2010a; Wildman et al., 2006). DeFelice in the year 1998 defined nutraceutical as a food or its component that offers any health benefits, which include the prevention or a targeted treatment of any disease. (Brower V., 1998) Both pharmaceutical and nutraceutical compounds might be utilized to cure or prevent diseases and their distinction is a bit erroneous but nutraceuticals generally lack patent protection in contrast to pharmaceuticals which have governmental sanction as they usually hold patent protection due to expensive testing which is involved (Keservani et al., 2010b; Chauhan et al., 2013; Zeisel, 1999).

Nutraceuticals (Keservani et al., 2017, 2020) offer a proactive health-care approach with enormous therapeutic effects on human body. It may include a variety of products originated from isolated nutrients, nutritional supplements, processed foods, and beverages, genetically tailored foods and products of herbal nature (Das et al., 2012; Bagchi et al., 2015). Mostly nutraceuticals are isolated from natural nutritional sources, and thus are considered to be relatively less toxic and offers few side effects as compared to drugs which are generally used to treat similar symptoms (Ting et al., 2014).

Nutraceuticals are potent means which help towards maintaining and improving human health and acting against nutritionally caused acute and chronic diseases and thereby advocating a higher quality of life. Nutraceuticals can be considered as a food supplement but not every food supplement is a nutraceutical because a dietary supplement should exhibit some function of prevention and/or disease treatment and if a dietary supplement is unable to prevent or treat any disease then it could not be considered as a nutraceutical. (Kalra, 2003; Ross, 2000) Nowadays, in addition to the therapeutic potentiality, nutraceuticals also offer an effective nutritional and safety profile which has brought various pharmaceutical and nutritional companies being aware and abreast of the changing trends in the healthcare industry. (Kalra, 2003).

11.2 RICKETS

Impaired chondrocyte differentiation, mineralization of the growth plate and osteoid mineralization leads to nutritional rickets in adolescents (Munns et al., 2016; Uday & Högler, 2018). Globally, nutritional rickets continues to remain the most widespread type as it is the most frequent triggering root of pediatric bone disease in our world (Ozkan, 2010).

11.2.1 ETIOLOGY OF RICKETS

In comparison with the prevalence of nutritional rickets a century ago, now the prevalence of nutritional rickets has significantly declined, but even in well-resourced countries the condition can be seen to be re-developing. (Goldacre et al., 2014; Thacher et al., 2013) In the early 2000s the incidence of nutritional rickets estimated in United States was 24.1 per 100,000 compared with 2.2 per 100,000 in the early 1980s. However, in the present times the incident rate reported ranges from 2.9–27 per 100,000 individuals in both United States and Europe (Thacher et al., 2013).

Mostly the incidence of nutritional rickets is highly prevalent in low and middle income countries. Hence, this brings out the reason for significant cause of deformities occurring in kids living especially in India, Africa, and Middle East (Creo et al., 2016; Prentice, 2016) The incidence of nutritional rickets is directly associated with the causative factors which can be listed as solar vitamin D deficiency and/or low nutritional calcium dietary intake in children. (Munns et al., 2016; Uday and Högler. 2019) Hence, one can conclude that incidence of rickets is found globally but its etiology varies geographically.

11.2.2 CAUSES OF RICKETS

In low- and middle-income countries of Asia and Africa, low dietary intake of calcium plays a critical role (Jones et al., 2018) whereas in temperate countries it is the vitamin D deficiency which causes rickets. (Creo et al., 2016) In the past, it was considered that the main causative issues pertaining with high incidence of nutritional rickets is vitamin D deficiency and existence of simultaneous calcium deficiency. But, this was revealed that only dietary calcium insufficiency can likewise develop nutritional rickets even in the non-existence of vitamin D insufficiency (Thacher et al., 1999). Through various studies it was found that in the absence of administration of vitamin D supplementation and with just exclusively breastfeeding, infants developed nutritional rickets. Also, infants of the mothers who were deficient in vitamin D during their pregnancy and dark-skinned infants are mostly the ones who are highly affected (Misra et al., 2008; Pettifor, 2004).

11.2.3 CLINICAL SIGNS OF RICKETS

In children, nutritional rickets is triggered by the failure of calcification of the developing bone which further leads to impaired growth, various skeletal deformities, and few significant clinical characteristics which differ depending on the subject of the age of the patient and the severity of the disease. In many of the cases nutritional rickets affects a child within the span of first 18 months of a child's life (Greenbaum, 2011; Pettifor, 2004). The prevalent clinical features of skeletal deformities linked to rickets includes development of asymmetrical, deformable skull, bowing of legs, knock knees, rachitic rosary of the ribcage, pigeon chest, and widening of the wrists and ankles and prolonged motor movement development (Agarwal et al., 2009; Greenbaum, 2011; Ozkan, 2010). Some other clinical signs associated with nutritional rickets includes bone pain or tenderness in spine, pelvis and limbs; loss of muscle strength; deferred formation of teeth; weakened growth; higher incidence of fractures; skeletal deformities and abnormal spine deformity. Some of the skeletal deformities may necessitate corrective surgery and positioning (Greenbaum, 2011).

11.2.4 PREVENTION AND TREATMENT OF RICKETS

Prevention of rickets is possible by either consuming vitamin D supplements or food enrichment and proper exposure to sunlight. In cases where prevention has not been successful, there is recommendation of treatment with calcium and/or vitamin D supplementation (Allgrove & Shaw, 2015; Elder & Bishop, 2014; Munns et al., 2016). In the case of management of rickets which are heritable wherein there is some defect in the vitamin D metabolic pathway or in the pathway pertaining to its activation, administration of vitamin D metabolites were found to be effective (Levine, 2020).

11.2.4.1 FOOD FORTIFICATION

Food fortification is the most effective and feasible solution which can be available to the population groups which are not able to maintain and have a balanced diet. Food fortification can be focussed upon the poor, the sick, the elderly, pregnant women, young children, and even the general population who are not meeting the adequate nutrient intake through their dietary choices (De Lourdes Samaniego-Vaesken et al., 2012) According to WHO, categorization of food fortification strategies has been categorized into three potential approaches which are the following:

1. Through mass fortification approach which included fortifying of wheat, salt and sugar.
2. Targeted approach wherein targeted foods which are mostly consumed by the target population is fortified such as infant feeds and complementary infant food products.
3. Market-driven approach wherein specific brands are fortified which are targeted towards a particular consumer group. (Allen et al., 2006; Das et al., 2013)

Nowadays supplementation with vitamins and minerals and foods enriched with them are available in the market but there is need to ensure that the integrity of the vitamins and minerals are well-maintained and their bioavailability in the gastrointestinal tract is retained. Hence, to improve their stability in the food products they are added into, for their bioavailability maintenance and for upkeeping with the health benefits they offer the strategy of micro- and nanoencapsulation is needed (Barba et al., 2015; Bender, 2002).

Following the variety of patterns in food supplementation, the year 1932 witnessed milk fortified with vitamin D, as a measure to prevent the rickets in children in the community (Backstrand, 2002).

Due to the instability related to many environmental conditions Vitamin D encapsulation was also found to be of great benefit. Enclosed in a shell-core element made up of alginate guluronic and mannuronic acid, fabricated by ionic gelation to ultrasonic atomization and finally drying of particles by microwave, ergocalciferol (vitamin D2) was encapsulated successfully to have a sustained release with its minimal degradation (Barba et al., 2015). Prepared by ionic gelation method vitamin D3 was encapsulated in a polymeric nanoparticles formulated by carboxymethyl chitosan and isolate of soy protein. Both the above-mentioned methodologies have been found to be suitable to encapsulate Vitamin D nutraceuticals to supplement and fortify foods (Teng et al., 2013).

For beverages, the nutraceuticals to be introduced needs to undergo an encapsulation technique of Nanoemulsions which prevents any alteration of its natural appearance, taste, and flavor. Nanoemulsions while increasing the bioavailability also offers small droplet size and thereby providing stability against any precipitation. Vitamin D encapsulated in nanoemulsions, resulted in enhanced bioavailability and increased shelf life i.e., more than a month at room temperature without any particle precipitation (Guttoff, 2015).

11.3 NUTRACEUTICALS FOR RICKETS

Low calcium intake and/or Vitamin D deficiency in children causes' defective chondrocyte differentiation and mineralization of the growth plate and defective mineralization of osteoid which leads to nutritional ricket (NR). Complete diagnosis of NR is based on the following (Paxton et al., 2013):

1. History identification;
2. Physical examination;
3. Biochemical testing; and
4. Radiography.

11.3.1 VITAMIN D STATUS RECOMMENDATION

Assuming 25OHD 2 (vitamin D2) and 25OHD 3 (vitamin D3) are of equal biological value, total of 25OHD is assessed. By measuring blood levels of total 25OHD Vitamin D status is assessed. (Institute of Medicine, 2011) According to Global Rickets Consensus high quality evidence has supported the Vitamin D status classification based on serum 25-hydroxyvitamin D (25OHD) levels depicted in table 11.1 (Munns et al., 2016).

TABLE 11.1 Vitamin D status classification according to global rickets consensus.

Vitamin D Status Classification	Levels
Deficiency	<30 nmol/l
Insufficiency	30–50 nmol/l
Sufficiency	>50 nmol/l

11.3.2 DIETARY CALCIUM INTAKE TO PREVENT RICKETS

Dietary Reference Intakes for Calcium stated by Institute of Medicine has recommended adequate dietary calcium intakes for infants, children as well as adults. (Institute of Medicine, 2011) In Consensus Statement of Global Consensus Recommendations of Nutritional Rickets recommendations have been stated and depicted in table 11.2 for prevention of rickets for different population groups. (Munns et al., 2016).

TABLE 11.2 Dietary Calcium Recommendation for Different Population Groups According to Global Rickets Consensus

Dietary Calcium Recommendation for Different Population Groups	
For infants 0-12 months of age	
For infants 0-6 months of age	200 mg/day
For infants 6-12 months of age	260 mg/day
For children over 12 months of age	
Deficiency	<300 mg/day
Insufficiency	300–500 mg/day
Sufficiency	>500 mg/day

According to Institute of Medicine for children aged 1–18 years, the daily calcium intake has been recommended as 700–1, 300 mg/day depending on the age (Ross et al., 2011).

11.3.2.1 CASE STUDIES SUPPORTING AS EVIDENCE

There have been various studies conducted to support the dietary calcium recommendation for different populations groups which are stated above. The studies include case control studies focusing on role of calcium deficiency in Indian children for nutritional rickets (Aggarwal et al., 2012; Balasubramanian et al., 2003), study conducted in Nigerian children focusing on rickets development due to malnutrition of calcium (Okonofua et al., 1991) and a robust study focusing on children of rural South Africa- on the relationship between calcium intake and calcium nutritional status (Eyberg et al., 1986).

11.3.3 RECOMMENDATION FOR VITAMIN D SUPPLEMENTATION FOR THE PREVENTION OF RICKETS

Institute of Medicine has also stated the recommendation regarding Vitamin D supplementation which is depicted in the table 11.3 (Institute of Medicine, 2011).

TABLE 11.3 Vitamin D Supplementation for Different Population Groups according to Institute of Medicine

Vitamin D Supplementation for Different Population Groups	
For infants from birth to 12 months of age	
For infants 0-12 months of age	400 IU/day (10 μg)
(Independently of their mode of feeding)	
For infants over 12 months of age, all children and adults	
Through diet and/ or supplementation	At least 600 IU/day (15 μg)

11.3.3.1 CASE STUDIES SUPPORTING AS EVIDENCE

Gallo et al. in their randomized trial study published in JAMA in 2013 assessed the effect of different oral Vitamin D supplementation doses in breastfed infants. This study has indicated that during infancy 400 IU/day of vitamin D as a supplement has been found to be adequate for the prevention of radiographic signs of rickets (Gallo et al., 2013). A randomized controlled trial study conducted in China by Specker et al. in 1992 revealed that even in infants with vitamin D deficiency, 400 IU/day of vitamin D supplement was necessary for the infants of 6 months of age to avert any radiographic signs of nutritional rickets. (Specker et al., 1992) Also, in a two-year surveillance Canadian study on infants who received 400 IU/day of vitamin D no incident reporting was recorded radiographically which confirmed rickets (Ward et al., 2007).

In the absence of NR, prevention of vitamin D deficiency was also studied in a double-blind RCT where in infants with no vitamin D deficiency effect of 400, 800, 1,200, and 1,600 IU of vitamin D 3 per day was measured. In 97% of infants after a year of administering 400 IU/day 25OHD levels >50 nmol/l levels were maintained. No additional benefit was found with the doses of 800 and 1,200 IU/day. Whereas there was a concern raised for potential toxicity when 1,600 IU/day was administered (Gallo et al., 2013).

11.3.4 DOSE OF VITAMIN D AND CALCIUM FOR THE TREATMENT

As per the Global Rickets Consensus, for treating nutritional rickets high quality evidence has been provided supporting the treatment of minimum recommended dose of vitamin D which is administering 2,000 IU/day (50 μg) for at least three months. Also, in combination of the above treatment strong evidence also supported the recommendation of routine administration of 500 mg/day of oral calcium to be taken as a supplement or through dietary source which has been established for any population regardless of the age and weight (Munns et al., 2016).

11.3.4.1 CASE STUDIES SUPPORTING AS EVIDENCE

In a study focusing on small group of children aged 2–36 months with nutritional rickets (n = 19), within three weeks administration of 5,000–10,000 IU of oral vitamin D 3 and 0.5–1.0g of daily calcium normalized serum calcium and phosphate levels. (Kruse, 1995) Also, several studies have advocated the combined administration of vitamin D supplement with calcium which was found to be adequate and recommendable for treatment of nutritional rickets in children (Kutluk et al., 2002; Oginni et al., 2003).

Thacher et al. in their study on 123 Nigerian children with nutritional rickets due to calcium deficiency, found radiographic evidence of almost healed rickets to be lower in children who received vitamin D alone (19%) and higher for children who were administered a combination of vitamin D supplement and calcium (58%) (Thacher et al., 1999).

Similarly, Aggarwal V et al. in their study on 67 Indian children with nutritional rickets attained full recovery at 12 weeks in children who received combined calcium and vitamin D therapy (50%) than with treatment done by vitamin D (15.7%) or calcium alone (11.7%). (Aggarwal et al., 2012)

Furthermore, various studies focused on India and Nigeria have rationalized the combined therapy for nutritional rickets because these studies have depicted that children and adolescent suffering with nutritional rickets consumed diet low in both vitamin D and calcium sources (Aggarwal et al., 2012; Balasubramanian et al., 2003; Thacher et al., 1999).

11.3.5 RECOMMENDATIONS FOR ROUTE OF ADMINISTRATION AND DURATION OF THERAPY

As per the Global Rickets Consensus, through various case studies strong evidence has supported the route of administration and duration for the therapy. This has led to the recommendation regarding oral treatment, which could more quickly rebuild 25OHD levels than intramuscular (IM) treatment.

Also, for routine daily treatment, both D2 and D3 depicted likewise effectiveness. In treatment where single large doses are received by the patients, D3 appears to be better compared to D2 because of the longer half-life which D3 has. Regarding duration of the therapy to be administered a minimum of 12 weeks is recommended as depending on the severity of the cases some children may have the need of prolonged treatment duration (Munns et al., 2016).

11.3.5.1 CASE STUDIES SUPPORTING AS EVIDENCE

Some studies mostly conducted on adults had compared intramuscular treatment and oral administration of vitamin D. Cipriani C et al. in their study published in 2013 on bioavailability factor after oral and intramuscular administration of 600,000 IU of D 2 or D 3 received by 24 volunteers aged 50–78 years of age. The study findings revealed that highest level of 25OHD were recorded within 30 days in the adults who were administered oral treatment and 120 days in those who received intramuscular treatment (Cipriani et al., 2013).

Another open labeled Randomized Controlled Trials study involving 92 adults with 25OHD <75 nmol/l evaluated the administration of 300,000 IU vitamin D 3 intramuscular treatment to 50,000 IU D 3 orally given over 3 months 6 times. The study revealed that a higher percentage of subjects who received oral treatment 50,000 IU D 3 attained the level of 25OHD >75 nmol/l at 3 months whereas it took 6 months for the ones who received intramuscular treatment subjects (Zabihiyeganeh et al., 2013).

Tripkovic et al. in their meta-analysis of studies had analyzed and compared the administration of vitamin D 2 and D 3 and their findings stated that, when administered as bolus doses, vitamin D 3 was more effective than vitamin D 2 at raising 25OHD concentrations, but when given as daily doses no significant differences were seen (Tripkovic et al., 2012).

Currently the literature consists of mostly the review articles and there is lack of robust randomized controlled trials which has assessed the duration of treatment for children with nutritional rickets. But the expert review committee of Pediatric Endocrine Society (PES) has recommended that for administration of daily oral therapy the duration should range from 8–12 weeks. (Misra et al., 2008). The recommendation regarding 8 and 12 weeks of daily oral treatment could also be found in review studies conducted by Pearce SH et al. in 2010 and Shaw NJ et al. 2013 in United Kingdom (Pearce et al., 2010; Shaw et al., 2013).

11.3.6 PREVENTION OF NUTRITIONAL RICKETS: RECOMMENDATION FOR DIETARY PRACTICES AND NUTRIENT INTAKES AMONG MOTHERS

As per Institute of Medicine there are high quality evidence available which advocates that a mother should avoid vitamin D deficiency by making sure that all women of reproductive age should meet the intakes of 600 IU/day of Vitamin D. Also, for pregnant mothers in addition to other essential micronutrients like folic acid and iron, it is recommended that they also receive 600 IU/day of vitamin D (Institute of Medicine, 2011).

11.3.6.1 CASE STUDIES SUPPORTING AS EVIDENCE

In the available literature maternal dietary intake has not been considered as a predictor for infant rickets but the data assembled for the above recommendation have been sourced from the various vitamin D intervention studies or case studies focusing on maternal health during pregnancy. Many studies focusing on nutritional rickets included information on the dietary record of mothers and the maternal 25OHD levels (Bhakhri & Debata, 2010; Brinsmead et al., 2011; Elidrissy et al., 1984). Also, in a study conducted by Soliman A et al. neonates and infants not supplemented with Vitamin D were found to have developed early biochemical signs of nutritional rickets (Soliman et al., 2013).

Studies have also supported that majority of infants with signs and symptoms of vitamin D deficiency belonged to mothers who were vitamin D deficient and exclusively breast-fed their child (Al-Atawi et al., 2009; Dawodu et al., 2005; Maiya et al., 2008; Robinson et al., 2006; Soliman et al., 2013).

11.3.7 RECOMMENDATION REGARDING NUTRIENT INTAKE ASSOCIATED WITH NUTRITIONAL RICKETS IN INFANTS

For infants along with the intake of 400 IU/day of vitamin D, from 26th week complementary foods sources rich in calcium should be introduced. During the childhood and adolescence period it is recommended that at least 500 mg/day of elemental calcium is taken (Munns et al., 2016).

11.3.7.1 CASE STUDIES SUPPORTING AS EVIDENCE

Many studies have advocated the finding that extended breast-feeding in infants with delayed introduction of complementary feeding is related to the incidence of cases of nutritional rickets who are not receiving vitamin D supplements (Akpede et al., 1999; Kaper et al., 2000; Majid et al., 2000; Siddiqui & Rai, 2005). In various studies there have been record of observational data suggestive of the recommendation that those infants who were introduced with the vitamin D supplementation in the first year of life successfully prevented the development of nutritional rickets (Lerch et al., 2007; Ozkan et al., 2009; Pedersen et al., 2003; Strand et al., 2003; Tezer et al., 2009).

11.3.8 TREATMENT / MANAGEMENT OF RICKETS DUE TO NUTRITIONAL DEFICIENCY OF VITAMIN D

Treatment therapies are carried out in two phases namely- early intensive and late maintenance phase. Several treatments are applied to treat rickets due to nutritional deficiency of vitamin D (Gordon et al., 2008). All the therapies comprise of some kind of vitamin D administration- vitamin D2 i.e., ergocalciferol or vitamin D3 i.e., cholecalciferol following monitoring for healing process (Mughal, 2011).

The intensive phase of vitamin D treatment is given for two to three months in conjunction with calcium supplementation (500 mg either through diet or by supplements) for children who have insufficient dietary calcium (Mughal, 2011; Munns et al., 2016).

1. **Single-dose therapy:** Here, a specific large dose of vitamin D is used which in which for infants > 1 month it is administered orally or intramuscularly at a dose of 100,000 to 600,000 international units (IU) (Ozkan, 2010). However oral treatment is generally recommended as this route reestablishes vitamin D more rapidly (Munns et al., 2016). The U.S. Endocrine Society recommends a dose of 50,000 IU of vitamin D once weekly for six weeks, followed by a maintenance phase with 400–600 IU of vitamin D daily (Holick et al., 2011).
2. **Multiple Doses of Vitamin D:** Depending on the age of the patient the doses are decided which are administered in this regimen. The recommended doses depicted in the table 11.4 are laid for infants for a span of 2–3 months (Ozkan, 2010).

TABLE 11.4 Recommended Dose of Vitamin D for Infants of Different Age Given for 2-3 Months

Age of infant	Recommended dose of vitamin D given for 2-3 months
infants < 1 month	1000 IU
1-12 months	1000 – 5000 IU
> 12 months	5000 IU

Successively, in the maintenance phase 400 IU of vitamin D is recommended. (Ozkan, 2010) Following the treatment, it was found that there was improvement in bone pain within two weeks after initiation of the therapy. Whereas by six months the metaphyseal swelling improves, and for the recovery of the bowlegs and knock knee it can take up to two years for the toddlers. But, in the case of adolescents there are some residuals which may ultimately require orthopedic surgical correction (Mughal, 2011).

11.4 REGULATORY CONSIDERATIONS FOR NUTRACEUTICALS FOR RICKETS

According to the core principles of universal health coverage the fundamental elements constitutes of the universal access to effective, quality and safe health products. The urgent need to expand the accessibility to health products is also highlighted in the Sustainable Development Goal 3 (Sustainable Development Goals) Hence for any countries' development and progress towards universal health coverage there is a need to work upon the accessibility of health products (World Health Organization Medicines Strategy, 2010; WHO Model List of Essential Medicines, 2019).

Since Colecalciferol-Vitamin D3, and ergocalciferol- Vitamin D2 and calcium is required as a highly significant health-care need of various population groups like-infants, children, adolescents and pregnant women they are considered to be of essential nature (WHO Model List of Essential Medicines, 2019; WHO Model List of Essential Medicines for Children, 2019). Thus, at all times vitamin D3, vitamin D2 and calcium qualifies to be available in adequate amounts with standard expected quality within the setting of operating health systems that too in the correct dose and at a cost that the individual and the community can manage to pay for (World Health Organization. Essential medicines and health products).

Colecalciferol is incorporated in both the listing of WHO Model List of Essential Medicines for Children and the WHO Model List of Essential Medicines and available in the dose of 400 IU/mL oral liquid and 400 IU; 1000 IU – solid oral dosage form. Ergocalciferol is found in the listing of the WHO Model List of Essential Medicines and can be taken as an alternative. It is accessible in the dosage forms (283) of 250 μg/mL (10 000 IU/mL) of oral liquid and 1.25 mg (50 000 IU) of solid oral dosage form.

In the WHO Model List of Essential Medicines Calcium is included in two different dosage forms- Calcium tablets in 500 mg (elemental) dose and in the form of Calcium gluconate injections which are available in 10- mL ampoule (100 mg/ mL). Whereas the WHO Model List of Essential Medicines for Children only includes calcium gluconate injections in 10 mL ampoules (100 mg/ mL).

For maintaining the quality assurance and regulating it at all stages the WHO Expert Committee on Specifications for Pharmaceutical Preparations has made several recommendations which cover the initial step of production, then ensures the quality control aspect, followed by inspection and finally the stage of distribution of colecalciferol, ergocalciferol and calcium (World Health Organization; 2015).

KEYWORDS

- **fortified foods**
- **nutraceuticals**
- **rickets**
- **supplement**
- **vitamin D 2**

REFERENCES

Agarwal, A., Gulati, D., Rath, S., & Walia, M., (2009). Rickets: A cause of delayed walking in toddlers. *Indian J. Pedia., 76*(3), 269–272.

Aggarwal, V., Seth, A., Aneja, S., et al., (2012). Role of calcium deficiency in development of nutritional rickets in Indian children: A case control study. *J. Clin. Endocrinol. Metab., 97*, 3461–3466.

Akpede, G. O., Omotara, B. A., & Ambe, J. P., (1999). Rickets and deprivation: A Nigerian study. *J. R. Soc. Promot. Health, 119*, 216–222.

Al-Atawi, M. S., Al-Alwan, I. A., Al-Mutair, A. N., Tamim, H. M., & Al-Jurayyan, N. A., (2009). Epidemiology of nutritional rickets in children. *Saudi J. Kidney Dis. Transpl., 20*, 260–265.

Allen, L. D., De Benoist, B., Dary, O., & Hurrell, R. E., (2006). *Guidelines on Food Fortification with Micronutrients.* World Health Organization/Food and Agriculture Organization, Geneva.

Allgrove, J., & Shaw, N. J., (2015). A practical approach to vitamin D deficiency and rickets. *Endocr. Dev., 28*, 119–133. doi: 10.1159/000381000.

Backstrand, J. R., (2002). The history and future of food fortification in the United States: A public health perspective. *Nutr. Rev., 60*, 15–26.

Bagchi, D., Preuss, H. G., & Swaroop, A., (2015). *Nutraceuticals and Functional Foods in Human Health and Disease Prevention.* Taylor & Francis, USA.

Balasubramanian, K., Rajeswari, J., Gulab, et al., (2003). Varying role of vitamin D deficiency in the etiology of rickets in young children vs adolescents in northern *India. J. Trop. Pediatr., 49*, 201–206.

Barba, A. A., Dalmoro, A., D'Amore, M., & Lamberti, G., (2015). Liposoluble vitamin encapsulation in shell–core microparticles produced by ultrasonic atomization and microwave stabilization. *LWT—Food Sci. Technol., 64*, 149–156.

Bender, D. A., (2002). Micronutrients—The vitamins and minerals. In: Bender, D. A., (ed.), *Introduction to Nutrition and Metabolism* (3rd edn.). CRC Press, London.

Bhakhri, B. K., & Debata, P. K., (2010). Nutritional rickets presenting with myelofibrosis. *Indian J. Pediatr., 77*, 1437–1439.

Brinsmead, T., Frawley, K., & Conwell, L. S., (2011). Images in pediatric endocrinology: Vitamin D deficiency rickets and other nutritional deficiencies in a 12-month-old infant. *J. Pediatr. Endocrinol. Metab., 24*, 13, 14.

Brower, V., (1998). Nutraceuticals: Poised for a healthy slice of the healthcare market? *Nat. Biotechnol., 16*, 728–731.

Chauhan, B., Kumar, G., Kalam, N., & Ansari, S. H., (2013). Current concepts and prospects of herbal nutraceutical: A review. *J. Adv. Pharm. Technol. Res., 4*, 4–8.

Cipriani, C., Romagnoli, E., Pepe, J., et al., (2013). Longterm bioavailability after a single oral or intramuscular administration of 600,000 IU of ergocalciferol or cholecalciferol: Implications for treatment and prophylaxis. *J. Clin. Endocrinol. Metab., 98*, 2709–2715.

Continuity and Change, (2010). *Implementing the third WHO Medicines Strategy 2008–2013.* Geneva: World Health Organization. (WHO/EMP/2010.2). https://apps.who.int/iris/rest/bitstreams/66520/retrieve (accessed on 25 July 2023).

Creo, A. L., Thacher, T. D., Pettifor, J. M., Strand, M. A., & Fischer, P. R., (2016). Nutritional rickets around the world: An update. *Paediatr. Int. Child. Health., 37*, 84–98.

Das, J. K., Salam, R. A., Kumar, R., & Bhutta, Z. A., (2013). Micronutrient fortification of food and its impact on woman and child health: A systematic review. *Systematic Rev., 2*, 67–90.

Das, L., Bhaumik, E., Raychaudhuri, U., & Chakraborty, R., (2012). Role of nutraceuticals in human health. *J. Food Sci. Technol., 49*, 173–183.

Dawodu, A., Agarwal, M., Sankarankutty, M., Hardy, D., Kochiyil, J., & Badrinath, P., (2005). Higher prevalence of vitamin D deficiency in mothers of rachitic than nonrachitic children. *J. Pediatr, 147*, 109–111.

De Samaniego-Vaesken, L., Alonso-Aperte, M. E., & Varela-Moreiras, G., (2012). Vitamin food fortification today. *Food Nutr. Res., 56*, 5459–5467.

Elder, C. J., & Bishop, N. J., (2014). Rickets. *Lancet, 383*, 1665–1676.

Elidrissy, A. T., Sedrani, S. H., & Lawson, D. E., (1984). Vitamin D deficiency in mothers of rachitic infants. *Calcif. Tissue Int., 36*, 266–268.

Eyberg, C. J., Pettifor, J. M., & Moodley, G., (1986). Dietary calcium intake in rural black South African children. The relationship between calcium intake and calcium nutritional status. *Hum. Nutr. Clin. Nutr., 40*, 69–74.

Gallo, S., Comeau, K., Vanstone, C., et al., (2013). Effect of different dosages of oral vitamin D supplementation on vitamin D status in healthy, breastfed infants: A randomized trial. *JAMA, 309*, 1785–1792.

Goldacre, M., Hall, N., & Yeates, D. G., (2014). Hospitalization for children with rickets in England: A historical perspective. *Lancet, 383*, 597, 598.

Gordon, C. M., Williams, A. L., Feldman, H. A., May, J., Sinclair, L., Vasquez, A., & Cox, J. E., (2008). Treatment of hypovitaminosis D in infants and toddlers. *J, Clic, Endo. Meta.*

Greenbaum, L. A., (2011). Rickets and hypervitaminosis D. In: Kliegman, R. M., Behrman, R. E., Jenson, H. B., & Stanton, B. F., (eds.), *Nelson Textbook of Pediatrics* (19th edn.). Philadelphia (PA): Elsevier Saunders.

Guttoff, M., Saberi, A. H., & McClements, D. J., (2015). Formation of vitamin D nanoemulsion-based delivery systems by spontaneous emulsification: Factors affecting particle size and stability. *Food Chem., 171*, 117–122.
Holick, M. F., Binkley, N. C., Bischoff-Ferrari, H. A., Gordon, C. M., Hanley, D. A., Heaney, R. P., Murad, M. H., & Weaver, C. M., (2011). Evaluation, treatment, and prevention of vitamin D deficiency: An endocrine society clinical practice guideline. *J. Clic. Endo Meta.*
Institute of Medicine, (2011). *Dietary Reference Intakes for Calcium and Vitamin D.* Washington, National Academies Press.
Jones, K. D. J., Hachmeister, C. U., Khasira, M., Cox, L., Schoenmakers, I., Munyi, C., Nassir, H. S., et al., (2018). Vitamin D deficiency causes rickets in an urban informal settlement in Kenya and is associated with malnutrition. *Matern. Child. Nutr., 14*, e12452.
Kalra, E. K., (2003). Nutraceutical-definition and introduction. *AAPS Pharmsci., 5*(3), 27–28.
Kaper, B. P., Romness, M. J., & Urbanek, P. J., (2000). Nutritional rickets: Report of four cases diagnosed at orthopedic evaluation. *Am. J. Orthop., 29*, 214–218.
Keservani, R. K., Kesharwani, R. K., Sharma, A. K., Vyas, N., & Chadoker, A., (2010b). Nutritional supplements: An overview. *International Journal of Current Pharmaceutical Review and Research, 1*(1), 59–75.
Keservani, R. K., Kesharwani, R. K., Vyas, N., Jain, S., Raghuvanshi, R., & Sharma, A. K., (2010a). Nutraceutical and functional food as future food: A review. *Der Pharmacia Lettre, 2*(1), 106–116.
Keservani, R. K., Sharma, A. K., & Kesharwani, R. K., (2017). An overview and therapeutic applications of nutraceutical and functional foods. *Recent Advances in Drug Delivery Technology*, 160–201. ISBN:9781522507543.
Keservani, R. K., Sharma, A. K., & Kesharwani, R. K., (2020). *Nutraceuticals and Dietary Supplements: Applications in Health Improvement and Disease Management.* CRC Press. ISBN: 9781771888738.
Kruse, K., (1995). Pathophysiology of calcium metabolism in children with vitamin D-deficiency rickets. *J. Pediatr., 126*, 736–741.
Kutluk, G., Cetinkaya, F., & Başak, M., (2002). Comparisons of oral calcium, high dose vitamin D and a combination of these in the treatment of nutritional rickets in children. *J. Trop. Pediatr., 48*, 351–353.
Lerch, C., & Meissner, T., (2007). Interventions for the prevention of nutritional rickets in term born children. *Cochrane Database Syst Rev, 4*, CD006164.
Levine, M. A., (2020). Diagnosis and management of vitamin D dependent rickets. *Front. Pediatr., 8*, 315.
Maiya, S., Sullivan, I., Allgrove, J., et al., (2008). Hypocalcemia and vitamin D deficiency: An important, but preventable, cause of life-threatening infant heart failure. *Heart, 94*, 581–584.
Majid, M. A., Badawi, M. H., al-Yaish, S., Sharma, P., el-Salam, R. S., & Molla, A. M., (2000). Risk factors for nutritional rickets among children in Kuwait. *Pediatr. Int., 42*, 280–284.
Misra, M., Pacaud, D., Petryk, A., Collett-Solberg, P. F., & Kappy, M., (2008). Vitamin D deficiency in children and its management: Review of current knowledge and recommendations. *Paediatrics, 122*(2), 398–417.
Mughal, M. Z., (2011). *Rickets.* Current osteoporosis reports. 2011 [PubMed PMID: 21968816].

Munns, C. F., Shaw, N., Kiely, M., Specker, B. L., Thacher, T. D., Ozono, K., et al., (2016). Global consensus recommendations on prevention and management of nutritional rickets. *Horm. Res. Paediatr., 85*, 83–106.

Oginni, L. M., Sharp, C. A., Badru, O. S., Risteli, J., Davie, M. W., & Worsfold, M., (2003). Radiological and biochemical resolution of nutritional rickets with calcium. *Arch. Dis. Child, 88*, 812–817.

Okonofua, F., Gill, D. S., Alabi, Z. O., Thomas, M., Bell, J. L., & Dandona, P., (1991). Rickets in Nigerian children: A consequence of calcium malnutrition. *Metabolism, 40*, 209–213.

Ozkan, B., (2010). Nutritional rickets. *J. Clic. Res. Pedia. Endo., 2*(4), 137–143.

Ozkan, B., Doneray, H., Karacan, M., et al., (2009). Prevalence of vitamin D deficiency rickets in the eastern part of Turkey. *Eur. J. Pediatr., 168*, 95–100.

Paxton, G. A., Teale, G. R., Nowson, C. A., et al., (2013). Vitamin D and health in pregnancy, infants, children and adolescents in Australia and New Zealand: A position statement. *Med. J. Aust., 198*, 142–143.

Pearce, S. H., & Cheetham, T. D., (2010). Diagnosis and management of vitamin D deficiency. *BMJ, 340*, b5664.

Pedersen, P., Michaelsen, K. F., & Molgaard, C., (2003). Children with nutritional rickets referred to hospitals in Copenhagen during a 10-year period. *Acta Paediatr., 92*, 87–90.

Pettifor, J. M., (2004). Nutritional rickets: Deficiency of vitamin D, calcium, or both? *Am. J. Clic. Nutri., 80*(6 Suppl), 1725S–1729S.

Prentice, A., (2008). Vitamin D deficiency: A global perspective. *Nutri. Rev., 66*(Suppl 2), S153-164.

Robinson, P. D., Hogler, W., Craig, M. E., et al., (2006). The re-emerging burden of rickets: A decade of experience from Sydney. *Arch Dis. Child, 91*, 564–568.

Ross, A. C., Manson, J. E., Abrams, S. A., et al., (2011). The 2011 report on dietary reference intakes for calcium and vitamin D from the institute of medicine: What clinicians need to know. *J. Clin. Endocrinol. Metab., 96*, 53–58.

Ross, S., (2000). Functional foods: The food and drug administration perspective. *Am. J. Clin. Nutr., 71*, 1735S–1738S.

Shaw, N. J., & Mughal, M. Z., (2013). Vitamin D and child health. 2. Extra skeletal and other aspects. *Arch. Dis. Child, 98*, 368–372.

Siddiqui, T. S., & Rai, M. I., (2005). Presentation and predisposing factors of nutritional rickets in children of Hazara division. *J. Ayub. Med. Coll. Abbottabad, 17*, 29–32.

Soliman, A., Salama, H., Alomar, S., Shatla, E., Ellithy, K., & Bedair, E., (2013). Clinical, biochemical, and radiological manifestations of vitamin D deficiency in newborns presented with hypocalcemia. *Indian J. Endocrinol. Metab., 17*, 697–703.

Specker, B. L., Ho, M. L., Oestreich, A., et al., (1992). Prospective study of vitamin D supplementation and rickets in China. *J. Pediatr., 120*, 733–739.

Strand, M. A., Peng, G., Zhang, P., & Lee, G., (2003). Preventing rickets in locally appropriate ways: A case report from North China. *Int. Q Community Health Educ., 21*, 297–322.

Sustainable Development Goals Knowledge Platform. Sustainable Development Goal 3. *Ensure Healthy Lives and Promote Well-being for All at All Ages* https://sustainabledevelopment.un.org/sdg3 (accessed on 25 July 2023).

Teng, Z., Luo, Y., & Wang, Q., (2013). Carboxymethyl chitosan-soy protein complex nanoparticles for the encapsulation and controlled release of vitamin D3. *Food Chem., 141*, 524–532.

Tezer, H., Siklar, Z., Dallar, Y., & Doğankoc, S., (2009). Early and severe presentation of vitamin D deficiency and nutritional rickets among hospitalized infants and the effective factors. *Turk. J. Pediatr., 51*, 110–115.

Thacher, T. D., Fischer, P. R., Pettifor, J. M., Lawson, J. O., Isichei, C. O., Reading, J. C., et al., (1999). A comparison of calcium, vitamin D, or both for nutritional rickets in Nigerian children. *N. Engl. J. Med., 341*, 563–568.

Thacher, T. D., Fischer, P. R., Tebben, P. J., Singh, R. J., Cha, S. S., Maxson, J. A., & Yawn, B. P., (2013). Increasing incidence of nutritional rickets: A population-based study in Olmsted County, Minnesota. *Mayo. Clin. Proc., 88*, 176–183. doi: 10.1016/j.mayocp.2012.10.018.

Ting, Y., Jiang, Y., Ho, C. T., & Huang, Q., (2014). Common delivery systems for enhancing *in vivo* bioavailability and biological efficacy of nutraceuticals. *J. Funct. Foods, 7*, 112–128.

Tripkovic, L., Lambert, H., Hart, K., et al., (2012). Comparison of vitamin D2 and vitamin D3 supplementation in raising serum 25-hydroxyvitamin D status: A systematic review and metaanalysis. *Am. J. Clin. Nutr., 95*, 1357–1364.

Uday, S., & Högler, W., (2018). Rickets and osteomalacia. In: Huhtaniem, I., & Martini, L., (eds.), *Reference Module in Biomedical Sciences* (2nd edn.). Amsterdam: Elsevier Inc.

Uday, S., & Högler, W., (2019). Spot the silent sufferers: A call for clinical diagnostic criteria for solar and nutritional osteomalacia. *J. Steroid Biochem. Mol. Biol., 188*, 141–146.

Ward, L. M., Gaboury, I., Ladhani, M., & Zlotkin, S., (2007). Vitamin D-deficiency rickets among children in Canada. *CMAJ, 177*, 161–166.

WHO Model List of Essential Medicines for Children, 7th List. Geneva: World Health Organization; 2019 (WHO/MVP/EMP/IAU/2019.07; https://apps.who.int/iris/bitstream/handle/10665/325772/WHO-MVPEMP-IAU-2019.07-eng.pdf?sequence=1&isAllowed=y (accessed on 25 July 2023).

WHO Model List of Essential Medicines, 21st List. Geneva: World Health Organization; 2019 (WHO/MVP/ EMP/IAU/2019.06; https://apps.who.int/iris/bitstream/handle/10665/325771/WHO-MVP-EMP-IAU2019.06-eng.pdf?sequence=1&isAllowed=y (accessed on 25 July 2023).

Wildman, R. E. C., Wildman, R., & Wallace, T. C., (2006). *Handbook of Nutraceuticals and Functional Foods* (2nd edn.). CRC Press, Boca Raton, FL.

World Health Organization, (2015). WHO Expert Committee on Specifications for Pharmaceutical Preparations. Forty-ninth report. *World Health Organ Tech Rep. Ser.*, (992), 1–210. Back cover. PMID: 26118121.

World Health Organization. *Essential Medicines and Health Products*. Essential medicines https://www.emro.who.int/health-topics/essential-medicines/about-essential-medicines.html (accessed on 25 July 2023).

Zabihiyeganeh, M., Jahed, A., & Nojomi, M., (2013). Treatment of hypovitaminosis D with pharmacologic doses of cholecalciferol, oral vs intramuscular; an open labeled RCT. *Clin. Endocrinol. (Oxf), 78*, 210–216.

Zeisel, S. H., (1999). Regulation of nutraceuticals. *Science, 285*, 1853–1855.

INDEX

β

β-carotene, 77
β-catenin, 7, 247, 248, 262, 263
β-cryptoxanthin, 77
β-estrogen binding, 75
β-lymphocyte, 140

γ

γ-carotene, 77
γ-carboxyglutamate, 94
γ-tocotrienol, 13, 199

A

Abaloparatide, 12
Aberrant growth pattern, 145
Abnormal
  curvature, 83, 84
  spine deformity, 296
Absorption, 3, 9, 54, 73–75, 89, 98, 113, 131, 165, 241, 268
Acidic, 32, 48, 132
Acquaints, 73
Active catechol, 16
Acute respiratory distress syndrome, 130
Adaptogenic herbal supplement aids, 99
Adenosine triphosphate, 34, 36
Adenylyl cyclase activity, 91
Adherence, 250
Adipocytes, 27, 98, 263
Adjuvant, 78, 94, 132, 135, 138
  induced arthritis (AIA), 138
Administration, 92
Adolescence, 84, 86, 100, 161, 295, 305, 306
  idiopathic scoliosis (AIS), 84, 86, 91, 92, 95, 96
  osteosarcomas, 155
  period, 304
Adrenal, 99
Adult
  male, 99
  person, 9
  scoliosis, 100
  skeletal remodeling, 247
  stage, 59, 65
Advantageous
  dietary, 52
  microscopic organisms, 52
Age attributable, 31
Aggravation, 44, 50, 195, 196, 198–200, 202
Aglycone, 75
Agonism, 75
Ailment, 33, 54, 60, 210, 214
Albeit, 35, 42
Alcoholism, 60, 67
Alendronate, 12
Alfalfa, 43, 54, 72, 75
Algae, 129
Alkaline
  environment, 76
  phosphatase, 72, 155
  substances, 77
Alkaloid compound, 242
Allergic rhinitis, 131
Alleviation, 100, 211
Allopurinol hypersensitivity syndrome (AHS), 180
Alluded, 26, 30, 45, 52
Alpha, 14, 130, 215
  tocopherol, 130
Alternative
  nutritional supplements, 4
  therapies, 83
Amalgamation, 36–39, 41, 44, 50
Amino
  acids, 43, 49, 89, 93, 94, 96, 164
  corrosive, 45
Ampicillin, 242
Analgesic, 83, 85, 89, 99–101, 105, 113, 132, 205, 207, 211, 213
  activities, 99
  properties, 100, 101, 113

Ananas comosus, 98
Anatomy, 79, 205, 208
Androgens, 11, 76
Anecdotal evidence, 154
Angina pectoris, 98
Angiogenesis, 136, 137
Ankle, 59
Anorexia, 33, 60
  nervosa, 33
Antagonistic effects, 14
Antartic krill, 139
Antecedents, 50, 51
Anthocyanidins, 46, 72
Anthocyanin, 16, 25, 26, 46, 76, 133, 181
  compounds, 16
  rich food, 46
Anti
  aging, 11, 76
  allergic
    activity, 103
    properties, 103
  arthritic
    effect, 138, 139
    potential, 135
  bacterial activity, 233, 238, 241, 242
  biotic, 225, 235, 238, 241, 242, 268
    resistance microbes, 242
  bodies, 126, 127, 232, 236
  cancer effects, 98
  citrullinated protein antibody, 128
  convulsive, 101
  diabetic activities, 3, 4
  immunoglobin, 104
  inflammatory, 3, 4, 13, 25, 72, 83, 89, 98, 100–103, 105, 113, 125, 130–134, 139, 158, 159, 175, 179–182, 193, 199, 201, 202, 205, 212, 214, 237, 253, 258–260
  osteoporotic, 25, 26, 47, 48, 51, 53, 54
    compounds, 26
    mechanism, 15
  oxidant, 3, 4, 13, 71, 72, 74, 76, 77, 87, 94, 95, 99, 102, 103, 113, 130, 132, 139, 163, 164, 181, 182, 211, 213, 214, 237, 264
    defense, 95
  oxidative effects, 100, 101
  resorptive (AR), 13, 16
  rheumatic
    agents, 128
    drugs, 129
    potential, 135
  thetical, 247
  thrombotic, 98, 130
Antigen, 103, 127, 133, 135
  induced arthritis (AIA), 133, 138, 139
  specific IgE antibodies, 103
Anxiolytic, 101
Apatite, 36, 70, 95
Apigenin, 13, 44, 247
Apoptosis, 28, 130, 133, 135, 231, 233, 245
  cell, 38
Appendicular, 65, 249
Appetite, 126
Arachidonate, 131
Arachidonic acid (AA), 15, 105, 131, 258, 259
  metabolism, 15
Archaeological record, 147
Armoracia rusticana, 242
Arthritic, 4, 73, 126, 128, 129, 133–135, 137, 138, 145, 178, 199, 206, 210, 214
  scores, 139
Articular cartilage, 135
Ascends, 29, 31, 36
Ascorbic corrosive, 39
Ascribed, 44
Asparagus, 44, 52, 97
Assessment, 128, 134, 155, 192, 194
Assimilated, 34, 35, 38, 40, 44, 47, 51, 53
Assist, 5, 33, 42, 43, 45, 49, 51, 70, 77, 158, 199, 212, 214
Association, 13, 160, 161, 256
Assortment, 27, 36–38, 42, 43, 46, 48, 51, 194
Assurance, 4, 191, 306
Asthma, 130
  annotations, 60
Asymmetrical mechanical strain, 100
Atomic component, 38
Attention deficit-hyperactivity disorder (ADHD), 131
Attributable, 34, 48, 49, 52
Augmentations, 35
Autoimmune diseases, 132
Autosomal dominant pattern, 148
Auxiliary hyperparathyroidism, 32, 51
Avenue, 62

Avocado soyabean unsaponifiables (ASU), 131, 132
products, 131, 132
Axial
part, 63
skeleton, 64, 146

B

Bacteria, 15, 61, 74, 126, 137, 223–227, 229, 232, 233, 241, 260, 268
biofilm, 223, 264
Baking soda, 77
Balanced
diet, 3, 4, 26, 54, 61, 159, 297
nutrition, 4
Bending, 4, 85, 129
Beneficial, 3, 14, 15, 61, 62, 92, 93, 98, 105, 129, 133, 137, 158, 160, 164, 198, 199, 201, 206, 224, 234–236, 253
Berberine, 247, 249, 252
Berberis species, 241
Berries, 45, 46, 72
Beta
carotene, 13
glucan, 52
Betulinic corrosive, 44
Biased practices, 78
Bicarbonates, 77
Bifidobacteria (BiFi), 51, 52, 74, 138
Bile acid metabolism, 15
Billion-dollar, 8, 12
Bioactive
compounds, 98, 182, 198
filaments, 52
substances, 45, 164
synthetic, 48
Bioavailable sulfur compounds, 44
Biochemical, 28, 90, 91, 93, 95, 113, 125, 236, 304
pathways, 90, 91, 93, 95
signs, 28, 304
testing, 298
Biological
membranes, 130
properties, 3
response modifiers, 129
Biomechanical theories, 85
Biophosphates, 13
Biosynthesis, 89, 94, 103
Biotic elements, 43
Bisphosphonates, 12, 157
Black cohosh, 75
Blastic stages, 151
Blend, 42, 43
Blood
brain barrier, 99
calcium, 29
cancer, 46
cells, 6, 149
coagulation, 73
magnesium level, 36
vessels, 8, 40, 67, 151
B-lymphocytes, 134
Body
fluids, 260
postures, 4
triggers osteoblasts, 67
Bone, 3–15, 25–40, 42–54, 59–79, 83, 85–87, 89, 91–96, 102, 104, 113, 126, 128, 135, 138, 145, 146, 149, 151–153, 155–165, 167, 197, 199, 200, 223–228, 230–235, 237, 238, 243–252, 261–264, 268, 269, 296, 305
abnormalities, 26, 79, 155, 158
aches, 145
amicable, 47
arrangement, 27, 37, 43
breakdown, 27, 29, 35, 44
calcification, 40, 53
calcium, 69, 70, 75
catabolism, 72
cell, 7, 28, 36, 38, 43, 64, 94, 149, 244, 247
collagen hydroxylation, 72
cracks, 59
creation, 26, 29, 30, 32, 37, 38, 43, 51, 151
cycle, 27
debilitating, 31
decay, 3, 249
defensive, 47, 48
deficiency, 61, 71
deformation, 67, 126
degeneration, 145
demineralization, 27, 29, 30, 36, 70
density, 9, 10, 32, 59, 60, 66, 67, 69, 70, 72–78, 160, 247, 262
depletion, 61, 77

deterioration, 29, 48, 164
development, 26, 33, 38, 43, 47, 50, 51, 60, 73, 247
digestion, 37–39, 47, 49, 50
enhancement, 73
erosion, 126
factures, 10
formation (BF), 7, 8, 10, 12, 13, 26, 28, 40, 60, 66, 72–75, 77, 78, 92, 94–96, 113, 160, 231, 244, 251, 263
fracture, 4, 11, 60, 61, 67, 244
fragility, 62, 66, 71, 77, 95
fragments, 151
framework, 27, 28, 35, 36, 70
protein, 35
growth, 3, 5, 8, 9, 59
healing, 71
health, 1, 3, 10, 13, 25, 26, 31, 46, 50, 53, 59, 60, 62, 69, 70, 74, 76, 78, 83, 92–94, 104, 125, 145, 159, 175, 191, 205, 223, 224, 238, 243, 244, 261, 264, 265, 268, 293
homeostasis, 10, 29, 48, 245, 249
infection, 51, 224, 227, 230, 232–235, 237, 238, 268
loss, 10–12, 14, 15, 59, 61, 72, 76, 83, 85–87, 96, 113, 161–164, 231, 233, 249–252, 262–264
magnesium, 36
marrow, 11, 27, 38, 65, 66, 72, 79, 135, 149, 244, 251
density (BMD), 45, 49, 70, 71, 73–77, 79, 86, 92–94, 96, 97, 102–105
mass, 26, 29, 39, 44, 47, 49, 59, 60, 66, 69–71, 75, 160–162, 247, 251
matrix, 5, 61, 64, 74, 93, 95, 149, 165, 226, 231, 244, 245
metabolism, 13, 62, 91, 95, 102, 113, 162, 163, 248, 264
microarchitecture, 25
mineral, 13, 15, 39, 48, 49, 53, 54, 60, 69, 70, 72, 86, 89, 92, 160, 165, 234
mineralization, 39, 53, 72, 74, 89, 92
misfortune, 27, 31, 33, 38, 39, 44, 45, 47–51
morphogenetic protein (BMP), 7, 13, 245, 246
network, 35
osteolysis, 42
pain, 12, 296, 305
physiology, 60, 78
porosity, 37
pressure, 28
production, 32
protection, 70, 102
protein, 25, 31, 38, 64
reconstruction, 26
redesigning, 27, 30, 32, 38
related messes, 38
remodeling, 32, 67, 68, 94, 146, 244, 245, 249
removal, 151
resorption, 13, 15, 50, 51, 69, 72, 73, 75, 77, 94, 102, 146, 151, 163, 226, 230, 231, 233, 244, 252
rigidity, 59
shaping cells, 27
strength, 47, 53, 62, 76, 96
structural integrity, 60
tears, 67
thickness, 32, 33, 43, 44
tissue, 31, 60, 69, 149, 232, 245
attachments, 60
turnover, 30, 32, 62, 156, 245
wall, 5
wedges, 67
well-being, 29, 35, 36, 38, 39, 43, 44, 46, 48, 49, 51–53
Borage oil, 131
Boron, 9, 38, 42, 70
Bosom, 43
orthosis, 86
Bountiful, 28, 34, 46, 52
mineral, 34
Bovine type II collagen (BIIC), 136
Bowels, 39
Brachydactyly, 246, 249
Bracing, 83, 86
Bradykinin, 99
Brittle, 126, 146, 149
Broad range, 89, 113, 163, 201, 206
Bromelain, 98, 99, 200, 201, 205, 212
Bronchitis, 98
Brussel sprouts, 39
Butein, 12, 252
Buttercup, 75

C

Calcification, 7, 30, 40, 192, 194, 296
Calcinuria excretion, 76
Calcitonin, 11, 12, 27, 30, 35, 66, 96, 158
  invigorates, 30
  salmon, 11
  therapy, 11
Calcium, 3–5, 7–10, 12, 26, 27, 29, 30, 32–38, 40–44, 48, 49, 51, 53, 54, 59, 60, 62, 64–67, 69–71, 73, 74, 76–78, 84, 89, 92, 94, 96, 97, 102, 113, 114, 145, 150, 158, 160, 161, 165, 192, 196, 200, 231, 248, 293–302, 304–307
  absorption, 54, 73, 74, 76, 92, 165, 200
  adversary, 36, 37
  assimilation, 29, 35, 44, 49, 53
  blend utilized, 44
  homeostasis, 29, 30, 49, 51, 73
    jigsaw, 29
  reuptake, 74
  supplements, 3, 9, 78, 92, 96
Callus, 67
Camellia sinensis plant, 45
Cancellous, 6, 64, 65
Cancer, 4, 12, 16, 37, 39, 43–46, 102, 131, 163, 164, 175, 179, 180, 198, 206, 211, 228, 236, 244, 263
Canine distemper virus, 149
Cannabidiol, 84, 101, 113
Capsule, 75, 128, 235
Carbohydrate, 4, 73
Carbon atoms, 98
Carbonate, 64, 96, 102, 158
Carboxylated osteocalcin levels, 94
Carcinogenicity, 25
Cardamonin, 12
Cardiac
  stroke, 154
  surgery, 85
Cardioprotective properties, 102
Cardiovascular, 4, 9, 37, 40, 43, 51, 94, 98, 99, 102, 104, 154, 163, 180
  disease, 9, 102
    management, 4
  disorders, 98
  function, 99
  illness, 51
  infection, 37, 43
  well-being, 40
Carotenoids, 13, 76, 77, 139
Carpal, 206–215
  tunnel syndrome (CTS), 205–215
Cartilage, 8, 73, 128, 130, 134–136, 226, 247
  degeneration, 130
Casein protein, 49
Catabolism, 135, 175, 177
Catalase, 47, 136
Catechins, 16, 45, 181
Cathepsin G (CTSG), 13
Cation fixations, 36
Cavities, 5
Cefepime, 242
Cell, 3, 5, 8, 27–29, 34, 36–38, 41, 43, 44, 47, 60, 64, 126, 127, 132, 134, 135, 137, 149, 191, 193, 198, 200, 226, 229–233, 237, 243–246, 251, 252, 256, 259–262
  debris, 67
  digestion, 30
  elicit, 127
  endurance, 38
  flagging, 28
  genealogies, 27
  growth, 131
  inflammatory response, 100
  layers, 34, 36
  level, 37, 96, 149
  membrane, 74, 105, 113, 179, 230, 238
  metabolism, 130
  multiplication, 50
  pannus, 127
  record, 39
  signaling pathways, 12
  stimulation, 133
  to-cell associations, 50
  walls, 260
Centralization, 35, 40, 42–44, 48, 53
Cephalgia, 153
Cerebral circulatory frameworks, 47
Chain breaking antioxidant, 130
Chelation, 53
Chemical, 27, 34, 36–38, 46, 47, 131, 132, 164, 237
  composition, 64
  inadequacy, 48
  messengers, 90
  substitution treatment, 47
Chicory, 52

Chiropractic care, 86
Cholecalciferol, 3, 9, 40, 41, 305
Cholesterol assimilation, 42
Choline, 43
Chondrocyte, 8, 27, 30, 127, 130, 246, 249, 293, 295, 298
  multiplication, 30
Chronic
  diseases, 4, 130, 293, 295
  disorder, 146
  fatigue syndrome (CFS), 131
  illness, 78
  pain, 101, 113, 126
Ciprofloxacin, 242
Citrus aurantium, 99
Citrus fruit, 72, 134, 136, 211, 213
C-Jun N-terminal kinase (JNK), 261, 246
Clinical
  case, 11, 94, 101
  signs, 296
Cluster, 148
Coadjunct, 130
Cobb angle, 85, 94
Coccyx, 63
Cocktails, 33
Coconut oil, 89
Cocos nucifera, 98
Cod liver oil, 105, 113
Cola beverages, 35
Colitis, 130
Collagen, 5, 13, 27, 28, 37–39, 44, 60, 64, 66, 70–72, 93, 96, 104, 105, 132, 137, 149–151, 177, 199, 226, 228, 263
  amalgamation, 37, 44
  induced arthritis (CIA), 136, 137
  molecule, 93
  type I α1 (COL1A1), 13
Collagenase, 149
  activity, 132
Collagenolytic effect, 132
Collard greens, 13
Colleagues, 11, 15, 196
Colon, 53
  corrosiveness, 53
Combat osteoporosis, 10
Combinational therapies, 79
Comminuted fracture, 67
Common lymphoid progenitors (CLPs), 245
Compact, 6, 64–66, 151
  bone, 6, 64–66
Compartment syndrome, 67
Complete
  blood count (CBC), 128
  Freund's adjuvant arthritis (CFAA), 136
Complex
  disorder, 86, 87
  etiology, 10
Complications, 67, 88
Composition, 15, 25, 59, 62, 65, 79, 98, 258
Compound, 3, 4, 12–16, 37, 52, 65, 72, 75, 89, 92, 100, 103, 105, 130, 131, 139, 163, 166, 182, 199, 201, 202, 206, 214, 224, 235, 241, 243, 260, 268, 294
  annual growth rate (CAGR), 159, 167
  fracture, 67
Concentration, 9, 10, 64, 69, 71, 73, 77, 94, 102, 154, 177, 205, 260
Concurrent therapy, 84
Condensation process, 7
Confirmatory diagnosis, 85
Congenital, 84, 85
  predispositions, 84
Connective tissue, 5, 37, 39
  membrane, 7
Consequent deficiency, 92
Conservative treatment, 86
Constituents, 71, 99
Consumers, 16, 206
Contamination, 41, 191, 192
Continual cycle, 26
Convenience, 15, 48
Convex side, 84
Copper (Cu), 9–11, 26, 37, 38, 42, 54, 70, 71, 73, 89, 96, 97, 139
  deficiency, 37, 71
  inadequacy, 37
  particle, 37
  supplementation, 97
Coronarin D, 12, 13
Coronary
  illness, 46
  supply route infection, 33
Correlation, 10, 137
Corrosive, 34, 36, 37, 39, 50, 198
Cortical, 35, 36, 64–66, 157, 247
  bone, 35, 36, 64–66, 157, 247
Corticosteroids, 33, 175, 179
Coumestans, 43, 75

Coumestrol, 43, 75
Counteraction, 38, 50
Cramps, 12
Craniosynostosis, 247
C-reactive protein (CRP), 128, 139, 197, 202, 226, 234
Creature
food, 40
meats, 39
Crepitation, 100
Critical influence, 37
Crosslink lysine, 71
Cruciferous vegetables, 39
Crystal structures, 7
Curcuma longa, 100, 101, 241, 253
Curcumin, 7, 12, 13, 25, 26, 84, 89, 100, 101, 113, 162, 182, 201, 205, 212, 214, 224, 247, 249–251, 262, 263
supplements, 101
Curve, 84–87, 91, 92, 195
advancement, 85
bone, 64
progression, 84, 86, 87, 91, 92
Cushing syndrome, 67
Cutaneous, 41, 207
Cutting recurrence, 43
Cyanidin, 46
Cyanocobalamin tend, 77
Cyclase activity, 91
Cycloxegenases, 131
Cylindrical reabsorption, 34
Cytokine, 14, 50, 70, 72, 133, 134, 138, 149, 162, 164, 215, 231–233, 236, 245, 251, 258, 260, 261, 263
production, 137
secretions, 134
Cytotoxic, 37, 230, 237

D

Dairy
cattle feed, 42
items, 39–41, 49, 51
products, 26, 97
Debasement, 38
Debilitate, 32, 33
bone wellbeing, 50
Decarboxylation, 72
Decaying, 5, 10
bone, 4
Declining proinflammation, 15
Deemed negligible, 85
Defense immune system, 128
Deficiency, 8, 29, 32, 34, 37, 38, 49, 61, 70, 73, 74, 91, 92, 94, 95, 155, 161, 162, 178, 200, 251, 293, 295, 296, 298, 300, 301, 303, 304
Degeneration, 3, 130, 209
disease, 154
lumbar scoliosis (DLS), 105
Degradation, 26, 70, 72, 100, 132, 198, 251, 298
Dehydrocholesterol, 9, 40, 41
Dehydroepiandrosterone (DHEA), 11, 16, 76
sulfate (DHEAS), 11, 16
Delay
puberty, 85
repair, 67
Delphinidin, 46
Dementia, 153
Demineralization, 27, 29, 30, 36, 70, 149
Dendritic cells (DCs), 127, 135, 137, 237, 243, 259–261
maturation, 135
Density, 13, 15, 25, 54, 59, 69–74, 86, 102, 157, 160, 165
Depicted, 61, 298–300, 302, 305
Depletion, 61, 67, 70, 72, 73, 75, 77, 79, 92
Destructive microorganisms, 52
Deterioration, 128, 132, 199
Devastating tragic, 3, 4
Development hormone, 30
Dexamethasone (DMS), 135
Diabetes, 33, 42, 51, 67, 73, 197, 198, 201, 205, 207, 210, 226, 228, 234
mellitus (DM), 67, 226, 228, 236
nerve pain, 131
Diagnosis, 227
Diameter, 63
Diaphysis, 5, 63, 155
Diarrhea, 12, 75, 241, 268
Diclofenac, 99
Dietary, 4, 8–10, 12, 14, 15, 25, 33–35, 37, 41, 42, 48, 49, 59, 62, 73, 75, 78, 87, 130, 139, 159, 160, 164–166, 176, 177, 192, 198, 202, 206, 235, 253, 261, 263, 293, 295–297, 299, 301, 303, 305
calcium, 9, 35, 49, 296, 299, 305

constituents, 59, 261
copper inadequacy, 37
guideline, 15
phenolic compounds, 130
Supplement, 63, 105
Health Education Act (DSHEA), 166
Dietician, 5
Diffuse scleroderma, 132
Digestion, 29, 32, 36–38, 41, 83, 87, 95, 104, 206, 241, 268
aid, 98
organs, 29
probiotic, 52
system, 44, 53, 98, 131
Dihomogamma lineic acid, 131
Dihomo-γ-linoleic acid (DGLA), 258
Diminish, 33, 37, 43, 51, 134, 138, 251
bone, 44
Dimorphandra mollis Benth, 13
Diosgenin, 12, 25
Diphysis, 8
osteoblast, 8
Disaccharides, 52
Disease
modifying antirheumatic drugs (DMARDs), 129
prevention, 4
Disintegration, 31, 42
Disorders, 3, 59, 62, 68, 100, 101, 105, 132, 159, 167, 175, 179, 198, 210, 214, 236, 259, 261
Display estrogenic, 43
Disruption, 26, 90, 237, 246
Dissolution, 5, 135
Distal
ends, 63
epiphysis, 5
lower homeobox 5 (DLX5), 246
radius, 59
fracture, 67
Diverse chronic, 132
Docosahexaenoic acid (DHA), 26, 50, 71, 74, 76, 105, 130, 139, 182, 198, 205, 212, 258, 259
Dopamine, 93, 99
Dormant photoproducts, 41
Double-blind clinical trial, 96, 102
Drastic improvement, 224, 243, 268
Drawn-out, 33, 40
Dried plum, 48
Drug, 12, 59, 60, 67, 69, 84, 129, 157, 166, 176, 200, 205, 223, 225, 238, 241, 268, 294
firms, 52
molecules, 79
Dual therapy, 129
Dubosiella, 15
Dull green verdant food, 39
Dynamic, 5, 125, 129
Dyslipidemic individuals, 131
Dysregulated, 10
bone, 146

E

Early-onset scoliosis, 86
Ease, 10, 129, 157
Eczema, 131
Edema, 99, 208
Efficacy, 4, 78, 79, 87, 104, 125, 135, 139, 157, 200, 237
Ehlers-Danlos syndrome, 83
Eicosanoids, 50, 131, 258, 259
Eicosapentaenoic acid (EPA), 26, 50, 71, 74, 105, 130, 139, 182, 198, 205, 212, 258, 259
Eight missense mutations, 247
Elastin, 37, 71, 96
Electron, 37, 46, 47, 92, 96, 179
insufficient, 46
particles, 47
Elemental calcium, 304
Elicit, 103
autoimmune inflammation, 127
Embelin, 12, 25, 251
Embellishment, 42
Embryonic
skeletal, 247
stage, 60
Emerging potential, 87, 113
Emodin, 13, 247, 249
Empowering, 36
Endochondral ossification, 6, 8
Endogenous
acid, 73
substances, 71
Endometriosis, 16

Endosteum, 5, 65
Endurance, 36
Enhanced quantity, 15
Enviro, 146
Enzyme, 37, 38, 92, 95, 113, 125, 131, 136, 175, 176, 178–180, 212, 233, 242
Eosinophil, 103
Eph signaling, 249
Ephausia superba, 139
Ephrin, 248, 249
Epicatechin (EC), 45, 166
  gallate (ECG), 45
Epidemiological, 205
  research, 154
Epidermis, 9, 40
Epidiosbulbin, 242
Epigallocatechin gallate (EGCG), 16, 45, 181
Epigenetic modifications, 127
Epinephrine, 99
Epiphyseal
  direction, 8
  line, 5
  plate, 8, 30
Epiphysis, 5, 8, 63, 155
Equilibrium, 33, 34, 37, 162
Equisetum arvense, 44
Equivalent, 27, 165
  nutrient, 62
Ergocalciferol, 41, 294, 298, 305–307
Eroded bone, 63
Erythrocyte sedimentation rate (ESR), 128, 139, 197, 202, 226, 234
Estrogen, 10–12, 14, 16, 26, 29–31, 42, 43, 47, 48, 51, 66, 70, 73–76, 84, 92, 102, 113, 177, 182, 263
  action, 75
  activities, 43, 75
  derivatives, 12
  hardship, 47
  inadequacy, 47, 51
  mimic agents, 12
  receptor (ERs), 10, 11, 14, 16, 43, 47, 75, 102, 113, 263
    beta, 14
Estrogenicity, 16
Etiology, 85, 86, 90, 91, 127, 148, 149, 153, 195, 208, 295
European Food Safety Authority (EFSA), 166
Evening primrose, 131
  oil, 131
Evidence grounded, 4
Exceptional natural mixtures, 43
Exercise, 33
Expansion, 35, 38, 39, 44, 200, 210
Experimental data, 91
Exploration, 43, 48, 49
Exposure, 4, 9, 161, 227, 293, 296
External
  force, 67
  shock, 60
Extracellular
  centralizations, 36
  fluids, 260
  liquid, 34
  matrix, 7, 8, 13, 135, 136, 228
  space, 36, 40, 245
Extractable, 45, 132

F

Fabaceae, 42
Fabricated nano micelles, 135
Fat
  deposition, 98
  dissolvable nutrient, 40
  embolism, 67
Fatality, 223, 237, 238, 268
Fatigue, 126, 128
Fatty acids (FA), 50, 54, 74, 113, 131, 137, 182, 205, 213, 224, 253, 256, 258, 259
Feasible solution, 297
Fed skim-milk, 96
Feminine cycles, 31, 33
Femoral neck, 47, 70, 103
  fracture, 67
Fetal, 27
Feverish, 128
Fiber, 48, 53, 54, 215
Fibrinogen (FN), 99, 138, 229
Fibrinolytic, 98
Fibroblast, 7, 127, 133, 134, 136, 247
  growth factors (FGFs), 7, 247
    tyrosine kinase receptors (FGFRs), 247
  like synovial cells (FLSC), 133
Fibromyalgia, 101
Fibrosis, 86, 208, 209
Fibrous, 4, 5, 7
  collagen, 4

Fibulae, 63
Filaments, 52
First line
drug, 129
management, 128
Fish
bone peptides (FBP), 165
head lipid (FHL), 15
Fitness club, 129
Flagging
particles, 50
pathways, 49
Flares, 5, 126, 180, 182
Flat bones, 7, 63, 64
Flavan-3-ols, 76
Flavonoids, 15, 44, 46, 72, 76, 133, 140, 163, 181, 199, 241, 242, 263, 264
Flavonols, 16
Flourish, 52
Fluctuates, 33, 47
Folate, 74, 77
Folic
acid, 72, 77, 303
corrosive, 43
Food
Drug Administration (FDA), 69, 159, 166
fortification, 62, 293, 297
Safety Standards
Act (FSSA), 165
Authority of India (FSSAI), 166
Supplements Europe (FSE), 166
Forerunner cholesterol, 41
Forestalling, 29, 34, 37, 38, 41, 42, 45, 47, 48
Formaldehyde, 136
Forskolin, 91
Fortified milk, 10
Foster osteoporosis, 33
Founder mutations, 146
Fox Osin mouse model, 262
Fracture, 3, 4, 10, 25, 26, 59–61, 66, 67, 69–78, 145, 146, 154–156, 158, 161, 226, 228, 246, 296
bone, 31, 67
risk cutoff, 71
Fragile, 145, 192
Free
radicals, 70, 71, 100, 130, 237
revolutionaries, 44, 45
Frequency, 40, 41, 79, 137, 147, 148, 162, 167, 209
Freund's adjuvent arthritis (FAA), 135
Fricasseeing, 45
Frontal part, 5
Fructo-oligosaccharides, 52
Fucogalactoxyloglucan, 260
Functional foods, 4, 54, 63, 165, 166, 237
Fungi, 61, 223–225, 227, 260

G

Galacto-oligosaccharide, 52, 53, 74
Gamble, 32, 33, 42, 45, 46, 51
Gambogic acid, 12, 13, 25, 26, 251
Gamma
carboxylation, 94
carotene, 13
linolenic acid (GLA), 131
Gastrointestinal
framework, 35
issues, 85
maturation, 51
microbiota, 52
microflora advancement, 52
tract, 293, 297
Gelatin, 89, 104
Gene, 13, 102, 127, 210, 242, 244, 246, 248, 258
expression, 72, 260
library, 13
Generation, 3, 5, 71, 180, 228, 233, 237, 260
Genetic, 10, 85, 146, 148, 176–178, 210
Genistein, 7, 12, 14, 25, 26, 43, 72, 75, 84, 89, 102, 164, 247, 249, 263
Genomic studies, 148
Gentamicin, 242
Global, 3, 159
impression, 100
occurrence rate, 84
regulatory criteria, 145
Glucocorticoids, 67
Glucosamine, 100, 139, 198
Glutamic corrosive, 43
Glutathione (GSH), 44, 47, 136, 237
Glycine, 14, 94
Glycitein, 14, 75
Glycolipids, 15
Glycoproteins, 27, 71, 132, 253

Glycoside, 75, 133, 242
Glycosyl
phosphatidylinositol (GPI), 248
transferase, 95
Glycyrrhizaglabra L, 13
Goiter, 61
Good
health, 62, 87
manufacturing practices (GMP), 166, 181
Gorlin syndrome, 249
Gossypin, 13
Gradual worsening, 86
Granuloma
composition, 132
extracts, 132
Gravity, 33
Greenery, 44
Greenstick fracture, 67
Growth
plate, 5, 293, 295, 298
rods, 86
Guggulsterone, 13
Guidance, 5, 268
Guideline, 8, 10, 39
Gulping green tea, 45
Gut, 15, 54, 137, 138, 237, 253, 259
microbiota (Gm), 137, 259
Gyrase, 242

H

Habitat, 61
Half-life enhancement, 70
Hampered, 94
protein level, 73
Hard boney structure, 7
Harmed bone, 27
Harpagoside, 13
Haversian canals, 151
Hazard, 50, 197
side effects, 12
Health
body tissue, 126
bone development, 4
care, 10, 59, 206, 293–295
crisis, 61
life style, 3
risks, 16, 78
specialists, 79
Heart
beats, 52
disease, 85, 131, 163, 206
Hedgehog (Hh), 249
signaling pathways, 249
Height loss, 26
Hematopoietic
cells, 250
genealogy, 27
Hematoxyphilic, 151
Heparan sulfate
proteoglycan 2 (HSPG2), 13
Hepatitis B, 128
Hepatitis C, 128
Hepatocyte growth factor (HGF), 244
Herbal, 26, 70, 213
dietary, 75
extract, 62
Hereditary, 34, 43, 85, 146, 148, 178, 210
neuropathy with pressure palsy (HNPP), 210
Hesperidin (HPD), 133, 134
Heterozygous, 247
High
bone mass, 247
dose, 15
fixations, 39
mobility group box 1 (HMGB1), 162, 262
phosphorus consumption, 35
plasma concentration, 16
protein diet, 32, 54
silica fixation, 44
vascularized, 146
Hindered development, 34
Hip, 11, 42, 61, 71, 73, 76, 77, 132, 156, 157, 192, 194, 197
fracture, 61, 71, 73, 77
Histamine, 98, 103, 104
concentration, 98
Holistic, 87
Hollow region, 63
Homeostasis, 29, 32, 35, 65, 90, 191, 236, 245, 249, 261
Homocysteine, 45
development, 45
level, 77, 94
serum, 77
Homocystinuria, 94

Homozygous
frameshift, 247
mutation, 249
Honeycomb, 63
Hormonal, 9–11, 16, 74, 90, 113, 161, 179
cancers, 75
imbalances, 83
replacement therapy, 11, 16
signs, 27
Horse
feed, 42, 43
tail, 44, 72
Host, 74, 191, 206, 224, 230, 232, 237, 245, 253, 268
Hot flashes, 12, 132
Hub, 62
Human
being, 4
body, 3, 5, 26, 160, 192, 201, 236, 237, 253, 293, 294
bone, 63
lung pathogens, 242
species, 63
Humerus, 154
Humoral B cell, 127
Humpback, 26
Hyaline cartilage, 5
Hyaluronoglucosaminidase 1 (HYAL1), 13
Hydration shell, 36
Hydrochloric acid, 149
Hydrolysis, 53, 102, 164
Hydroxide, 7
Hydroxyapatite, 35, 36, 64, 69, 70, 160
needle, 7
Hydroxyl group, 93, 133
Hydroxylysine, 37, 96
Hydroxyproline, 154
Hyoid bone, 63
Hypercalcemia, 12, 73, 161
Hypercalciuria, 73
Hyperglycemia, 46
Hyperlipidemia, 46
Hyperparathyroidism, 33
Hyperprolactinemia, 67
Hypersecretion, 73
Hypersensitivity, 60
Hypertension, 33, 177, 207
Hyperthyroidism, 30, 33
Hypervascularization, 151
Hypervitaminosis D, 41
Hypnotic, 101
Hypomagnesemia, 95
Hypothalamic norepinephrine, 99
Hypothyroidism, 30, 205, 210
Hypotriglyceridemic, 130

I

Iatrogenic scoliosis, 84
Ibandronate, 12
Ideal, 4, 40, 42, 61, 260
structured adult, 4
Idiopathic, 83, 84, 87, 105, 161, 192, 194
Illness, 147, 148, 154, 156, 158, 195, 200, 206, 208, 214, 223, 235, 238, 268
Immature thoracic cage, 85
Immersed, 50
Immobility, 3, 4
Immune, 76, 89, 103, 127, 137, 200, 236, 237, 243, 245, 253, 256, 258–261, 269
cell, 67, 101, 229, 230, 232, 236, 243, 244, 256, 258, 259
response, 95, 135, 137, 227, 232, 236, 237, 253, 258, 259, 261
system, 126, 128, 137, 193, 199, 237, 238, 244, 253, 256, 259, 261
Immunological
progenitors, 146
reaction, 37
Immunomodulatory, 125, 132, 133, 236, 237, 253, 258–261
Immunosorbent assay, 136
Impacted fracture, 67
Impairment, 32, 94, 154, 178
Imparts, 29
Impedes, 153, 163
Impetuses, 38
Import-export certificate (IEC), 166
In vitro, 37, 47, 49, 83, 98, 102, 162, 164, 230, 250, 253, 262, 263
In vivo, 47, 83, 102, 162, 164, 230, 238, 251, 253, 258, 262, 263
Inadequate, 33, 37, 148, 160, 241
Incidence, 84, 147, 148, 154, 162, 224, 227, 228, 235, 295, 296, 304
detection, 146
Incorporate, 34, 35, 39, 40, 44–49, 191, 192, 197, 235, 259
Indirect mechanisms, 260

Individual hormonal level, 10
Indomethacin, 135
Inducible nitric oxide synthase (iNOS), 113, 233, 260
Industrial scale, 62
Inexpensive, 13, 86
Infant stage, 59
Inflammation, 15, 100, 105, 113, 126, 130–132, 134–137, 146, 175, 176, 178, 180, 181, 191, 195, 197–199, 201, 205, 207, 208, 211–213, 215, 224, 231, 232, 236, 237, 253, 258–261
  arthritis, 128, 129, 175, 176
  disease, 126, 130
  factors, 133, 134
  infiltration, 136, 137
  mediators, 98, 100, 104, 105, 113, 134
  phase, 67
  scoliosis, 84
  variables, 146
Influence creation, 41
Infuses procollagen, 74
Ingestion, 29, 35, 75
Inhalation, 11
Inhibit, 3, 16, 76, 100, 133, 135, 137, 164, 200, 233, 242, 250, 256, 259, 262–264
Initial
  osseous lesion, 149
  surgery, 86
Initiate osteocalcin, 43
Injectable routs, 11
Inner
  layer, 6
  surface, 5
Innovation, 61
Inorganic, 4, 27, 35, 64
  materials, 4
  mixtures, 27
  phosphate, 35
  phosphorus consumption, 35
Instability, 67, 297
Instigate, 30, 40, 113
Insulin, 30
Insusceptible framework, 39, 51
Integrated hip, 70
Interaction, 27–30, 90, 96, 225, 231
  workshops, 129
Interchangeable magnesium, 35, 36
Interferon (IFN), 134, 246, 261
  nonsupervisory factor 8 (ICSBP), 246
Interleukin (IL), 29, 71, 133, 136, 139, 149, 167, 244, 260, 263
Interleukin-1β (IL-1β), 100, 104, 133, 135, 136, 182
Interleukin-6 (IL-6), 29, 72, 105, 133–136, 149, 167, 231, 244, 261
International units (IU), 9, 10, 26, 42, 92, 94, 96, 102, 300–306
Intestine, 73, 74, 76, 92, 135
Intracellular
  cancer, 44
  cell flagging, 34
  flagging, 50
Intracortical area, 65
Intracranial pressure, 153
Intramembranous ossification, 6, 7
Intraperitoneal injections, 91
Intricately, 53
Intrinsic capacity, 12
Inulin, 52, 53
  type fructans, 52
Invigorate, 29, 30, 38, 43, 47, 50
  osteoblastic, 38
  substance, 29
Ions, 7, 66, 165
Iron, 38
  assimilation, 38
  over-burden people, 38
Irregular bones, 63, 64
Isobutyrate, 53
Isodeoxyelephantopin, 13
Isoflavone, 14–16, 42, 43, 47, 48, 75, 76, 102, 164, 263
  compounds, 16
  nutraceuticals, 16
Isolated, 14–16, 75, 76, 133, 226, 235, 241, 242, 253, 294
Isoleucine, 43
Isomerization, 40
Isothiocyanates, 242, 243
Isovalerate, 53
IκB kinase (IKK), 251, 263

J

Jerusalem artichokes, 52
Jigsaw, 151
Joint
  hypermobility, 85

pain, 128, 129, 134, 192, 194, 197, 201
redness, 128
stiffness, 67
swelling, 128
tenderness, 128
warmth, 128
Juvenile, 84, 178

K

Kaempferol, 25, 26, 43, 44, 72, 224, 242
Kale, 13, 97
Kidney, 32
Klinefelter syndrome, 67
Kneading, 29
Knee, 63, 126, 192, 193, 296
osteoarthritis, 132
Krill oil, 139
Kynurenic acid, 103

L

Laboratory evaluation, 128
Lactitol, 52
Lactobacillus, 51, 52, 74, 138
Lactoferrin, 71
Lactose subordinates, 53
Lactulose, 52
Ladies foster osteoporosis, 31
L-ascorbic acid, 42, 43, 199
Late blossom, 43
LDL cholesterol, 131
Lean creature, 40
Least therapeutic effect, 130
Leg cramps, 12
Legumes, 15, 75, 175
Leptin, 92, 113, 179
Lesions, 146, 147, 149, 155, 208, 234
Lessen morphine, 95
Leucine, 43
peptidase, 132
Leukotriene, 131, 182, 258
generation, 103
Ligament, 37, 44, 65, 67, 192, 195, 196, 198, 199, 207, 209
Lighter less dense, 6
Lignans, 43
Limbs, 8, 64, 65, 296
Limonoids, 72
Line therapies, 87, 105
Linoleic acid, 74, 76, 182
Linolenic acid (LA), 130, 131, 212, 256
Lipid, 15, 36, 50, 64, 71, 73, 98, 102, 130–132, 181, 182, 198, 202, 233, 256
extracted, 15
soluble, 130
Lipopolysaccharides (LP), 133
Lipoprotein, 71, 102
Lipoxygenase, 103, 131
Liquid stream, 28
Liquor utilization, 33
Little marrow, 64
Liver infection, 51
Living
microscopic organisms, 51
tissues, 5
Long
bone, 5, 8, 30, 63–65, 67, 126, 227
chain, 74
fatty acid (LCFA), 98
fructo-oligosaccharides, 52
glucose polymer, 52
run, 28, 29, 40, 158, 162
term adverse effects, 26
Low
benchmark admissions, 49
bone
mineral, 25
protein, 25
boundaries, 63
cylindrical maximal phosphorus reabsorption, 34
density, 71
fat eating regimen, 40
gastrointestinal framework, 39
lycopene, 77
plasma phosphate, 34
prevalence, 154
L-tryptophan (TRP), 103
Lucerne, 42
Lumbar spine, 47, 49, 70, 76, 92, 146, 156
Lumisterol, 41
Luteolin, 13, 44, 72
Lycopene, 13, 77
Lysine, 37, 43, 93, 96
Lysosomal enzymes, 132, 149
Lysyl oxidase, 37, 71, 96
Lytic, 151, 155

M

Macromolecules, 132
Macronutrients, 86
Macrophage, 127, 133–135, 137, 230, 232, 236, 237, 244, 253, 256, 260
colony stimulating factors (M-CSF), 162, 244, 250, 251
genealogy, 27
inflammatory proteins (MIP), 261
phagocytic activity, 260
Magnesium (Mg), 9, 11, 26, 35–37, 42, 49, 62, 64, 65, 70, 72–74, 84, 89, 95, 97, 113, 264
homeostasis, 36, 37
oxide supplement, 70
Magnetic resonance imaging (MRI), 85, 197, 202, 234
Magnificent calcium sources, 34
Maize, 53
Malabsorption, 34, 67
Malformation, 16, 145
Malfunctions, 92
Malignant growth, 46
Mammoth potential, 125
Mandible, 63
Manganese (Mn), 9, 89, 95–97
supplements, 96
Manifestation, 91, 196
Mannose receptor (MR), 261
Marine sources, 129
Marrow, 65, 149, 244
microenvironment, 149
Masculinizing, 76
Matrix ossification, 32
Mature, 7, 38, 42, 46, 48, 52, 64, 135, 252
dendritic cells, 135
Measles, 149
Mechanical
powers, 28
pressure, 28
strength, 37, 69, 76
stress, 130
Mechanism, 12, 102, 104, 135, 138, 162, 182, 211, 213, 224, 229, 230, 241, 242, 245, 251, 262
Mechanotransduction, 28
Mediated bidirectional communication, 248
Medicago sativa, 42
Medications, 10, 25, 86, 128, 129, 145, 154, 157–159, 176, 179, 198
Medicinal
food, 63
forms, 62
plant, 98, 99, 182, 237, 241, 253
prescription, 10
Medieval period, 147
Medium chain fatty acid (MCFA), 98, 256
Medullary, 5, 63
Melanin, 9, 41
Melatonin, 74, 84, 86, 89, 91, 92, 103, 113
deficiencies, 86
supplements, 91
Membrane, 128, 136, 179, 259
Menaquinone, 39, 40, 94
pills, 40
rich creature, 40
Menarche, 85, 86
Menopausal, 10, 33, 42, 43, 71, 75, 102, 132, 177, 248
chemical treatment, 42
side effects, 43
Mesenchymal
cells, 7
foundational microorganism, 27
improvement, 37
stem cells (MSCs), 134, 244, 246, 250, 251, 263
chondrogenesis, 134
tissue, 27
Metabolic, 9, 15, 32, 73, 74, 91, 94–96, 98, 102, 113, 132, 165, 178, 248
bone, 25, 161
course, 37
cycles, 36, 38
exercises, 34, 37, 38
factors, 85
pathways, 15
state, 30
syndrome, 131, 177
Metabolites, 15, 44, 49, 133, 297
Metacarpal, 63
bones, 59
Metacarpophalangeal joints, 128
Meta-investigations, 41, 49
Metalloproteinase, 72, 133, 136, 231
Metaphyseal swelling, 305

Metaphysis, 5
Metatarsals, 63
Methicillin-resistant, 226, 241, 242, 263
  S. aureus (MRSA), 226, 230, 235, 241, 242
Methionine, 43
Methotrexate, 129, 262
Microarray analysis, 13
Microbial
  communities, 15
  maturation, 53
Micronutrients, 49, 62, 86, 303
Microorganisms, 39, 52, 191, 225, 242
Microscopic, 51, 63
Microstructure, 15, 262
Microwave, 215, 298
Middle-centered tubular section, 5
Mild
  cases, 91
  curves, 86
Milk based protein (MBP), 71
Mimic, 12
Mineral, 4, 5, 7, 9, 10, 26, 28, 32, 34, 35, 38, 42, 44, 48, 49, 51, 54, 60, 61, 64, 65, 72, 74, 78, 84, 87, 89, 91, 92, 95–97, 105, 113, 149, 165, 198, 214, 215, 231, 297
  homeostasis, 5, 65
  nutrients, 10
Mineralization, 5, 9, 27, 60
  interaction, 28, 35, 37, 38
  period, 32
  phase, 27, 32
Minor element, 38
Minuscule
  casing, 33
  extent, 38
Misfortune, 47, 48, 51
Miso, 40
Missense mutations, 247
Mitigated, 4, 179
  impacts, 46
  normal physiology, 62
  properties, 50
  stress, 130
  substance, 44
Mixed lytic-blastic phase, 151
Moderate physical activity, 129
Molecular, 54, 71, 100, 228, 248, 256, 259, 261
  biology, 148
  level, 131, 155
  pathways, 92
Monocyte, 134
  chemoattractant protein (MCP), 261
Monomers, 52
Mononuclear, 27, 72
Monosodium urate (MSU), 175, 176, 181, 185
Monostotic illness, 147
Moral, 129
Morphine, 95
Mosaic pattern, 151
Mouse model, 135, 163, 164, 251, 252, 264
Mucopolysaccharide, 60
Multi
  drug resistance (MDR), 223, 225, 226, 235, 237, 238, 241–243, 264, 268
  functional element, 71
  nucleated osteoclast, 27
  stage process, 67
  vitamins, 78
Multiple
  beverages, 33
  doses, 305
  factors, 10
  fragments, 67
Muscle, 11, 33, 65, 67, 76, 105, 191, 192, 195, 196, 200, 207, 208, 296
Myeloid related protein 8 (MRP8), 232
Myeloperoxidase (MPO), 233, 258
Myocytes, 27

N

Nano
  drug delivery system, 135
  emulsions, 298
  micelles, 135
Naringenin (NAG), 133–136
  oral bioavailability, 135
Native cultures, 98
Natto, 39, 40
Natural
  dietary supplements (NPS), 175
  killer (NK), 102, 137, 236, 243, 256, 260, 261
    cells (NKC), 137, 236, 260
  sources, 12, 13, 74, 101, 103, 104, 134, 198

Nausea, 12
Nerve damage, 67
Neurofibromatosis, 101
Neurologic
impairments, 85, 154
theories, 85
Neuromuscular, 84, 161
Neuronal regulation, 95
Neuropathic, 84, 101
Neuroprotective, 102, 214
Neurotransmitter, 89, 93, 94, 98, 99
imbalance, 83, 85–87, 113
levels, 98, 99, 113
Neutral
lipids, 15
lysosomal proteases, 132
Niacin, 43, 72
Nicotinic cholinergic, 99
Nitric oxide (NO), 135, 233, 237, 258, 260
Nobiletin (NOB), 136
Non
digestible oligosaccharides, 74
mineralized osteoid, 32
receptor tyrosine kinase (NRTK), 252
steroidal
anti-inflammatory medications (NSAIDs), 85, 99, 176, 179, 182
phenolic compounds, 43
structural scoliosis, 84
Norepinephrine, 99, 103
Normal marrow cells, 149
Notch protein, 47
Noteworthy, 30, 35, 41, 42, 44, 49, 51, 133, 135–139
Noticeable rib hump, 85
Nourishment, 48, 238
Novel nutraceutical, 129
Nuclear
acids, 36
factor kappa-B (NF-κB), 14, 100, 104, 251, 252, 260–264
receptor, 10
Nucleocapsid, 148, 149
protein, 149
antigens, 149
Numbness, 128
Nutraceuticals, 3–5, 7, 10, 12–15, 23, 25, 26, 53, 54, 59–63, 68, 77–79, 81, 83, 84, 86, 87, 89, 90, 92, 104, 105, 110, 113, 114, 125, 130, 132, 133, 136, 139, 140, 145, 147, 159, 165–167, 175, 185, 191–193, 197, 198, 202, 203, 205–207, 215, 223–225, 235–238, 241–243, 250–253, 261, 262, 264, 265, 268, 269, 293–295, 298, 307
therapies, 84
Nutrient, 72
menaquinone, 40
Nutrition, 4, 5, 10, 25, 26, 59, 62, 86, 130, 131, 163, 165, 166, 181, 191, 197, 205, 206, 223, 243, 293–296, 298, 300–304
deficiencies, 59
rickets (NR), 293–296, 298–304
supplement, 5, 131, 294

O

Obovatol, 13, 251
Obscure skin, 41
Observational, 3–7, 11, 12, 47, 84, 85, 90, 92, 93, 97, 98, 102–105, 133–136, 138, 139, 146, 147, 154, 163, 179, 208, 224, 226, 229–231, 241–243, 259, 261, 263, 268
data, 304
investigations, 38
Obvious, 36, 37, 157, 164
Oleanolic
acid, 13, 247, 249
corrosive, 44
Oligo
fructose, 52
saccharides, 52
Omega-3 fatty acids, 71, 105, 129, 130, 182, 212–214, 259
Oncorhynchus keta, 103
Onset, 84–86, 130, 146, 148, 162, 175, 178, 224, 225, 233
Optic
atrophy, 153
foramen, 153
Optimum reduction, 129
Oral, 99, 302, 305
administration, 105, 135, 302
melatonin supplements, 91

pathogens, 242
phosphate supplements, 34
Orchestral
fashion, 4
substances, 52
Organic
anion transporter 1 (OAT1), 176, 179
bone matrix, 9
treatment aids, 78
Organs, 4, 65, 126, 135, 155, 199
Oriental medicine, 101
Oscillospira, 15
Ossification, 6–8, 247
Osteitis deformans, 146
Osteoarthritis (OA), 98–100, 104, 129, 130, 132, 156, 158, 198, 200, 201, 246
disorders, 132
pain, 99
mixture, 132
Osteoblast (OBs), 5, 7, 8, 10, 11, 27–30, 32, 36–39, 47, 51, 54, 60, 64, 66, 71–73, 75, 76, 91, 93, 95, 102, 113, 149–151, 162–164, 226, 230–232, 244–249, 251, 261–263
activity, 32, 151, 162
apoptosis, 102
cells, 27, 37, 246
movement, 30, 32
Osteoblastogeneis, 72
Osteocalcin, 71, 72, 74, 94, 246, 263
Osteoclast (OCs), 5, 8, 11, 12, 27, 29, 30, 36, 38, 44, 51, 54, 60, 64, 66, 71–76, 91, 93, 95, 102, 103, 146, 148, 149, 151, 157, 162, 163, 165, 230, 231, 244, 245, 248, 250–252, 261
apoptosis, 12, 102
bone, 11, 30, 50
breakdown, 37
cell, 27, 29, 39, 71
death, 72
development, 11, 12
forerunner cells, 27
movement, 29, 38
precursors, 149
Osteoclastogenesis, 13, 154, 162–164, 230, 231, 244, 249–252, 262–264
Osteoclastosis interaction, 44
Osteocytes, 5, 7, 27, 28, 36, 64, 65, 76, 229–232, 244, 249
Osteods, 7
Osteogenesis, 6, 67, 155, 244, 245, 248
cells, 7
tissue, 155
Osteoid, 7, 35, 64, 150, 162, 293, 295, 298
seam, 150
Osteolysis process, 71
Osteomalacia, 34, 161
Osteomyelitis, 223–237, 241–243, 245, 252, 253, 258, 262–264, 268, 269
treatment, 223, 224, 235, 237, 264, 268
Osteon, 6, 10
Osteonecrosis, 39
Osteopenia, 92, 94, 95, 102, 114, 244
Osteoporosis, 4, 5, 8, 10–15, 25, 26, 31–33, 36, 42–45, 47–51, 54, 59–61, 66, 67, 69–71, 73–75, 78, 86, 102, 146, 163, 206, 236, 244, 247, 248, 251, 262
breaks, 45
counteraction, 42, 50
patients, 26, 95
rats, 15
symptoms, 26
Osteoprogenitor cells, 7
Osteoprotegerin (OPG), 74, 91, 102, 103, 149, 167, 244, 249–251
ligand (OPGL), 249
Osteosarcoma, 154, 155
Osteosil calcium, 44
Outrageous non-veggie, 32
Ovariectomization, 15
rats, 104, 165
rodent models, 47
Ovariectomy, 95, 263, 264
Ovaries, 29, 33
Overabundance, 39
Overcast cover, 41
Overdosage, 39
Oversaw parenteral nourishment, 34
Oxalates, 53
Oxidation
catalysts, 45
collagen, 45
metabolism, 131
pressure increments, 47
renders, 76

stress, 16, 72, 87, 100, 130, 135, 139, 180, 233, 237, 238
Ozone, 41
assimilation, 41

P

Paget's disease of bone (PDB), 146–148, 151, 153–159, 161, 162, 165, 167
Pain
persists, 85
score, 101
Paleozoic age, 44
Pantothenic corrosive, 43
Papain injection, 104
Papilledema, 153
Parafollicular cells, 12
Paramyxoviruses, 148
Paraparesis, 153
Parathyroid
chemical, 27, 34–36
invigorates, 35
hormone (PTH), 12, 29, 34, 66, 69, 70, 74, 95, 96, 248
chemical, 29
related protein (PTHrP), 248
organ, 29, 35, 36
fundamental cells, 29
Parcel, 53
Parsley, 76, 213, 243
Partial collapse, 67
Pasteurization, 61
Patella, 63, 64, 194
Pathogenesis, 10, 83, 85, 91, 92, 127, 135, 140, 193, 246, 262
Pathologic
alterations comparable, 147
inflammatory process, 130
Pathophysiology, 205
Pattern recognition receptors (PPRs), 137, 259
Paw swelling, 135, 138
Peak, 60, 160, 161
Pediatric
bone, 295
Endocrine Society (PES), 303
Pelargonidin, 46
Pelvic, 64
accounting, 146
girdle, 63, 65
Penetrability, 34
Penetrance, 148
Pentothenic corrosive, 43
Peptide, 12, 30, 75, 127, 128
chemical, 30
citrullination, 127
hormone, 12
molecules, 12
Perception, 61
Periodontitis, 242
Periosteum, 5, 8, 65, 247
Peritoneal mast cells, 104
Permeability, 32, 99, 113
Peroxidase, 104
Peroxisome proliferator-activated receptor (PPAR), 258, 259
Perpetual
herbaceous plant, 42
weed, 44
Persistent, 33, 36, 62, 197, 199, 208, 223, 225, 229, 231, 233, 238, 268
Personal exercise programs, 129
Petunidin, 46
Pharmaceutical, 4, 59, 61, 63, 78, 205, 206, 235, 293, 294, 306
industry, 4
products, 59
Pharmacogenomic implications, 10
Pharmacokinetics, 79
Pharmacological
activities, 98
drug response, 79
effects, 102, 104
function, 14
potentials, 90
properties, 4, 62, 89
therapy, 26
Pharmacotherapeutic, 83, 85, 86, 105
intervention, 86
Phenolic cancer, 46
Phenols, 76, 253
Phenomenon, 3, 4, 233
Phenylalanine, 43
Phosphate, 7, 12, 27, 29, 35, 60, 64–66, 71, 92, 96, 150, 301
Phospholipids, 15, 34, 139, 256
Phosphoric corrosive, 35

Phosphorus, 3, 4, 8, 9, 27, 34, 35, 37, 42, 49, 64, 72
inadequacy, 34
statement, 27
Phosphorylation, 100, 136, 137, 258
Phylloquinone, 39, 40
Physical
aberrant, 145
activity, 32, 53, 129
examination, 85, 298
fitness, 62
therapy, 84–87, 105
Physician excessive, 5
Physiologic
capacities, 34
cycles, 36
demands, 155
factors, 41
function, 60
negative criticism instrument, 29
ramifications, 37
Physiotherapeutic, 129, 158
scoliosis, 86
specific rehabitation (PSSR), 86
Physis, 5
Phytates, 51, 53
Phytochemicals, 26, 54, 62, 70, 98, 99, 125, 130, 139, 145, 224, 237, 238, 241–243, 253, 268
Phytoconstituents, 76, 84, 87, 100, 105
Phytoestrogen, 14, 15, 43, 47, 48, 54, 74, 75, 102, 113, 164, 248, 262
activity, 43
Phytonutrients, 13, 42, 213, 237
Phytosterols, 131, 139
foods, 131
Pinealectomy, 91
Pinnacle point, 41
Piper
longum, 242
nigrum, 242
Piperine, 242
Pituitary organ, 30
Placebo, 71, 96, 97, 100, 102, 139, 181
pill, 97
Placental growth factor (PGF), 244
Plant
cell membranes, 131
extracts, 89
genotype, 43
origin, 62, 78, 256
pigments, 13
seeds, 131
sterol, 131
Plasma
calcium, 29, 96
concentration, 77
layer, 40
Plasmid, 242
Plastic distortion, 154
Platelets, 27, 102
Plum, 48, 72
Plumbagin, 12, 13, 250, 252
Plummeted, 27
Poisonousness, 39
Poly unsaturated fatty acids, 26
Polymerization degrees (DP), 52
Polymers, 76
Polymorphism, 10
Polymorphonuclear cells (PMN), 230, 231
Polyostotic, 148, 154
Polypeptide, 7
chemical, 29, 30
Polyphenol, 13, 47, 100, 103, 133
group, 15
Polyphenolic
compounds, 45
phytoalexin, 101
Polyps, 131
Polysaccharides, 52, 229, 253, 260, 261
Polyunsaturated
fatty acids (PUFA), 25, 49, 50, 74, 76, 139, 256, 258, 259
Population, 5, 54, 61, 62, 83, 84, 145–147, 151, 154, 162, 209, 294, 297, 299, 301, 306
Porosis, 10, 66
Porous, 10, 26, 65, 66
Portal circulation, 98
Post menopausal
menopausal, 5, 8, 10–13, 16, 25, 26, 45, 47, 48, 51, 54, 62, 67, 70, 75, 77, 93, 96, 102, 132, 178, 182, 262
bone, 47
females, 48, 54, 70
ladies, 45, 47, 48, 51
patients, 10

operative, 95
care, 86
morphine, 95
osteoporosis, 11
patients, 11
Postural, 60, 84, 85, 208
control, 85
Potassium, 9, 42, 49, 64, 70, 72, 77
Potent
anti-inflammatory, 99
anti-oxidant, 100
Potential, 9, 13, 36, 48, 51, 53, 62, 76, 78, 84, 87, 90, 92, 98, 104, 105, 113, 114, 125, 134–139, 145, 149, 164, 167, 175, 181, 182, 198, 213, 241, 248, 250, 258, 262, 264, 297, 301
Pragmatic, 145, 194
Prebiotic, 25, 26, 52–54, 74, 237
activity, 52
filaments, 52, 53
inulin (PRE), 138, 191, 296, 300, 303
research, 52
rich food, 52
substances, 52
Precious stone, 27, 35, 36
hydration shell, 35
Pre-clinical, 87, 90, 95, 96, 99, 103–105
studies, 95
Precursor, 9, 11, 93, 99, 103, 113, 244, 250, 251
Predictors, 86
Predisposition, 85, 176, 210
Predominant
catechin, 45
magnesium, 9
phytoestrogens, 43
Preliminaries, 41
Premenopausal women, 70, 182
Pre-osteoblast, 7
Prepubertal, 49
Prerequisites, 41
Prevails, 6
Prevalent, 26, 91, 146, 147, 195, 210, 223, 226–228, 230, 295, 296
Prevented curve, 86, 96
Pre-vitamin D3, 9
Primary
bile acid metabolism, 15
ossification center, 8
regions, 63
Pro inflammatory
cytokines, 134–137
pathways, 130
Probiotic (PRO), 25, 26, 51, 54, 74, 137, 138, 140, 237
forestall, 51
Proficient
method, 28
prebiotics, 53
Progressive scoliosis, 91
Pro-inflammation
cytokine, 10, 103, 263
regulations, 260
factors, 15
Proliferate, 91, 133, 134, 137, 208, 209, 237, 250, 256, 259, 260
bone tissue, 151
synovitis, 135
Proline, 93
Prolonged, 92, 99, 195, 210, 223, 230, 238, 268, 296, 302
Promensil, 42
Prominence, 52
Prone, 59, 61, 72, 92, 126, 129, 162, 177, 210
Pro-oxidant, 130
Propionate, 53
Prostacyclin, 134
Prostaglandins, 49, 50, 71, 99, 134, 182, 209, 258
Prostate malignant growth, 43
Protectins, 49, 198, 258
Protective nutrients, 25
Protein, 4, 15, 16, 36, 38, 42, 64, 71, 73, 79, 94, 135, 198, 215, 228–230, 233, 245
antibody, 128
modifications, 127
rich, 32
Proteoglycans, 27, 64, 95, 177
Protruding shoulder blade, 85
Proven therapeutic, 129
Provocative
entrail illness, 51
prostanoids, 50
Proximal, 5, 59, 63, 209, 210
epiphysis, 5
femur, 59
humerus, 59
Prunus domestica L, 48

Pseudo, 126, 247
Psoriasis, 130, 131
P-tyrosol, 99
Puberty, 86
Punica granatum L, 13
Pyridoxal phosphate, 43

Q

Quadriparesis, 153
Quality, 4, 25, 36, 38, 39, 62, 73, 78, 101, 104, 113, 129, 132, 158, 166, 293–295, 298, 301, 303, 306
  control, 15
Quantity, 4, 32, 79, 151, 160, 214, 215
Queasiness, 39
Quercetin, 7, 12–14, 25, 26, 44, 72, 76, 84, 89, 98, 103, 104, 163, 164, 181, 224, 246, 247, 249–251, 263, 264
Quinolinic acid, 103

R

Radiation, 9, 40, 194
Radicles, 46, 47
Radii, 63
Radiographic, 298
  damage, 129
Radiological
  appearance, 151
  examination, 85
Raloxifene, 12
Randomized, 62, 70, 92, 95–97, 99, 102, 161, 181, 300, 303
Rapid, 9, 62, 71, 85, 145, 160, 161, 233, 252
Rat synovium, 137
Ratio, 33, 61, 79, 132, 191, 209, 261
Rationale, 87
Reabsorbed, 29, 35, 67, 73, 176
  bone tissue, 149
Reactive oxygen species (ROS), 105, 136, 180, 233, 256, 258–260
Reality bones, 5
Receptor, 38, 102, 136, 149, 167, 231, 244, 245, 247, 248, 259, 261
  activator for NFkappaB ligand (RANKL), 72, 149, 162, 163, 244, 249–251, 262, 264
    receptors, 72
Recommended daily allowances (RDAs), 166
Reconstitution, 61
Reconstruction, 27
Rectifying, 48
Red
  clover, 42, 72, 75
  wine, 76, 163
Reduce
  fertility, 16
  testis size, 16
Reformation, 145
Regimen, 32, 35, 37, 41, 78, 158, 305
Reimbursing, 5
Reinforcements, 45
Relative remission, 126
Reliant, 27
Reluctant, 42
Remedy, 11, 201
Remodeling, 54, 60, 66, 67, 146, 149, 226, 231, 244
Renal, 29, 34–36, 128, 158, 176–180, 228, 229
  cylindrical phosphate reabsorption, 34
  problems, 36
  tubules, 29, 35
Render nutritional, 61
Reparative phase, 67
Repository, 35, 36
Repress, 35
Reproductive
  hormone, 25
  organs, 16
Research group, 13
Reseveratrol, 12, 13, 25, 26
Resolvins, 49, 50, 258
Resorption, 11, 50, 51, 60, 66, 71–73, 75, 231
Respiratory syncytial virus, 149
Restorative plants, 43
Restrain, 29, 35, 37, 38, 42, 50, 53
  vein calcification, 40
Resveratrol, 7, 14, 15, 47, 72, 84, 89, 99, 101, 102, 113, 163, 224, 247, 249–251, 262
  supplementation, 262
Retinol, 73
Rheumatoid arthritis (RA), 105, 125–130, 132–140, 205, 208–210, 212, 244, 250
  disorder, 125
  pathogenesis, 127, 135, 137

Rheumatoid factor, 128
Rhodiola rosea, 99, 113
Rhodioloside, 99
Rhythmic gymnastics, 85
Riboflavin, 43
Ribs, 63, 64, 85
Rickets, 34, 161, 162, 221, 293–304, 307
Rigidity, 60, 96
Risedronate, 12, 158
Rosavins, 99
Roseburia, 15
Rosmarinic acid, 13, 251
Rotational
  component, 84
  prominence, 84
Rutin, 99, 200

S

Sacrum, 63, 156
Safeguard, 38, 40, 42–44, 47–49, 51
  veins, 40
Safety, 4, 78, 79, 87, 104, 125, 139, 165, 166, 237, 295
Salidroside, 99
Sarcopenia, 105
Saturated
  fatty acid (SFA), 256
  oil, 98
Sauerkraut, 40
Scaffold, 150
Scapula, 63, 64
Scattering, 9
Scavenger receptor (SR), 261
Schroth therapy, 86
Sciatic, 84
Scientific
  knowledge, 4
  studies, 4, 145
Scleroderma, 132
Sclerosis, 131, 151, 157
Scoliometer, 85
Scoliosis, 83–87, 90–92, 94–96, 98–101, 103–105, 110, 113, 114, 156
  patients, 85, 92, 100
  progression, 91
  trials, 96
Scope, 27, 34, 35, 41, 52, 86, 155, 194
Scull bones, 7
Scurvy, 39
Secondary
  hyperparathyroidism, 32, 62
  line
    agents, 129
    management, 129
  ossification center, 8
Secretion, 72, 134, 176, 182, 229, 231, 259, 260
Sedative, 101
Segment, 59, 159
Selenium, 9, 42
Self
  esteem, 128, 129
  management skills, 129
Senile population, 25
Serine, 94, 245
Serotonin, 99, 103, 113
Serum, 11, 15, 77, 92, 94, 98, 104, 132, 136, 138, 155, 176, 298, 301
  amyloid A (Saa), 138
  uric acid (SUA), 176–182
Sesamoid bones, 63, 64
Severe, 86, 94, 126, 138, 146, 178, 200, 237, 296, 302
  curves, 85, 86
  pain, 126
Sex steroid, 11
Shark skin gelatin, 104
Shelf life, 62, 298
Short
  bones, 64
  chain, 51, 53, 54, 137, 260
    fatty acids (SCFAc), 137, 260
    unsaturated fats, 51
  term solution, 86
Shoulder girdle, 63
Sickliness, 38
Silicon, 42, 44
Silymarin, 13, 246, 247, 252
Simmering, 45
Simple fracture, 67
Single
  bone, 146
  living cell, 36
Sinusitis, 98
Skeletal, 4, 26, 30, 33, 36, 37, 59–61, 65, 67, 70, 73–79, 83, 85, 96, 146, 147, 155, 157, 160–162, 192, 248, 249, 252, 256, 296
  bones, 4

  deformities, 249, 296
  health management, 79
  system, 59, 70, 74, 79, 146, 157, 160
Skin
  cancer, 9
  diminishes, 41
  pigmentation, 41
Skull, 63–65, 146, 147, 151, 154, 156, 157, 296
Slender, 33
Slick fish, 41, 198
Slow
  acting symptomatic approach, 132
  progressing skeletal, 145
Smother proinflammatory cytokines, 50
Sodium bicarbonate, 77
Soft
  connective tissue, 6
  tissues, 4, 100
Sole
  mineral, 38
  precursor, 103
Solid
  bone mineralization, 38
  gums, 39
  hormonal pathway, 29
  neurons, 33
Soluble phosphatase, 45
Soy
  bean, 15, 42, 52, 97, 213
    isoflavones, 48
  food, 16, 47, 48, 102
  product, 15, 75, 164, 263
  protein, 48, 75, 298
Specialized cells, 5
Species, 15, 138, 164, 243
Spherical bone, 64
Spice, 44, 212, 214, 262
Spinach, 13, 39, 69, 70, 73, 76, 77, 97, 199, 205, 211, 213–215, 243
Spinal, 11, 42, 67, 73, 76, 77, 83–85, 154, 156, 157, 296
  bracing, 86
  cord, 65, 86, 153, 156
  curvature, 85, 114
  deformity, 83, 84, 105
  fractures, 11
  fusion surgery, 95
  health, 114
Spiral
  fracture, 67
  projection, 67
Spongy, 5, 6, 8, 65, 66
  bone, 6, 8, 66
  layer, 6
Sporadic cases, 148
Sprague Dawley rats, 99
Stability, 70, 91, 93, 157, 297, 298
Stage trabecular, 8
Stanol esters, 131
Staphylococcal abscess communities (SACs), 229
Starches, 48, 215
Starvation, 34
Static, 5
Stem cells, 7, 65
Steric structure, 43
Sternum, 63
Steroid
  estrogens, 43
  hormone, 76
  synthetic, 47
Stiffness, 128
Stomach microscopic organisms, 51, 52
Stooped posture, 26
Strain, 98, 138, 194, 197, 201, 215, 242
Streamlining, 38
Stroke, 11
Strolling, 33, 194, 202
Stromal
  lesions, 155
  related enzymes, 132
Structural, 3–6, 8, 40, 62–64, 77, 79, 93, 96, 133, 198, 229, 248
  organs, 4
  scoliosis, 84
Sturdiest organ, 5
Sub-atomic oxygen, 37
Substantial health benefits, 129
Substrate, 52
Subtypes, 39, 40, 84
Sugar monomers, 52
Suggestive diet, 61
Sunscreen, 41
Superfamily, 10, 245, 249
Superoxide-dimutase, 47
Supersede, 35

Supplanted, 32
Supplementation, 14, 45, 47, 49, 51, 97, 101, 103, 137, 160–162, 213, 263, 293, 296, 297, 300, 304, 305
Supplements, 9, 15, 40, 42, 43, 52, 59, 62, 71–73, 75, 76, 78, 87, 94–97, 101, 103–105, 165, 166, 192, 198, 212, 214, 235, 237, 264, 296, 304, 305
Surgical
  intervention, 85, 86, 105
  lengthening, 86
  trauma, 98
Swollen, 126, 128, 176, 208, 214
Symbiotic (SYM), 138
Symptomatic effects, 139
Symptoms, 10, 25, 26, 59, 62, 126, 128, 131, 132, 139, 146, 155, 156, 178, 196, 206, 207, 210–214, 224, 226, 227, 233, 234, 294, 304
Syndromes, 83, 156
Syndrome-X, 131
Synergy, 53
Synovial
  cells, 133
  fluid, 130, 177
  lipids, 130
  macrophages, 133
  membranes, 127
  tissue, 134–136
Synoviocytes, 127, 133, 136
Synthetic
  drugs, 83, 84, 87
  hormones, 11
  substances, 46, 50

T

T-cell, 127, 135, 232, 236, 238, 243, 256, 259, 261
Tachykinin receptor 1 (TACR1), 13
Tachysterol, 41
Tarsals, 63, 64
Tart Cherry Juice (TCJ), 181
Tea polyphenols, 45, 46
Tear, 67, 196
Temporal bone involvement, 154
Tendons, 64, 65, 67, 207, 208, 214
Teriparatide, 12
  abaloparatide (TYMLOS), 12, 16
Testosterone creation, 31
Tetraterpenoids, 13
Theaceae family, 45
Therapeutic, 10, 11, 45, 59, 78, 84, 86, 89, 90, 94, 96, 98, 100, 104, 105, 113, 130, 133, 135, 137, 139, 145, 154, 157–159, 162, 165, 167, 180, 192, 198, 201, 207, 213, 214, 223–226, 235–238, 241, 243, 253, 268, 293–295, 301–303, 305
  agents, 59, 90, 104, 130, 159
  benefits, 113, 130
  effect, 89, 96, 130, 135, 139, 293, 294
  treatment, 137
Thiamin, 43
Thigh region, 63
Thiocolchicoside, 13
Thoracic cavity, 63
Threonine, 43, 245
Thrombin, 71
Thrombophlebitis, 98
Thromboxane, 134, 258
Thyroid
  invigorating chemical (TSH), 30
  para-follicular cells, 30
Thyroxine (T4), 30, 67
Tibia, 63, 74, 146, 154, 156
Tilapia nilotica, 15
Timeframe, 33
Tingling, 128, 207, 214
Tissue, 5, 6, 34, 36, 37, 60, 65–67, 86, 96, 126, 132, 134, 135, 149, 151, 157, 175, 176, 178, 191, 192, 195, 196, 198, 199, 205, 208–210, 224, 226–229, 233, 235, 258, 259
  constituents, 132
  exercises, 36
  irregularities, 37
  proteases, 132
T-lymphocytes, 134, 140
Tocopherol, 130, 199
Tolerance, 127
Tomatidine, 242
Topoisomerase, 72, 102, 242
Torque, 71
Tough, 6, 128
Toxicity, 5, 12, 15, 16, 180, 301
Trabecular, 65
  bone, 65, 157

Trace minerals, 5, 9, 96
Traditional growing rods, 86
Transducer, 36, 37
Trans-resveratrol, 47
Transversion, 41
Trauma, 67, 84, 155, 224, 225, 227, 228, 234
  location, 67
Triglycerides, 131
Triiodothyronine (T3), 30
Triterpenoids, 44
Trochanter region, 76
Trypsin, 99, 164, 200
Tryptamine, 103
Tryptophan, 43, 89, 103
Tuberculosis, 128
Tubular structure, 63
Tumor, 154, 158
  growth, 16
  necrosis factor (TNF), 72, 104, 113, 129, 133–138, 162, 215, 231–233, 244, 249–251, 258–263
    alpha (TNFα), 215
    related activation-induced cytokine (TRANCE), 249
Tyrosine
  kinase, 7, 72, 102, 248
  metabolism, 93

U

Ubiquitin binding site, 148
Ulnar deviation, 128
Ultraviolet
  radiation, 9, 16
  radiation B (UVB), 9, 41
    openness, 41
    radiation, 9
Unavoidable factors, 67
Underlying capacities, 34
Unequal hip, 85
Unequivocally, 51, 247
Unfermented, 45
United Kingdom (UK), 8, 62, 146, 147, 162, 155, 303
Unmineralized bone matrix, 64
Unmistakable, 27, 40, 47
  cell types, 27
Unpalatable, 52
Unprecedented, 34
Unsaponifiable, 132
  extracts, 132
  products, 131
Unsaturated
  fats, 49–51, 53, 198
  polyhydroxy
    alcohols, 39
    liquor, 39
Upregulate, 14
Upsurge, 73
Uptake, 5, 69
Urate
  transporter 1 (URAT1), 176, 179
  transporter-channel (URAT), 179
Urged, 38
Urinary calcium, 70, 77
Ursolic corrosive, 44
Uterine hypertrophy, 16
Utilitarian, 48
Utilization, 33, 35, 44, 45, 47, 49, 50, 139, 192, 197

V

Valerate, 53
Valine, 43
Valosin-containing protein (VCP), 155
Vascular, 5, 40, 71, 94, 99, 151, 153, 154, 209, 226–228, 234, 244, 259
  calcification, 40, 94
  endothelial growth factor (VEGF), 71, 209, 244
  primitive connective tissues, 7
Vegetative protein, 73
Vertebrae, 63, 64, 84, 92, 153, 156, 157, 228
  column, 63, 65, 86, 153
  fracture, 67
  trabecular BMD (VTBMD), 97
Viral infection, 148
Visualized nutraceutical, 62
Vital organs, 60, 65, 126
Vitamins, 4, 26, 61, 62, 77, 84, 87, 89, 91, 92, 94, 97, 105, 113, 181, 198, 212, 214, 237, 293, 297
Vitamin C, 26, 54, 72–74, 89, 92, 93, 97, 180, 199, 213
  proteins, 72
  supplements, 93

Vitamin D, 3, 4, 9, 10, 26, 29, 32, 38, 40–43, 49, 51, 53, 59, 62, 66, 67, 70, 71, 78, 84, 86, 89, 92–97, 102, 105, 113, 114, 145, 155, 161, 162, 200, 236, 293–307
  amalgamation, 41
Vitamin D3, 3, 9, 41, 84, 94, 96, 294, 306
Vitamin E, 73, 130, 132
Vitamin K, 39, 40, 62, 73, 89, 94, 97
Vitamin K2, 26, 54
Vitis vinifera L, 13
Volatile organosulfur compound, 242
Vulnerable host, 148

W

Water-dissolvable, 46
Wavelength, 9
WBC count depletion, 75
Weak bones, 54, 67
Wedelolactone, 13, 247, 249, 251, 252
Weight-bearing
  bones, 26
  exercises, 33
Western blotting, 136
Wide
  diameter, 66
  range, 92, 102, 105, 225, 230
Wistar rats, 74, 95
Withanolides, 13
Worldwide market, 62, 159
Worsen, 84, 87, 92, 95, 202
Woven bone, 8
Wrinkles, 73
Wrist, 63, 70, 207, 208, 210, 214

X

Xanthine oxidase (XOD), 175–177, 179–181, 185

Y

Yeast, 61, 97, 260
Yellow marrow, 63

Z

Zerumbone, 12, 13, 25, 252
Zinc (Zn), 9–11, 26, 38, 42, 49, 54, 71, 72, 96, 139
  subordinate chemicals, 38
  supplementation, 39
Zoledronic acid (ZA), 12, 157, 158, 167
Zyflamend, 13